Unlocking Artificial Intelligence

Christopher Mutschler • Christian Münzenmayer
Norman Uhlmann • Alexander Martin
Editors

Unlocking Artificial Intelligence

From Theory to Applications

 Springer

Editors
Christopher Mutschler
Division Positioning and Networks
Fraunhofer IIS, Fraunhofer Institute
for Integrated Circuits IIS
Nürnberg, Germany

Christian Münzenmayer
Division Smart Sensing and Electronics
Fraunhofer IIS, Fraunhofer Institute
for Integrated Circuits IIS
Erlangen, Germany

Norman Uhlmann
Division Development Center
X-Ray Technology
Fraunhofer IIS, Fraunhofer Institute
for Integrated Circuits IIS
Fürth, Germany

Alexander Martin
Fraunhofer IIS, Fraunhofer Institute
for Integrated Circuits IIS
Nürnberg, Germany

ISBN 978-3-031-64831-1   978-3-031-64832-8   (eBook)
https://doi.org/10.1007/978-3-031-64832-8

This work was supported by Fraunhofer Institut für Integrierte Schaltungen IIS   •

This Springer imprint is published by the registered company Springer Nature Switzerland AG
The registered company address is: Gewerbestrasse 11, 6330 Cham, Switzerland

If disposing of this product, please recycle the paper.

Preface

In recent years it has become apparent that the deep integration of artificial intelligence (AI) methods in product and services is essential for companies in Germany and world-wide to stay competitive. The use of AI allows large volumes of data to be analyzed, patterns and trends to be identified, and well-founded decisions to be made on an informative basis. It also enables the optimization of workflows, the automation of processes and the development of new services, thus creating potential for new business models and significant competitive advantages.

The use of AI in industry offers new opportunities to increase productivity, improve quality, reduce costs and generate new, innovative solutions. Customer satisfaction can also be increased through improved customer interaction and personalized offerings. The use of AI offers significant potential in terms of quality, efficiency and competitiveness - not only for multinational enterprises but also for the small and medium-sized enterprises (SME) which are the industrial backbone of the European economy. On the one hand, the quality of products and services can be increased through the use of suitable tools and methods, which minimizes the susceptibility to errors, optimizes processes and thus increases customer satisfaction. The automation of recurring tasks enables resources to be freed up and can lead to increased efficiency and productivity. On the other hand, the use of AI enables SMEs to better implement customer requirements, offer innovative solutions, stand out from the competition and remain competitive in an increasingly globalized and digitalized economy.

However, the use of AI in SMEs and industry also brings new requirements, such as building up specialist knowledge and mastering technological complexity. The rapid development and the in-depth knowledge required to implement and support suitable methods and tools currently pose major challenges for SMEs in particular.

To meet the above-mentioned challenges and support the adoption and integration of AI in industry and SMEs, structural measures are required. One suitable measure, for example, would be the financing of transfer structures such as the ADA Lovelace Center. Such a targeted development of transfer structures facilitates the transfer of knowledge between research institutions and companies and provides industry and

SMEs with low-threshold access to specialist knowledge and resources in order to exploit the full potential of these technologies.

The ADA Lovelace Center is a pioneering competence center for AI in Bavaria, the establishment of which was funded by the Bavarian State Ministry of Economic Affairs, Regional Development and Energy. A central focus of the ADA Lovelace Center is on the development of AI-based solutions for industrial applications in sectors of outstanding importance for Bavaria. These include transportation and traffic, production and Industry 4.0, rail transport, financial services and insurance, logistics and healthcare as well as sports. Concepts and solutions for specific issues are researched and implemented in close cooperation with the application partners. A wide range of AI skills are applied and further developed to promote the targeted and sustainable development of AI skills within partner companies. In addition to scientific research, particular attention is paid to the promotion of young scientists, who are integrated into industrial research at an early stage. The ADA Lovelace Center bundles and expands the AI expertise and infrastructure of the Friedrich-Alexander-University Erlangen-Nürnberg, Ludwig-Maximilians-University Munich, the Fraunhofer Institute for Integrated Circuits IIS, the Fraunhofer Institute for Integrated Systems and Device Technology IISB and the Fraunhofer Institute for Cognitive Systems IKS. Thus, the ADA Lovelace Center has significant expertise in all relevant AI processes.

The center has created an internationally visible network for the Bavarian economy, which is dedicated to the fundamental issues of data collection and analysis using AI methods, taking into account data protection and data security. The ADA Lovelace Center supports companies in the Bavarian economy by researching, developing and implementing concrete solutions for issues in the field of AI and enables them to transform their business processes and develop new data-driven business models. This book presents an excerpt from various application areas and methodologies and research areas of AI and explains how those methods and processes can be used successfully in practice.

Nuremberg, Fürth, Erlangen   *The Editors*

Acknowledgements

First of all the Editors want express their greatest thank to Nadine Chrobok-Pensky and her team for their excellent, professional and also human-focused project management, motivation, organization, and friendly reminders with a huge amount of commitment and patience. Your work and support of the whole team during the project was outstanding and highly appreciated.

We would like to express our sincere gratitude to all the authors for their valuable contribution to this book. The expertise and dedication have filled the book with high value content in the field of artificial intelligence and its applications, making it a valuable resource for readers in this field. Your thorough research before and within our joint research project, your insightful analysis, and clear writing style have undoubtedly played a crucial role in the success of this fantastic book. Your commitment to delivering high-quality content is commendable and greatly appreciated. We, the editors, thank you for your outstanding work. Your contribution will undoubtedly make a significant impact on the readers and researchers in the field.

On behalf of the entire team and authors, we would like to express our profound thank you to the Bavarian Ministry of Economic Affairs, Regional Development and Energy for your generous support of the ADA Lovelace Center project. Without the financial funding, it would not have been possible to successfully execute this project. Your support has allowed us to conduct important research and gain valuable insights. Through your funding, we were able to provide resources and materials that were essential to our work for scientific community and local, national and international industry. We say *thank you* to you for your trust in our project and your support across all its stages. Your financial support has not only contributed to the realization of this project but will also have a lasting impact on research in this field.

It is also very important to mention that such a piece of work can not be done in such excellent quality without the support and advice of an highly rated advisory board consisting of industry and scientific experts, who were always reachable and willing to give advice, support and direction of research and development. The complete ADA team says thank you for your contribution, enthusiasm and work in all phases of the project.

The editors also want to say thank you to all the people "behind the scenes" for management, calculations, administrative tasks, organization of meetings and some food, rooms, projectors, hot and cold drinks, good words of support, flexible and agile management. Thank you for been with the complete team. Your work was highly appreciated.

The entire team would like to express our sincere appreciation for the invaluable collaboration and support of our cooperation partners. The expertise and dedication have been instrumental in the success of this endeavor. The commitment to our shared goals and your willingness to work together have greatly contributed to the progress and achievements of the project. We are truly grateful for the opportunity to collaborate with such a dedicated group of cooperation partners.

We also want to thank Ralf Gerstner from Springer Verlag for his patience and continuous support, and the reviewer and proofreaders who helped us improve the book.

Last but not least, the complete team wants to say thank you to all the coffee machines all around which were able at all times during day and night to provide everybody in need with excellent coffee to keep the work and innovation up and running.

Nuremberg, Germany   *Christopher Mutschler*
Erlangen, Germany   *Christian Münzenmayer*
Fürth, Germany   *Norman Uhlmann*
Nuremberg, Germany   *Alexander Martin*

February 2024

Contents

8   Acquisition of Semantics for Machine-Learning and Deep-Learning based Applications

Part II Applications

Part I
Theory

Chapter 1
Automated Machine Learning

Florian Karl[1,2,4], Janek Thomas[2], Jannes Elstner[1], Ralf Gross[3], Bernd Bischl[1,2,4]

Abstract In the past few years automated machine learning (AutoML) has gained a lot of traction in the data science and machine learning community. AutoML aims at reducing the partly repetitive work of data scientists and enabling domain experts to construct machine learning pipelines without extensive knowledge in data science. This chapter presents a comprehensive review of the current leading AutoML methods and sets AutoML in an industrial context. To this extent we present the typical components of an AutoML system, give an overview over the state-of-the-art and highlight challenges to industrial application by presenting several important topics such as AutoML for time series data, AutoML in unsupervised settings, AutoML with multiple evaluation criteria, or interactive human-in-the-loop methods. Finally, the connection to Neural Architecture Search (NAS) is presented and a brief review with special emphasis on hardware-aware NAS is given.

Key words: AutoML, Neural Architecture Search, Black-box Optimization

1.1 Introduction

Machine learning (ML) has achieved remarkable results across a number of domains in many different applications. However, this success highly depends on the identification of a good model and its integration with suitable preprocessing procedures, feature engineering, and other stages within an ML pipeline. Furthermore, even upon

[1]Fraunhofer Institute for Integrated Circuits IIS, Fraunhofer IIS, Nuremberg, Germany
[2]Ludwig-Maximilians-Universität München, Munich, Germany
[3]Siemens Digital Industries, Nuremberg, Germany
[4]Munich Center for Machine Learning, Munich, Germany

Corresponding author: Florian Karl
e-mail: `florian.karl@iis.fraunhofer.de`

© The Author(s) 2024
C. Mutschler et al. (eds.), *Unlocking Artificial Intelligence*,
https://doi.org/10.1007/978-3-031-64832-8_1

identifying a suitable model, it still requires precise tuning, as model performance depends substantially on numerous hyperparameters. In short, ML experts must exert considerable manual effort and conduct extensive experimentation to achieve success in ML projects through hyperparameter optimization (HPO) and model selection.

Automated machine learning (AutoML) can help alleviate this issue by automatically identifying suitable models or even pipelines, which in turn frees experts up to devote themselves to more interesting and relevant work. However, ML projects should not be viewed merely as the search for an optimal model for a given dataset. The machine learning workflow CRISP-ML(Q), as outlined in [103], consists of six phases: (1) business and data understanding, (2) data engineering, (3) model engineering, (4) model evaluation, (5) model deployment, and (6) model monitoring and maintenance. While all of the described phases can profit from automation and reduction of manual effort, some are clearly better suited; in particular those centered around model development are most attainable as of now [70] and most of current AutoML research centers around this topic [62]. That is not to say AutoML does not

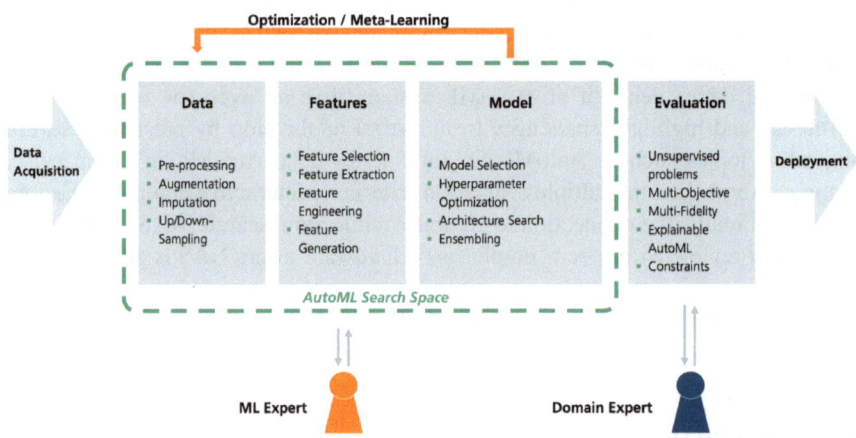

Fig. 1.1: An overview of what AutoML encompasses in the model engineering and evaluation of the ML workflow. Most elements can be found in the course of this chapter and core concepts are explained in the following subsections.

extend to other phases: monitoring of data and models, also a relevant topic in the area of machine learning operations (MLOps) for example, provides plenty of opportunities for automation and has also been considered as closely connected to model building, which led to research in the area of *online AutoML* [18]. Furthermore, automated data science (*AutoDS*) is a movement aiming at automation of stages in the ML workflow focused around data, like data acquisition or EDA [31]. Independent of the relevant part of the ML workflow, the overall goal of AutoML is to reduce tedious tasks to a minimum and make ML engineers increasingly efficient in their work. For the reasons outlined above we will in this work focus on automation in the context of model development. An overview of the steps that can be automated and the topics related to AutoML in this phase can be found in Figure 1.1.

In the model engineering phase it is usually assumed that a (mostly) clean dataset is available,[1] that a performance measure as well as a validation scheme have been defined, and that possible constraints for deployment are known.

The result of the model engineering phase is an optimized ML pipeline. An ML pipeline is a sequence of preprocessing, modeling and postprocessing operations that is trained on the available data and can be used to predict new observations. Since many ML algorithms are available, each with their unique hyperparameters, finding an optimal pipeline is a complex and – since training an ML model can require a lot of computational effort – expensive black-box optimization problem. This is further complicated by the fact, that (1) the search space can be mixed (numerical and categorical) and involve dependencies or hierarchies, (2) observations, i.e., measured performance of evaluated pipelines, can be inherently noisy and (3) dependencies between pipeline steps and certain hyperparameters are mostly unknown. A considerable amount of surveys on AutoML exist [121, 36, 126] as well as a textbook [62] to provide a general introduction into the topic. Additionally, recently surveys on subtopics of AutoML have been published, including e.g., neural architecture search (NAS) [115], AutoML for time series forecasting [6] and AutoML for unsupervised methods [8].

This chapter aims to provide a condensed overview over the domain of AutoML in the ML workflow phases related to modeling in a practical context, its current applications, existing solutions and limitations. After this brief introduction, Section 1.2 presents an evaluation on various AutoML components such as search space, optimization methods or ensembling. Section 1.3 highlights several selected topics with relevance for industrial applications such as AutoML for time series data, AutoML in low supervision scenarios, multi-objective AutoML and the integration of human experts. After Section 1.4, which gives a brief introduction to NAS with emphasis on the topic of hardware-aware NAS, the chapter concludes with a critical look at challenges and open research topics in the field of AutoML.

1.2 Components of AutoML Systems

In practice, a distinction is made within the field of AutoML between automatically searching for optimal ML pipelines for "traditional" non deep learning ML methods, and the search for network architectures in deep learning, NAS. Despite its success and relevance, NAS is in literature classified as a subfield of AutoML [62]. Most of this section concerns the former, while in Section 1.4 the use and adaption of various methods for NAS is discussed. In general, AutoML systems consist of several

---

[1] Data cleaning is technically not a part of AutoML and we will exempt this matter from discussion here, though automation in this phase is an interesting topic in itself. We assume that standard procedures (e.g., removal of constant and duplicate features) are always utilized.

different components: search space, optimization, ensembling, feature engineering and meta-learning. Different choices for all these components with their advantages and drawbacks are discussed in the following.

1.2.1 Search Space

The search space Λ defines which algorithms, preprocessing and postprocessing operations are considered as well as the ranges of required hyperparameters. Usually, these spaces are defined in a single ML framework, e.g., scikit-learn [90], WEKA [35] or H2O [53]. Λ can be written as a directed acyclic graph, where each node can be associated with discrete, continuous and conditional hyperparameters. Numeric hyperparameters have a finite range defined by a lower and upper bound. It may not be reasonable to optimize all numeric hyperparameters on a linear scale: The learning rate of neural networks, for example, is generally tuned on an exponential scale, e.g., $10^{-10}, \ldots, 10^{-1}$. Conditional hyperparameters are only active based on the values of other hyperparameters. For example, the γ hyperparameter of a Support Vector Machine (SVM) is only meaningful when the kernel is set to Radial Basis Functions. For an SVM with Linear Basis Functions, γ is inactive as it is not used by the model.

The resulting hierarchical structure of Λ makes optimization with standard techniques challenging. There is also a clear trade-off in the definition of Λ. On the one hand, a too restrictive search space may exclude the optimal pipeline and thus cannot be found. On the other hand, a large search space will likely result in difficult and expensive optimization. Ideally the search space should consist of complementing operations, i.e., a method that works well if another method does not and vice-versa. This prevents creation of an unnecessarily vast search space and allocation of too much budget on several methods that are expected to produce similar results. Unfortunately, it is very hard to learn such an optimal search space in a data-driven way. A large amount of meta-data is required to learn desirable behavior and general statements are difficult to make over all possible datasets. Notions of transfer learning [91] and adaptive search spaces [54] exist within black-box optimization, but to the best of our knowledge, they have not been successfully integrated in AutoML solutions. Currently, the search space is usually defined in an ad-hoc manner by the developer of the respective AutoML framework with respect to his or her domain knowledge, ML expertise and intended applications as well as some benchmark experiments.

1.2.2 Optimization

In addition to the challenges that the structure of Λ pose, despite best design efforts, the search space can become quite large if many different operations and model families are considered. The optimization problem is also a *black-box* optimization problem, which means that no derivatives of performance with respect to hyperparameters are available. In many settings the problem is *computationally expensive*, as a single evaluation of the black-box typically requires a full cross-validation. Lastly, the optimization is also *stochastic*, as evaluations are only estimates of the pipeline's true generalization performance. If those limitations are not taken into account, issues with overtuning can arise [17, 85].

Simple Optimization methods such as random or grid search can be quite competitive [47, 48], as they are hard to misspecify: These approaches forego the use of complex algorithms and there is minimal risk for human error. More sophisticated optimization techniques can break if assumptions, e.g., on the optimization surface, do not hold. In general, random search is always preferable to grid search, as irrelevant hyperparameters do not force identical evaluations [11]. This is also shown in a benchmark study by Zöller et al. [126].

Bayesian Optimization is a sequential global optimization method [86, 69] that was developed for expensive black-box problems and is now widely used in HPO and AutoML [99, 60]. The basic idea is to approximate the optimization surface with a probabilistic model, most commonly a Gaussian process; this surrogate is cheap to evaluate and analyze. Optimization is usually initiated by evaluating a certain number of points in a (pseudo-)random manner after which the surrogate model is fit. An infill criterion (or acquisition function) is optimized over the surrogate to select the next optimal point to evaluate; examples include expected improvement or probability of improvement [99]. Typically, an infill criterion balances exploitation of regions with high performance and exploration of regions with high uncertainty [99]. The point obtained through the optimization procedure is then evaluated and the surrogate model is retrained. Applying the method to AutoML poses a challenge, as Bayesian optimization in its original formulation requires a fully numeric configuration space. This is not the case for almost all AutoML systems and applications. A possible solution is to use an appropriate surrogate like a Random Forest instead of a Gaussian process and impute inactive hyperparameters [60]; Neural Networks have also shown promise in combination with a suitable Bayesian treatment like adding a Bayesian linear regressor to the last hidden layer [100]. Another approach is to use a Gaussian process surrogate, but learn lower dimensional numeric embeddings [88].

Bayesian optimization is quite flexible and can be extended to optimize multiple performance measures concurrently [58, 71] as well as to handle the stochasticity of the underlying AutoML problem [93]. Letham et al. [76] propose a way to handle noisy or unknown constraints [51], i.e., constraints where it is not (immediately) clear if the proposed ML pipeline is feasible or not. Local Bayesian optimization has been implemented by Eriksson et al. [39] to deal with challenging high dimensionality and large sample budgets. In its original sequential formulation Bayesian optimization is an iterative algorithm that evaluates one point at a time, which can be problematic when trying to parallelize it. It is therefore desirable to adapt Bayesian optimization to generate multi-point or batch proposals, which can be achieved through certain infill criteria [61, 24].

Evolutionary Algorithms are population-based and stochastic methods inspired by evolution in biology. Evolutionary algorithms generally follow the same procedure: After an initial population is sampled and fitness of each individual (an ML pipeline constitutes an individual) is assessed, a sub-population is chosen as parents for the next generation of offspring. Mutation (random perturbation of an individual) and crossover (combination of attributes of two individuals) operations are employed to generate offspring. These steps are repeated until a given stopping condition is met, when the best performing pipeline is returned as the final result. While requiring a substantial number of evaluations, evolutionary algorithms are a popular choice due to their ability to handle complex search spaces,[2] ease of implementation, straightforward possibilities for parallelization and the low probability of getting stuck in local optima [4]. Other population based approaches such as particle swarm optimization follow a similar idea. In particle swarm optimization, units of a population of candidate solutions (i.e., particles) traverse the search space based on information about their own respective known states and the behavior of the entire population. In general, these population based optimization techniques are not as efficient as Bayesian optimization and require a larger amount of iterations to find good solutions.

Multi-Fidelity Approaches aim to optimize budget allocation by stopping poor performing pipelines or models early, so as to not waste available budget. Instead of exploiting optimal selection of pipelines to evaluate like the previously presented methods, multi-fidelity methods attempt to find good solutions by allocating available budget in an optimal manner. Evaluating ML pipelines (and mainly deep learning architectures) can be expensive and the underlying assumption is, that with less budget, i.e., on a lower fidelity, one can already determine with confidence which pipelines will perform best. Suitable budget types are, for instance, epochs trained, size of training set, or image resolution.

The simplest implementation of this is successive halving [66]. With successive halving, an amount of randomly sampled configurations is trained on a low fidelity. The worse performing half of models is then discarded, while the better performing half of models is trained on a higher fidelity and the procedure is repeated. Hyperband [79] is an extension of successive halving to solve the problem of determining the amount of sampled pipelines and the initial fidelity by conducting several successive halving processes from different starting conditions. It is not entirely clear, which type of budget (if multiple are a sensible choice) is best for which application so that high rank correlation between fidelities is achieved [34]; the budget is generally chosen by the practitioner in an ad-hoc manner. Multi-fidelity approaches can be used in conjunction with many other optimization methods: The simple random sampling in Hyperband can, e.g., be upgraded to Bayesian optimization [40].

---

[2] One simply has to define suitable operations for mutation and crossover. While not always trivial, proper operations have been defined for several elements across the ML pipeline [89].

While these methods constitute arguably the most popular optimizers for AutoML, HPO, and NAS, other methods have been applied successfully to such tasks, including iterative racing, Monte-Carlo Tree Search (MCTS), and gradient-based optimization methods. The interested reader shall be referred to [121, 36, 126, 62, 115] among others.

1.2.3 Ensembling

An additional tool used by many AutoML frameworks is model ensembling (in the context of AutoML mostly done through *stacking*) [16]. In the optimization process, many candidate pipelines are proposed until the budget is used up or a different termination criterion is reached. Finally, the best k pipelines can be combined in a powerful ensemble. This can be achieved by simple (weighted) averaging of the pipeline predictions [16] or by training a model using the predictions of the pipelines as new features [117, 106]. Some successful AutoML tools include post-hoc stacking to further boost performance [43], others have ensemble methods as an integral part of the underlying algorithms [19]. Wistuba et al. [116] even use multiple levels of this stacking approach of hundreds of pipelines to achieve very strong predictive performance outperforming 3000 out of 3500 ML expert teams in 12 hours in an ML competition. Similarly, one of the best-performing AutoML tools [47], *AutoGluon*, relies largely on ensembling and stacking of models in multiple layers [38]. While in many cases ensembling helps boosting the performance of an AutoML tool [52], considerable drawbacks in terms of model size, inference time and general model complexity are apparent. The trade-off between complexity of the solution and its predictive performance needs to be quantified and a conscious decision has to be made [71].

1.2.4 Feature Selection and Engineering

Feature selection and feature engineering are two important preprocessing steps in an ML pipeline.

Feature Selection filters relevant features, that are then presented to the ML model. This serves the purpose of reducing model complexity, reducing costs of data acquisition and improving performance by eliminating noisy variables. Including feature selection into AutoML produces the search space $\{0, 1\}^p \times \Lambda$, where p is the number of initially available features. While often done in a separate step, there is a lot of merit in combining feature selection and the remaining parts of the ML pipeline into one optimization problem [13].

Feature Engineering is the process of extracting features from raw data, i.e., multiple data sources. Simple examples of such extractions are sums or averages with an $n - 1$ relation (combining n sources into one feature) to the target. The amount of possible extractions grows immensely with a growing number of relations and variety of extractor functions. This space will quickly become impossible to exhaustively search, so smart search heuristics need to be employed. Kanter et al. [72] propose a greedy exploration strategy to progressively optimize predictive accuracy. If the data is stored in entity-relation-entity format, Cheng et al. [23] propose an efficient way to extract features from the graph that can be used for general ML algorithms. It should be noted, that if automated feature engineering is included in the AutoML framework, we see the input not as a single cleaned data source as discussed in the introduction, but as multiple linked sources.

1.2.5 Meta-Learning

In meta-learning, the information on how machine learning models perform on many different datasets is used to approach new datasets more efficiently [108]. Many different tasks such as few-shot [9, 112] or multi-task [15] learning are closely related to meta-learning, but in the context of AutoML a distinction can be made. Meta-learning can be seen as a fundamental concept of AutoML and refers to methods that can leverage information from previous tasks and recommend pipelines or architectures and warm start the AutoML process, thus aiding model selection. In contrast, methods such as transfer learning or few-shot learning produce a fixed architecture and improve only model training (see Figure 1.2). To utilize meta-

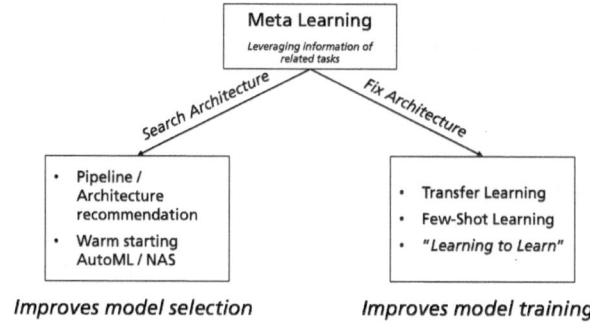

Fig. 1.2: Meta-learning can be divided into two approaches: Some methods support model selection, while others produce a fixed architecture and support model training.

learning for AutoML, it is possible to discretize the search space Λ and create a matrix of datasets $d \in D$ and ML pipelines $\lambda \in \Lambda$ where entries in the matrix correspond to the estimated generalization performance of λ on d; a technique often utilized in general algorithm selection and configuration. Such data can be collected from online ML databases like OpenML [109]. This matrix will most likely be very sparse, as not all combinations of datasets and pipelines are evaluated. The estimation of the performance of λ on a new dataset d_{new} corresponds to a recommender system in which datasets are users and ML pipelines are items. Recommender systems like collaborative filtering can be used as meta-learning AutoML tools to predict which pipelines to try out next [44, 120].

Additional information about the datasets such as size, number and type of features can help to improve the recommendation by calculating the similarity between datasets. Instead of a random or space filling initialization of the search, the best working configurations on similar datasets are used, which can speed up the optimization. Van Rijn et al. [107] have even used meta-learning to compute symbolic defaults that work well across a large number of datasets. Many meta-learning approaches are limited to examining and comparing configurations from the same search space. The inclusion of transformers however can allow for utilizing information across different search spaces for meta-learning in the context of HPO and AutoML [22]. Taking the notion of meta-learning even one step further, recent approaches to AutoML or NAS, that require a lot of computation on (sometimes synthetic) datasets beforehand, can produce models or architectures for a variety of datasets [75] or even do inference for new datasets out of the box and in a "one-shot manner" [57].

1.2.6 A Brief Note on AutoML in the Wild

Various AutoML tools have been developed and published over the last years - both open source and commercial. However, few success stories of AutoML in real-world applications and industry are known, even though AutoML often performs very well on the benchmark data frequently used in scientific publications and ML competitions. One possible reason for this could be that the application of AutoML is not necessarily made public by users. Nevertheless, the large amount of commercial solutions strongly indicates that a market for AutoML tools exists. AutoML has also been included in major ML platforms from companies such as Microsoft or Amazon.

In general, there is no single AutoML framework that is suitable for all use cases, since there exists a trade-off between the flexibility and stability of pipeline creation. Compared to frameworks developed by research groups, most frameworks developed in industry limit themselves to shorter pipelines as well as simpler search spaces and optimization strategies, c.f. H2O AutoML. This makes them quite robust and very desirable for users who are looking to harness the power of ML with limited ML knowledge, but not as flexible and powerful as some open source alternatives that offer more functionality. If the ML task at hand becomes more difficult, e.g., by a larger number of observations, higher dimensionality, or number of classes, some frameworks will experience crashes or produce unsatisfactory results. In particular, many AutoML frameworks have issues handling *high cardinality* categorical features, i.e., categorical features with an extensive amount of possible values. In Gijsberger et al. [48] H2O AutoML was found to be the most stable framework, whereas Zöller et al. [126] report crashes as well as memory and time constraint violations for all considered frameworks. The overall differences between frameworks can be marginal, and a more complex search space does not necessarily guarantee improved results. Both benchmarks [48, 126] are limited to specific hardware configurations. Gijsberger et al. [48] conduct their benchmark on machines with 32 GB memory and 8 vCPUs with time limits of 1 and 4 hours per run. Zöller et al. [126] conduct their benchmark on almost identical machines for 1.25 hours per run. A second iteration of the AutoML benchmark conducted in [48] has recently been released by Gijsbers et al. [47]. While this offers up additional insights into various time limits for AutoML, more benchmarking is required as it remains unclear how AutoML frameworks behave under different conditions, for example in large scale distributed systems.

All the referenced benchmarks give an overview over current open source frameworks and attempt a side-by-side comparison incorporating different criteria, which present a good resource for further reference. Finally, a recent trend towards *targeted AutoML solutions* offers a new trade-off between generalization and functionality by focusing on certain use cases that share common properties in the data or set-up. One example is an AutoML tool targeted at predictive maintenance of expensive machinery that has been introduced by German company Weidmüller [1]. The search space of such frameworks is designed to include operations that domain experts deem useful, e.g., certain feature extraction methods for sensor data or proven model classes. Meta-learning in such a focused domain is also considerably easier, as task similarity can be assumed to be much higher [62].

1.3 Selected Topics in AutoML

1.3.1 AutoML for Time Series Data

While AutoML approaches are often benchmarked on and targeted towards tabular data, NAS publications tend to highlight a lot of applications on image data. When moving to different data types like text data, time series data or even multi-modal data, AutoML methods are not as proven and may require substantial modifications or additions.

Time series data is especially prevalent in real-world industry applications, which is the reason we choose it to highlight why AutoML methods that work on tabular data can not simply be transferred to other types of data, unique scenarios, and ML tasks other than classification or regression. In terms of ML solutions for the typical time series tasks - time series forecasting and time series classification - we can divide approaches into three groups [6]: Feature engineering in combination with "traditional" ML approaches, deep learning, and solutions specific to time series. A more detailed taxonomy can be found in [84]. The sheer number of possible tasks on time series data and the abundance of available methods to solve them make creating an appropriate search space a complicated task.

Feature engineering can be also especially daunting for time series tasks, yet including it in the AutoML framework can be very helpful for users [26]. Several popular packages exist to automate feature construction, among the most used are tsfuse [32] and tsfresh [26].

Despite the popularity of deep learning methods and their successful application to both time series forecasting [81] and classification [65] it might be better to opt out of expensive deep learning solutions for several applications in industrial settings such as anomaly detection or predictive maintenance [7]. However, plenty of applications are suitable for deep learning; in terms of commercial deep learning tools for time series tasks, e.g., Amazon Web Services has launched GluonTS, which operates with probabilistic methods and deep learning [5]. Solutions specific to time series tasks are quite diverse. One such example, that has garnered some interest in the past years is the Matrix Profile, which was first introduced by Yeh et al. in [122]. The Matrix Profile is a metric – essentially achieved by folding a univariate time series with itself – that indicates for each step in the time series how far the most similar pattern is located in the series. This information can be leveraged for a variety of time series and pattern mining tasks like motif detection or anomaly detection among others. In terms of AutoML, many previously discussed techniques can be carried over to certain time series tasks. The need to tackle various tasks based on time series data has led to several tools adapting to support those tasks recently [6]. Specifically, the AutoML framework *Driverless AI* from H2O offers specialized time series tools [3] and *AutoGluon* from Amazon supports a number of time series applications [2]. NAS for time series tasks, in particular forecasting, has recently shown some success: Wu et al. in [118] discuss the importance of hyperparameter tuning in deep learning approaches for time series and Deng et al. in [34] have developed a NAS framework that can outperform "traditional" time series approaches on a variety of forecasting tasks from different application domains.

1.3.2 Unsupervised AutoML

While most research focuses on AutoML for supervised learning, there is a growing body of work that applies AutoML to unsupervised learning. Unsupervised learning is a type of ML where algorithms identify patterns and structures within data without the use of labeled examples. The methods for unsupervised ML are also often highly dependent on critical hyperparameters or even appropriate pre-processing. The biggest challenge in AutoML for unsupervised tasks is that performance evaluation is not as clear as in the supervised case, and defining a single, appropriate metric to evaluate performance of a model is not straightforward [8]. We aim to provide a brief introduction to AutoML for unsupervised ML and its challenges and highlight the important tasks of anomaly detection and clustering. Anomaly detection deals with detecting data points that differ significantly from the training distribution, and can be useful for tasks like identifying rare events, e.g., fraud detection or intrusion detection [8]. Unsupervised anomaly detection methods often use meta-learning to learn from other detection models and datasets to build their own detection model [125], or use human-in-the-loop strategies to train the model [124] (see also Section 1.3.4). Clustering is a task that especially struggles with appropriate evaluations; often, internal clustering validation criteria are used to ascertain the quality of a model. Many such metrics exist, it is often hard for users to interpret these metrics and using these criteria for model selection and tuning is a challenge. Meta-learning approaches have been shown to utilize these metrics in a meaningful way and have demonstrated good results in AutoML for clustering [33, 67] (c.f. Section 1.2.5). Some AutoML frameworks that rely heavily on meta-learning have been proposed as well [29, 95]. Another AutoML component, that has proven useful for clustering is ensembling [45, 49, 50] (c.f. Section 1.2.3). For a comprehensive overview of AutoML methods for unsupervised settings we refer to Bahri et al. [8].

1.3.3 AutoML Beyond a Single Objective

Existing AutoML approaches mainly find an optimal pipeline with respect to one evaluation criterion. In many real-world applications, however, practitioners are usually interested in other objectives as well. There can be multiple metrics that measure the performance of an ML model, and there may also be interest in including secondary objectives such as model complexity, energy efficiency, robustness, interpretability or sparseness, which makes the AutoML problem a multi-objective optimization problem [71, 87].

A simple solution to this is scalarization, i.e., turning the multi-objective problem into a single objective problem, for example by instead optimizing a weighted sum of the objectives [71]. This however not only requires extensive a priori knowledge about the optimization problem and an idea about possible trade-offs, but one solution cannot comprehensively describe a multi-objective optimization problem with conflicting objectives. Therefore, it may be sensible to use multi-objective optimization methods, which usually try to approximate the pareto-front, i.e., to find solutions where no single objective can be further improved without trade-offs in other objectives and thus present a decision maker (DM) with a set of non-dominated options from which a suitable solution can be chosen.

Several concepts and optimization methods presented in Section 1.2 can be adapted for the multi-objective case [71],[3] but multi-objective AutoML comes with its own set of challenges. In practice it can be hard to judge if an objective identified in the business understanding phase should be formulated as an objective or a constraint for the ML problem (e.g., should a model be as energy efficient as possible or should energy consumption lie under a certain threshold). The AutoML problem can also be alternatively formulated as a quality diversity optimization problem, i.e., identifying optimal configurations for each of a set of constrained regions at the same

---

[3] For an introduction to these methods for HPO and AutoML we refer to [71, 87].

time. This formulation has also shown a lot of promise and presents yet another way to tackle this issue [97].

It is often unclear which optimization method to choose for such a multi-objective AutoML problem as benchmarking and evaluation is notoriously hard [71, 42]. Another weakness of multi-objective methods is that approximating the pareto-front becomes increasingly difficult or even impossible as the number of objectives increases [77, 64], especially when their interactions are unknown. Yet another approach, that respects the complexity of real world problems, is to involve a DM or domain expert to interact with the optimization algorithm. This addresses some problems, such as objectives that may be hard to quantify or even unknown *a priori*. We will explore some of these interactive methods in the next section.

1.3.4 Human-In-The-Loop AutoML

The goal of automation in ML is not to exclude the – often extremely valuable – human influence altogether, but more so to reduce tedious, manual tasks as much as possible. How to include domain experts, DMs and ML experts to draw from their knowledge with minimal manual effort for them is very much a vision for AutoML and NAS [59]. How to integrate key human stakeholders in the ML process and specifically into AutoML during the model phase is a key topic in current AutoML research. Often, trade-offs cannot be specified a priori or it is hard to quantify certain objectives; there may also be some hidden objectives that are not formulated at all. Moreover, even if objectives can be specified, it may not be necessary to explore the entire Pareto-front as is often done in multi-objective optimization, but rather the search for pipelines should be focused around regions preferred by a DM. In such situations, including the preferences of the DM in the optimization process is most often beneficial [78]. Instead of specifying DM preferences in advance, preferences are best included interactively. This essentially puts humans back into the AutoML loop, but instead of configuring ML pipelines, they adjust the search process based on their preferences and expertise.

There are many ways by which to include DM preferences into the search process. Hakanen et al. [55] show intermediate solutions to the DM and adjust the search by asking for preferred ranges for the objective functions. Gibson et al. [46] allow the DM to set aspirational targets in the objective space that the DM would like to explore. Wang et al. [113] built an interactive visualization tool that allows the DM to modify search space, budget and model selection in an end-to-end workflow.

Overall, human-in-the-loop AutoML is in some sense contrary to the spirit of AutoML, and it is challenging to find the right amount of balance between automation and human intervention. Human-in-the-loop methods also rely on high quality feedback from the DM [78] and the preferences of DMs are always biased to some extent and can even change over time. This can be especially challenging with approaches like preferential Bayesian optimization, when a DM may pick one configuration over another and make a different decision under different circumstances.

1.4 Neural Architecture Search

As discussed previously, NAS is a sub-field of AutoML solely focusing on the architectural design of deep neural networks [127]. Deep learning – among other reasons – has celebrated great success, because it offers end-to-end solutions for ML, no longer requiring practitioners to build pipelines and carefully select pre-processing operations. However, the choice of architecture as well as hyperparameters like learning rate can have great influence on the overall performance of a deep learning model. One constraint in terms of applicability of NAS is available data for training and validation. Deep learning itself requires an abundance of data [98] and NAS even more so.[4] Shorten and Khoshgoftaar [98] allude to the lack of sufficient data in some domains, suggesting that the same may be true for several industrial use cases.

1.4.1 A Brief Overview of the Current State of NAS

The majority of NAS focuses on convolutional neural networks for computer vision, but approaches for audio [111], video [94], text [101], time series [34], and tabular [73] data exist. Various architectural designs, such as skip-connections [56], inception modules [104], and more generally multi-branch networks [37] are employed frequently. These design choices determine the search space just as the set of preprocessing operations and model families compose the search space for AutoML. Commonly, NAS works with cells, i.e., basic building blocks similar to inception modules that are stacked to make up the network [128], but search spaces are sometimes also hierarchical or chain-structured. The final composition of these cells can be predefined or searched with meta-architecture optimization to choose the number of cells and the configuration of their connections. This already creates an additional hierarchical structure in the optimization procedure by itself [82].

While design choices and their hyperparameters are optimized jointly in AutoML, hyperparameters of deep neural networks like regularization, learning rate and schedules are often not considered in NAS. Joint NAS and HPO exists [123], but further increases the computational complexity of NAS.

The optimization given the search space is very similar to AutoML and while early NAS approaches were centered around reinforcement learning, [127] almost all AutoML optimization techniques introduced so far have been successfully applied to NAS, including random search [80], Bayesian optimization [68, 114], gradient-based methods [83], multi-fidelity approaches [74], and evolutionary algorithms [63] among others. Some adaption in the optimizers is required, though. For Bayesian optimization, e.g., a suitable distance function or kernel is required to measure similarity of architectures. For evolutionary algorithms, mutation and crossing operations for architectures need to be defined.

---

[4] This is somewhat emphasized by Cui et al. in [30], who consider CIFAR10 a small dataset in the context of NAS.

A big issue in NAS is its enormous computational cost. NAS can require thousands of GPU hours to find state-of-the-art solutions, even for rather simple problems. With the trend to larger and larger models, e.g., transformers [110], NAS has become prohibitively expensive [102], which make multi-fidelity methods an important algorithmic component for utilized optimization procedures. In addition, many approaches to reduce the computational cost further exist, e.g., by sharing weights between proposed architectures [92] or by starting from small architectures and learning how to scale them [105]. Another issue with current NAS methods and research is that it is often hard to trust published results due to a lack of shared experimental protocols and missing ablation studies [119]. New results oftentimes only report little performance gain, that could very well be only due to certain "tricks" in the evaluation protocol [119].

1.4.2 Hardware-aware NAS

In practice, neural networks are deployed on different hardware platforms, including GPUs, CPUs, mobile phones and other edge devices and are therefore subject to hardware constraints. Energy efficiency and model size are of special concern when deploying deep learning models on edge devices. Hardware performance is also crucial for many real world applications, for example low latency for autonomous driving [10]. This has inspired the subfield of hardware-aware NAS, which involves taking into account hardware metrics such as latency, memory footprint, etc. in the NAS search process [10]. For this, the standard single-objective NAS problem can, e.g., be altered to a constrained optimization problem or a multi-objective optimization problem that includes appropriate hardware metrics.

Hardware metrics are either hardware-agnostic such as the number of model parameters or hardware-dependent such as latency. During the NAS search, hardware-dependent metrics are usually drawn from look-up tables or estimated using a prediction model [25]. Some hardware-aware NAS methods also jointly optimize hardware design and architecture [41], for example by including the buffer size of FPGA chips in their search [20]. One additional challenge for NAS is the fact that efficient architectures for one specific hardware are often not necessarily efficient on different hardware [14]. To solve this, approaches that guarantee optimal performance on a variety of different platforms have been proposed [27].

1.5 Conclusion and Outlook

While AutoML can be useful in many applications and could help increasing the efficiency of data scientists, its practical use arises not without its challenges. One common approach is to apply AutoML tools as a quick and automated baseline to see if, for a given dataset, any learning progress can be made [120]. However, in some benchmarks of open source AutoML systems, it was shown that for many publicly available datasets the improvement over a simple random forest might be very small [48] – especially on tasks, that are easy to solve. Furthermore, existing AutoML solutions are not necessarily usable out-of-the-box for several applications. For one, most AutoML tools optimize a single performance measure, but for many real-world problems multiple competing performance measures exist. While we discussed the current research in this area in Section 1.3.3, it is evident that a lot of work still needs to be done before these methods can be applied productively on a wider scale, as existing solutions lack in maturity compared to single-objective AutoML tools.

A related area of future work is human-in-the-loop AutoML (as discussed in Section 1.3.4); the utopian vision for (Automated) ML should not exclude human experts, but rather integrate them into ML processes as efficiently as possible. Along the same lines, when applying AutoML to real world problems, the flexibility of AutoML solutions is restricted in the type of problems to which they can be applied. Many AutoML tools can only solve fully supervised regression and classification tasks and other data types and ML tasks have only recently been explored. Similarly, AutoML for unsupervised or semi-supervised learning still holds a lot of potential for future work.

For many data science projects, the main difficulty is not the modeling itself but to properly map the underlying business question to a data science problem, acquire the required (labeled) data as well as the general readiness and capability for ML [12, 28]. Bringing automation into other stages of the ML workflow seems daunting as, e.g., business understanding or deployment seem hard to automate, but at the same time offer great potential for further research. Recent trends to combine AutoML with Large Language Models have sparked a lot of ideas in this direction, as these models could provide a good interface to help facilitate automation across the ML workflow.

Another open challenge for AutoML is budget selection. It is often unclear to users how much budget to allow for the AutoML process for finding a suitable pipeline. Stopping the search prematurely could result in a suboptimal pipeline, whereas prolonging the search excessively may waste resources or, in the worst-case scenario, lead to overfitting [85]. Some frameworks like *H2O AutoML* include rudimentary early stopping mechanisms which help alleviate this problem, but they have shown to not be optimal [85]. Finally it should be noted that AutoML does not aim to and for the foreseeable future will not be able to replace ML experts and researchers: AutoML is not intended to discover new methods or model types. While publications have played around this idea by showing that an AutoML system based on a search space only including basic operations can indeed "discover" deep learning architectures or methods like back-propagation [96], this is not really feasible for discovery of new methods at the moment – and furthermore is extremely expensive [96]. A follow-up publication showed that along the same lines, it is possible to discover new algorithmic components through this type of procedure like finding an improved version of the optimization algorithm *Adam*[5] [21].

References

1. Automated machine learning (industrial automl). `https://www.weidmueller.de/de/produkte/automatisierung_software/automated_machine_learning/index.jsp`. Accessed: 2023-12-11.
2. Time series forecasting - autoglion 1.0.0 documentation. `https://auto.gluon.ai/stable/tutorials/timeseries/index.html`. Accessed: 2023-12-11.
3. Time series in driverless ai. `https://docs.h2o.ai/driverless-ai/latest-stable/docs/userguide/time-series.html`. Accessed: 2023-12-11.
4. A. Abraham and L. Jain. *Evolutionary multiobjective optimization*. Springer, 2005.
5. A. Alexandrov, K. Benidis, M. Bohlke-Schneider, V. Flunkert, J. Gasthaus, T. Januschowski, D. C. Maddix, S. Rangapuram, D. Salinas, J. Schulz, L. Stella, A. C. Türkmen, and Y. Wang. Gluonts: Probabilistic time series models in python, 2019.
6. A. Alsharef, K. Aggarwal, M. Kumar, and A. Mishra. Review of ml and automl solutions to forecast time-series data. *Archives of Computational Methods in Engineering*, 29(7):5297–5311, 2022.
7. S. D. Anton, L. Ahrens, D. Fraunholz, and H. D. Schotten. Time is of the essence: Machine learning-based intrusion detection in industrial time series data. In *2018 IEEE International Conference on Data Mining Workshops (ICDMW)*, pages 1–6. IEEE, 2018.
8. M. Bahri, F. Salutari, A. Putina, and M. Sozio. Automl: state of the art with a focus on anomaly detection, challenges, and research directions. *International Journal of Data Science and Analytics*, 14(2):113–126, 2022.
9. E. Bart and S. Ullman. Cross-generalization: Learning novel classes from a single example by feature replacement. In *2005 IEEE Computer Society Conference on Computer Vision and Pattern Recognition (CVPR'05)*, volume 1, pages 672–679. IEEE, 2005.
10. H. Benmeziane, K. E. Maghraoui, H. Ouarnoughi, S. Niar, M. Wistuba, and N. Wang. A comprehensive survey on hardware-aware neural architecture search. *arXiv preprint arXiv:2101.09336*, 2021.
11. J. Bergstra and Y. Bengio. Random search for hyper-parameter optimization. *Journal of Machine Learning Research*, 13(Feb):281–305, 2012.
12. L. Bernardi, T. Mavridis, and P. Estevez. 150 successful machine learning models: 6 lessons learned at booking. com. In *Proceedings of the 25th ACM SIGKDD International Conference on Knowledge Discovery & Data Mining*, pages 1743–1751, 2019.
13. M. Binder, J. Moosbauer, J. Thomas, and B. Bischl. Multi-objective hyperparameter tuning and feature selection using filter ensembles. In *Proceedings of the 2020 Genetic and Evolutionary Computation Conference*, pages 471–479, 2020.
14. H. Cai, L. Zhu, and S. Han. Proxylessnas: Direct neural architecture search on target task and hardware. In *7th International Conference on Learning Representations, ICLR*, 2019.
15. R. Caruana. Multitask learning. *Machine learning*, 28(1):41–75, 1997.
16. R. Caruana, A. Niculescu-Mizil, G. Crew, and A. Ksikes. Ensemble selection from libraries of models. In *Proceedings of the twenty-first international conference on Machine learning*, page 18. ACM, 2004.
17. G. C. Cawley and N. L. Talbot. On over-fitting in model selection and subsequent selection bias in performance evaluation. *Journal of Machine Learning Research*, 11(Jul):2079–2107, 2010.

---

[5] *Adam* has been established as a standard optimizer when it comes to stochastic gradient descent in the context of training deep learning models.

18. B. Celik, P. Singh, and J. Vanschoren. Online automl: An adaptive automl framework for online learning. *Machine Learning*, 112(6):1897–1921, 2023.
19. B. Chen, H. Wu, W. Mo, I. Chattopadhyay, and H. Lipson. Autostacker: A compositional evolutionary learning system. In *Proceedings of the genetic and evolutionary computation conference*, pages 402–409, 2018.
20. W. Chen, Y. Wang, S. Yang, C. Liu, and L. Zhang. You only search once: A fast automation framework for single-stage dnn/accelerator co-design. In *2020 Design, Automation & Test in Europe Conference & Exhibition (DATE)*, pages 1283–1286. IEEE, 2020.
21. X. Chen, C. Liang, D. Huang, E. Real, K. Wang, Y. Liu, H. Pham, X. Dong, T. Luong, C.-J. Hsieh, et al. Symbolic discovery of optimization algorithms. *arXiv preprint arXiv:2302.06675*, 2023.
22. Y. Chen, X. Song, C. Lee, Z. Wang, R. Zhang, D. Dohan, K. Kawakami, G. Kochanski, A. Doucet, M. A. Ranzato, S. Perel, and N. de Freitas. Towards learning universal hyperparameter optimizers with transformers. In *Advances in Neural Information Processing Systems*, volume 35, pages 32053–32068. Curran Associates, Inc., 2022.
23. W. Cheng, G. Kasneci, T. Graepel, D. Stern, and R. Herbrich. Automated feature generation from structured knowledge. In *Proceedings of the 20th ACM international conference on Information and knowledge management*, pages 1395–1404. ACM, 2011.
24. C. Chevalier and D. Ginsbourger. Fast computation of the multi-points expected improvement with applications in batch selection. In *Learning and Intelligent Optimization - 7th International Conference*, volume 7997 of *Lecture Notes in Computer Science*, pages 59–69. Springer, 2013.
25. K. T. Chitty-Venkata and A. K. Somani. Neural architecture search survey: A hardware perspective. *ACM Computing Surveys*, 55(4):1–36, 2022.
26. M. Christ, A. W. Kempa-Liehr, and M. Feindt. Distributed and parallel time series feature extraction for industrial big data applications. *arXiv preprint arXiv:1610.07717*, 2016.
27. G. Chu, O. Arikan, G. Bender, W. Wang, A. Brighton, P.-J. Kindermans, H. Liu, B. Akin, S. Gupta, and A. Howard. Discovering multi-hardware mobile models via architecture search. In *Proceedings of the IEEE/CVF Conference on Computer Vision and Pattern Recognition*, pages 3022–3031, 2021.
28. M. Chui, J. Manyika, M. Miremadi, N. Henke, R. Chung, P. Nel, and S. Malhotra. Notes from the ai frontier: Insights from hundreds of use cases. *McKinsey Global Institute*, 2018.
29. N. Cohen-Shapira and L. Rokach. Automatic selection of clustering algorithms using supervised graph embedding. *Information Sciences*, 577:824–851, 2021.
30. J. Cui, P. Chen, R. Li, S. Liu, X. Shen, and J. Jia. Fast and practical neural architecture search. In *The IEEE International Conference on Computer Vision (ICCV)*, October 2019.
31. T. De Biel, L. De Raedt, H. H. Hoos, and P. S. Wu. Automating data science. *Report from Dagstuhl Seminar 18401*, 2019.
32. A. De Brabandere, P. Robberechts, T. Op De Beeck, and J. Davis. Automating feature construction for multi-view time series data. In *Proceedings of the ECML/PKDD Workshop on Automating Data Science*, pages 16–20, 2019.
33. M. C. De Souto, R. B. Prudencio, R. G. Soares, D. S. De Araujo, I. G. Costa, T. B. Ludermir, and A. Schliep. Ranking and selecting clustering algorithms using a meta-learning approach. In *2008 IEEE International Joint Conference on Neural Networks (IEEE World Congress on Computational Intelligence)*, pages 3729–3735. IEEE, 2008.
34. D. Deng, F. Karl, F. Hutter, B. Bischl, and M. Lindauer. Efficient automated deep learning for time series forecasting. In *Machine Learning and Knowledge Discovery in Databases: European Conference, ECML PKDD*, pages 664–680. Springer, 2023.
35. F. Eibe, M. A. Hall, and I. H. Witten. The weka workbench. online appendix for data mining: practical machine learning tools and techniques. In *Morgan Kaufmann*. Morgan Kaufmann Publishers, 2016.
36. a. Elshawi, M. Maher, and S. Sakr. Automated machine learning: State-of-the-art and open challenges. *arXiv preprint arXiv:1906.02287*, 2019.
37. T. Elsken, J. H. Metzen, and F. Hutter. Neural architecture search: A survey. *The Journal of Machine Learning Research*, 20(1):1997–2017, 2019.

38. N. Erickson, J. Mueller, A. Shirkov, H. Zhang, P. Larroy, M. Li, and A. Smola. Autogluon-tabular: Robust and accurate automl for structured data. *arXiv preprint arXiv:2003.06505*, 2020.
39. D. Eriksson, M. Pearce, J. Gardner, R. D. Turner, and M. Poloczek. Scalable global optimization via local bayesian optimization. *Advances in neural information processing systems*, 32, 2019.
40. S. Falkner, A. Klein, and F. Hutter. Bohb: Robust and efficient hyperparameter optimization at scale. *arXiv preprint arXiv:1807.01774*, 2018.
41. H. Fan, M. Ferianc, Z. Que, H. Li, S. Liu, X. Niu, and W. Luk. Algorithm and hardware co-design for reconfigurable cnn accelerator. In *2022 27th Asia and South Pacific Design Automation Conference (ASP-DAC)*, pages 250–255. IEEE, 2022.
42. M. Feurer, K. Eggensperger, E. Bergman, F. Pfisterer, B. Bischl, and F. Hutter. Mind the gap: Measuring generalization performance across multiple objectives. In *Advances in Intelligent Data Analysis XXI: 21st International Symposium on Intelligent Data Analysis*, pages 130–142. Springer, 2023.
43. M. Feurer, A. Klein, K. Eggensperger, J. Springenberg, M. Blum, and F. Hutter. Efficient and robust automated machine learning. *Advances in neural information processing systems*, pages 2962–2970, 2015.
44. N. Fusi, R. Sheth, and M. Elibol. Probabilistic matrix factorization for automated machine learning. In *Advances in Neural Information Processing Systems*, pages 3348–3357, 2018.
45. J. Ghosh and A. Acharya. Cluster ensembles. *Wiley interdisciplinary reviews: Data mining and knowledge discovery*, 1(4):305–315, 2011.
46. F. J. Gibson, R. M. Everson, and J. E. Fieldsend. Guiding surrogate-assisted multi-objective optimisation with decision maker preferences. In *Proceedings of the Genetic and Evolutionary Computation Conference*, pages 786–795, 2022.
47. P. Gijsbers, M. L. Bueno, S. Coors, E. LeDell, S. Poirier, J. Thomas, B. Bischl, and J. Vanschoren. Amlb: an automl benchmark. *arXiv preprint arXiv:2207.12560*, 2022.
48. P. Gijsbers, E. LeDell, J. Thomas, S. Poirier, B. Bischl, and J. Vanschoren. An open source automl benchmark. In *ICML AutoML workshop*, 2019.
49. A. Gionis, H. Mannila, and P. Tsaparas. Clustering aggregation. *ACM transactions on knowledge discovery from data*, 1(1), 2007.
50. K. Golalipour, E. Akbari, S. S. Hamidi, M. Lee, and R. Enayatifar. From clustering to clustering ensemble selection: A review. *Engineering Applications of Artificial Intelligence*, 104:104388, 2021.
51. R. B. Gramacy and H. K. Lee. Optimization under unknown constraints. In *Bayesian Statistics 9*, pages 229–256. Oxford University Press, 2011.
52. I. Guyon, A. Saffari, G. Dror, and G. Cawley. Model selection: Beyond the bayesian/frequentist divide. *Journal of Machine Learning Research*, 11(Jan):61–87, 2010.
53. H2O.ai. *H2O: Scalable Machine Learning Platform*, 2023. version 3.42.0.3.
54. H. Ha, S. Rana, S. Gupta, T. Nguyen, S. Venkatesh, et al. Bayesian optimization with unknown search space. *Advances in Neural Information Processing Systems*, 32, 2019.
55. J. Hakanen and J. D. Knowles. On using decision maker preferences with parego. In *Evolutionary Multi-Criterion Optimization: 9th International Conference, EMO 2017*, pages 282–297. Springer, 2017.
56. K. He, X. Zhang, S. Ren, and J. Sun. Deep residual learning for image recognition. In *Proceedings of the IEEE conference on computer vision and pattern recognition*, pages 770–778, 2016.
57. N. Hollmann, S. Müller, K. Eggensperger, and F. Hutter. Tabpfn: A transformer that solves small tabular classification problems in a second. In *The Eleventh International Conference on Learning Representations, ICLR 2023*, 2023.
58. D. Horn, T. Wagner, D. Biermann, C. Weihs, and B. Bischl. Model-based multi-objective optimization: taxonomy, multi-point proposal, toolbox and benchmark. In *International Conference on Evolutionary Multi-Criterion Optimization*, pages 64–78. Springer, 2015.
59. F. Hutter. Automl | deep learning 2.0: Extending the power of deep learning to the meta-level. https://www.automl.org/deep-learning-2-0-extending-the-power-of-deep-learning-to-the-meta-level/, March 2022. Accessed: 2023-05-12.

60. F. Hutter, H. H. Hoos, and K. Leyton-Brown. Sequential model-based optimization for general algorithm configuration. In *Learning and Intelligent Optimization*, pages 507–523. Springer, 2011.
61. F. Hutter, H. H. Hoos, and K. Leyton-Brown. Parallel algorithm configuration. In *Learning and Intelligent Optimization: 6th International Conference*, pages 55–70. Springer, 2012.
62. F. Hutter, L. Kotthoff, and J. Vanschoren. Automated machine learning-methods, systems, challenges, 2019.
63. W. Irwin-Harris, Y. Sun, B. Xue, and M. Zhang. A graph-based encoding for evolutionary convolutional neural network architecture design. In *2019 IEEE Congress on Evolutionary Computation (CEC)*, pages 546–553. IEEE, 2019.
64. H. Ishibuchi, N. Tsukamoto, and Y. Nojima. Evolutionary many-objective optimization: A short review. In *2008 IEEE congress on evolutionary computation (IEEE world congress on computational intelligence)*, pages 2419–2426. IEEE, 2008.
65. H. Ismail Fawaz, G. Forestier, J. Weber, L. Idoumghar, and P.-A. Muller. Deep learning for time series classification: a review. *Data mining and knowledge discovery*, 33(4):917–963, 2019.
66. K. Jamieson and A. Talwalkar. Non-stochastic best arm identification and hyperparameter optimization. In *Artificial Intelligence and Statistics*, pages 240–248, 2016.
67. Y. Jiang and N. Verma. Meta-learning to cluster. *arXiv preprint arXiv:1910.14134*, 2019.
68. H. Jin, Q. Song, and X. Hu. Auto-keras: An efficient neural architecture search system. In *Proceedings of the 25th ACM SIGKDD International Conference on Knowledge Discovery & Data Mining*, pages 1946–1956. ACM, 2019.
69. D. R. Jones, M. Schonlau, and W. J. Welch. Efficient global optimization of expensive black-box functions. *Journal of Global optimization*, 13(4):455–492, 1998.
70. S. R. Kaminwar, J. Goschenhofer, J. Thomas, I. Thon, and B. Bischl. Structured verification of machine learning models in industrial settings. *Big Data*, 2021.
71. F. Karl, T. Pielok, J. Moosbauer, F. Pfisterer, S. Coors, M. Binder, L. Schneider, J. Thomas, J. Richter, M. Lang, et al. Multi-objective hyperparameter optimization–an overview. *arXiv preprint arXiv:2206.07438*, 2022.
72. U. Khurana, D. Turaga, H. Samulowitz, and S. Parthasrathy. Cognito: Automated feature engineering for supervised learning. In *2016 IEEE 16th International Conference on Data Mining Workshops (ICDMW)*, pages 1304–1307. IEEE, 2016.
73. A. Klein and F. Hutter. Tabular benchmarks for joint architecture and hyperparameter optimization. *arXiv preprint arXiv:1905.04970*, 2019.
74. A. Klein, L. C. Tiao, T. Lienart, C. Archambeau, and M. Seeger. Model-based asynchronous hyperparameter and neural architecture search. *arXiv preprint arXiv:2003.10865*, 2020.
75. H. Lee, E. Hyung, and S. J. Hwang. Rapid neural architecture search by learning to generate graphs from datasets. In *9th International Conference on Learning Representations, ICLR 2021*, 2021.
76. B. Letham, B. Karrer, G. Ottoni, E. Bakshy, et al. Constrained bayesian optimization with noisy experiments. *Bayesian Analysis*, 14(2):495–519, 2019.
77. K. Li, K. Deb, and X. Yao. R-metric: Evaluating the performance of preference-based evolutionary multiobjective optimization using reference points. *IEEE Transactions on Evolutionary Computation*, 22(6):821–835, 2017.
78. K. Li, M. Liao, K. Deb, G. Min, and X. Yao. Does preference always help? a holistic study on preference-based evolutionary multiobjective optimization using reference points. *IEEE Transactions on Evolutionary Computation*, 24(6):1078–1096, 2020.
79. L. Li, K. Jamieson, G. DeSalvo, A. Rostamizadeh, and A. Talwalkar. Hyperband: A novel bandit-based approach to hyperparameter optimization. *arXiv preprint arXiv:1603.06560*, 2016.
80. L. Li and A. Talwalkar. Random search and reproducibility for neural architecture search. *arXiv preprint arXiv:1902.07638*, 2019.
81. B. Lim and S. Zohren. Time-series forecasting with deep learning: a survey. *Philosophical Transactions of the Royal Society A*, 379(2194), 2021.
82. H. Liu, K. Simonyan, O. Vinyals, C. Fernando, and K. Kavukcuoglu. Hierarchical representations for efficient architecture search. *arXiv preprint arXiv:1711.00436*, 2017.

83. H. Liu, K. Simonyan, and Y. Yang. Darts: Differentiable architecture search. *arXiv preprint arXiv:1806.09055*, 2018.

84. M. Löning, A. Bagnall, S. Ganesh, V. Kazakov, J. Lines, and F. J. Király. sktime: A unified interface for machine learning with time series. *arXiv preprint arXiv:1909.07872*, 2019.

85. A. Makarova, H. Shen, V. Perrone, A. Klein, J. B. Faddoul, A. Krause, M. Seeger, and C. Archambeau. Overfitting in bayesian optimization: an empirical study and early-stopping solution. In *2nd Workshop on Neural Architecture Search (NAS 2021 collocated with the 9th ICLR 2021)*, 2021.

86. J. Močkus. On bayesian methods for seeking the extremum. In *Optimization Techniques IFIP Technical Conference*, pages 400–404. Springer, 1975.

87. A. Morales-Hernández, I. Van Nieuwenhuyse, and S. Rojas Gonzalez. A survey on multi-objective hyperparameter optimization algorithms for machine learning. *Artificial Intelligence Review*, pages 1–51, 2022.

88. A. Nayebi, A. Munteanu, and M. Poloczek. A framework for Bayesian optimization in embedded subspaces. In *Proceedings of the 36th International Conference on Machine Learning*, volume 97 of *Proceedings of Machine Learning Research*, pages 4752–4761. PMLR, 2019.

89. R. S. Olson, N. Bartley, R. J. Urbanowicz, and J. H. Moore. Evaluation of a tree-based pipeline optimization tool for automating data science. *Proceedings of the Genetic and Evolutionary Computation Conference 2016*, pages 485–492, 2016.

90. F. Pedregosa, G. Varoquaux, A. Gramfort, V. Michel, B. Thirion, O. Grisel, M. Blondel, P. Prettenhofer, R. Weiss, V. Dubourg, J. Vanderplas, A. Passos, D. Cournapeau, M. Brucher, M. Perrot, and E. Duchesnay. Scikit-learn: Machine learning in Python. *Journal of Machine Learning Research*, 12:2825–2830, 2011.

91. V. Perrone, H. Shen, M. W. Seeger, C. Archambeau, and R. Jenatton. Learning search spaces for bayesian optimization: Another view of hyperparameter transfer learning. *Advances in Neural Information Processing Systems*, 32, 2019.

92. H. Pham, M. Y. Guan, B. Zoph, Q. V. Le, and J. Dean. Efficient neural architecture search via parameter sharing. *arXiv preprint arXiv:1802.03268*, 2018.

93. V. Picheny, D. Ginsbourger, Y. Richet, and G. Caplin. Quantile-based optimization of noisy computer experiments with tunable precision. *Technometrics*, 55(1):2–13, 2013.

94. A. Piergiovanni, A. Angelova, A. Toshev, and M. S. Ryoo. Evolving space-time neural architectures for videos. In *Proceedings of the IEEE International Conference on Computer Vision*, pages 1793–1802, 2019.

95. Y. Poulakis, C. Doulkeridis, and D. Kyriazis. Autoclust: A framework for automated clustering based on cluster validity indices. In *2020 IEEE International Conference on Data Mining (ICDM)*, pages 1220–1225. IEEE, 2020.

96. E. Real, C. Liang, D. So, and Q. Le. Automl-zero: Evolving machine learning algorithms from scratch. In *International conference on machine learning*, pages 8007–8019. PMLR, 2020.

97. L. Schneider, F. Pfisterer, P. Kent, J. Branke, B. Bischl, and J. Thomas. Tackling neural architecture search with quality diversity optimization. In *International Conference on Automated Machine Learning*, pages 9–1. PMLR, 2022.

98. C. Shorten and T. Khoshgoftaar. A survey on image data augmentation for deep learning. *Journal of Big Data*, 6, 12 2019.

99. J. Snoek, H. Larochelle, and R. P. Adams. Practical bayesian optimization of machine learning algorithms. *Advances in neural information processing systems*, 25, 2012.

100. J. Snoek, O. Rippel, K. Swersky, R. Kiros, N. Satish, N. Sundaram, M. Patwary, M. Prabhat, and R. Adams. Scalable bayesian optimization using deep neural networks. In *International conference on machine learning*, pages 2171–2180. PMLR, 2015.

101. D. So, Q. Le, and C. Liang. The evolved transformer. In *International conference on machine learning*, pages 5877–5886. PMLR, 2019.

102. E. Strubell, A. Ganesh, and A. McCallum. Energy and policy considerations for deep learning in NLP. In *Proceedings of the 57th Conference of the Association for Computational Linguistics, ACL 2019, Florence, Italy, July 28- August 2, 2019, Volume 1: Long Papers*, pages 3645–3650. Association for Computational Linguistics, 2019.

103. S. Studer, T. Bui, C. Drescher, A. Hanuschkin, L. Winkler, S. Peters, and K.-R. Müller. Towards crisp-ml(q): A machine learning process model with quality assurance methodology. *Machine Learning and Knowledge Extraction*, 3(2):392–413, 2021.

104. C. Szegedy, W. Liu, Y. Jia, P. Sermanet, S. Reed, D. Anguelov, D. Erhan, V. Vanhoucke, and A. Rabinovich. Going deeper with convolutions. In *Proceedings of the IEEE conference on computer vision and pattern recognition*, pages 1–9, 2015.

105. M. Tan and Q. Le. Efficientnet: Rethinking model scaling for convolutional neural networks. In *International conference on machine learning*, pages 6105–6114. PMLR, 2019.

106. M. J. Van der Laan, E. C. Polley, and A. E. Hubbard. Super learner. *Statistical applications in genetics and molecular biology*, 6(1), 2007.

107. J. N. van Rijn, F. Pfisterer, J. Thomas, A. Muller, B. Bischl, and J. Vanschoren. Meta learning for defaults: Symbolic defaults. In *Neural Information Processing Workshop on Meta-Learning*, 2018.

108. J. Vanschoren. Meta-learning: A survey. *arXiv preprint arXiv:1810.03548*, 2018.

109. J. Vanschoren, J. N. Van Rijn, B. Bischl, and L. Torgo. Openml: networked science in machine learning. *ACM SIGKDD Explorations Newsletter*, 15(2):49–60, 2014.

110. A. Vaswani, N. Shazeer, N. Parmar, J. Uszkoreit, L. Jones, A. N. Gomez, Ł. Kaiser, and I. Polosukhin. Attention is all you need. In *Advances in neural information processing systems*, pages 5998–6008, 2017.

111. T. Véniat, O. Schwander, and L. Denoyer. Stochastic adaptive neural architecture search for keyword spotting. In *ICASSP 2019-2019 IEEE International Conference on Acoustics, Speech and Signal Processing (ICASSP)*, pages 2842–2846. IEEE, 2019.

112. O. Vinyals, C. Blundell, T. Lillicrap, D. Wierstra, et al. Matching networks for one shot learning. In *Advances in neural information processing systems*, pages 3630–3638, 2016.

113. Q. Wang, Y. Ming, Z. Jin, Q. Shen, D. Liu, M. J. Smith, K. Veeramachaneni, and H. Qu. Atmseer: Increasing transparency and controllability in automated machine learning. In *Proceedings of the 2019 CHI conference on human factors in computing systems*, pages 1–12, 2019.

114. C. White, W. Neiswanger, and Y. Savani. Bananas: Bayesian optimization with neural architectures for neural architecture search. In *Proceedings of the AAAI Conference on Artificial Intelligence*, volume 35, pages 10293–10301, 2021.

115. C. White, M. Safari, R. Sukthanker, B. Ru, T. Elsken, A. Zela, D. Dey, and F. Hutter. Neural architecture search: Insights from 1000 papers. *arXiv preprint arXiv:2301.08727*, 2023.

116. M. Wistuba, N. Schilling, and L. Schmidt-Thieme. Automatic frankensteining: Creating complex ensembles autonomously. In *Proceedings of the 2017 SIAM International Conference on Data Mining*, pages 741–749. SIAM, 2017.

117. D. H. Wolpert. Stacked generalization. *Neural networks*, 5(2):241–259, 1992.

118. X. Wu, B. Shi, Y. Dong, C. Huang, L. Faust, and N. V. Chawla. Restful: Resolution-aware forecasting of behavioral time series data. In *Proceedings of the 27th ACM International Conference on Information and Knowledge Management*, CIKM '18, page 1073–1082, 2018.

119. A. Yang, P. M. Esperança, and F. M. Carlucci. Nas evaluation is frustratingly hard. *arXiv preprint arXiv:1912.12522*, 2019.

120. C. Yang, Y. Akimoto, D. W. Kim, and M. Udell. Oboe: Collaborative filtering for automl model selection. In *Proceedings of the 25th ACM SIGKDD international conference on knowledge discovery & data mining*, pages 1173–1183, 2019.

121. Q. Yao, M. Wang, H. J. Escalante, I. Guyon, Y. Hu, Y. Li, W. Tu, Q. Yang, and Y. Yu. Taking human out of learning applications: A survey on automated machine learning. *CoRR*, abs/1810.13306, 2018.

122. C.-C. M. Yeh, Y. Zhu, L. Ulanova, N. Begum, Y. Ding, H. A. Dau, D. F. Silva, A. Mueen, and E. Keogh. Matrix profile i: all pairs similarity joins for time series: a unifying view that includes motifs, discords and shapelets. In *2016 IEEE 16th international conference on data mining (ICDM)*, pages 1317–1322. Ieee, 2016.

123. A. Zela, A. Klein, S. Falkner, and F. Hutter. Towards automated deep learning: Efficient joint neural architecture and hyperparameter search. *arXiv preprint arXiv:1807.06906*, 2018.

124. D. Zha, K.-H. Lai, M. Wan, and X. Hu. Meta-aad: Active anomaly detection with deep reinforcement learning. In *2020 IEEE International Conference on Data Mining (ICDM)*, pages 771–780. IEEE, 2020.
125. Y. Zhao, R. A. Rossi, and L. Akoglu. Automating outlier detection via meta-learning. *arXiv preprint arXiv:2009.10606*, 2020.
126. M.-A. Zöller and M. F. Huber. Benchmark and survey of automated machine learning frameworks. *Journal of artificial intelligence research*, 70:409–472, 2021.
127. B. Zoph and Q. V. Le. Neural architecture search with reinforcement learning. In *5th International Conference on Learning Representations, ICLR 2017, Toulon, France, April 24- 26, 2017, Conference Track Proceedings*, 2017.
128. B. Zoph, V. Vasudevan, J. Shlens, and Q. V. Le. Learning transferable architectures for scalable image recognition. In *Proceedings of the IEEE conference on computer vision and pattern recognition*, pages 8697–8710, 2018.

Chapter 2
Sequence-based Learning

Christoffer Loeffler[1,2], Felix Ott[2], Jonathan Ott[2], Maximilian P. Oppelt[2], Tobias Feigl[2]

Abstract Learning from time series data is an essential component in the AI landscape given the ubiquitous time-dependent data in real-world applications. To motivate the necessity of learning from time series data, we first introduce different applications, data sources, and properties. These can be as diverse as irregular and (non-)continuous time series data as well as streaming and spatio-temporal data. To introduce the mechanics of learning from time series data, we elaborate on the most renowned convolutional, recurrent and transformer architectures for learning from time series. Then, we discuss essential characteristics of learning with time series. Therefore, we explain deep metric learning, which learns feature representations that capture the similarity between time series data. We further describe time series similarity learning to extract representations that allow comparison between sequences of spatio-temporal data. In addition, we discuss the interpretability of learning methods on time series data that target safety, non-discrimination, and fairness.

Key words: Time series data, classification, regression, forecasting, spatio-temporal networks, deep metric learning, time series similarity, model interpretability

2.1 Introduction

As humans, we live in a world where (any kind of) events and interactions occur in a specific causal order. For instance, when we talk to other people, words form sentences, and sentences form stories. When we walk, run or drive, our position in space changes depending on our current position and the change over time.

[1]Escuela de Ingeniería Informática, Pontificia Universidad Católica de Valparaíso, Valparaíso, Chile
[2]Fraunhofer Institute for Integrated Circuits IIS, Fraunhofer IIS, Nuremberg, Germany

Corresponding author: Tobias Feigl
e-mail: `tobias.feigl@iis.fraunhofer.de`

C. Mutschler et al. (eds.), *Unlocking Artificial Intelligence*,
https://doi.org/10.1007/978-3-031-64832-8_2

During physical exercise, our heart and breathing rates increase depending on the level of exertion. Even during mental stress or cognitive load, the conductivity of our skin, the variability of our heart rate, or the movement patterns of our pupils may change over time [44]. If all these sequential changes and past, current, and future observations were recorded, their data points would not be independent and identically distributed anymore, a fundamental statistical assumption that simplifies the application of machine learning. Thus, suitable learning paradigms would require specialized methods for handling such time series.

Hence, time series analysis describes patterns that occur over time. Research on this topic has gained momentum in recent years as sensor data streams become ubiquitously available. Many applications in health [44, 42], industry [39], education [49, 48] or entertainment [24, 40] process high-dimensional time series and raise new and interesting challenges for classical methods. Here, the tasks of classification, regression, forecasting or anomaly detection of sequential events (time series) are particularly noteworthy, as they contribute important information with real-world impact [59]. Historically, methods like autoregressive integrated moving average (ARIMA) [9] or the vector autoregressive (VAR) model [9] analyze lower-dimensional time series based on the principle of automatic regression. However, complex multivariate time series (MTS) with correlated random variables that are typical for signal processing or economics applications, pose an impossible challenge for such classic statistical approaches. By contrast, modern methods based on deep learning (DL) have demonstrated remarkable performance, especially with complex, high-dimensional data. They model natural stochastic noise to reduce the information complexity and can directly predict tasks. Alternatively, they can be combined with, e.g., Bayesian methods, to form more robust hybrid models [23]. The family of DL models for time series data include recurrent neural networks (RNNs) [21] or temporal convolutional networks (TCNs) [3] and are considered revolutionary [29]. Such models can, e.g., automatically learn time dependencies or handle temporal structures such as trends and seasonality directly from the data. Furthermore, they can extract patterns over very long periods of time and largely eliminate the need for manual feature engineering, data scaling and stationary data. Thereby, they provide more abstract, high-level features for downstream tasks. This holds even if the MTS is irregular or complex [49].

Throughout this chapter, we use the following notation from [46] to describe an MTS $\mathbf{X} = \{\mathbf{x}_1, \ldots, \mathbf{x}_i\} \in \mathbb{R}^{i \times j}$, which is an ordered sequence of $j \in \mathbb{N}$ channels. Each channel $\mathbf{x}_i = (x_{t,1}, \ldots, x_{t,j})$, where $t \in \{1, \ldots, i\}$, represents a series of observations with $i \in \mathbb{N}$ being the length of the time series. The training and test sets for MTS are a subset of the array $\mathcal{X} = \{\mathbf{X}_1, \ldots, \mathbf{X}_n\} \in \mathbb{R}^{n \times i \times j}$, where n represents the number of time series. We extend our annotation system for the classification task. We pre-define a label set Θ that contains H classes and that is associated with a label l. The objective of this task is to determine the unknown class label $l \in \Theta$ for a given MTS. The training labels $\mathcal{L} = \{\mathbf{l}_1, \ldots, \mathbf{l}_n\} \in \Theta^{n \times H}$ correspond to the training MTS set $\mathcal{X}$. Supervised sequential problems are formulated using input-label pairs $\{\mathbf{x}_i, \mathbf{y}_i\}$.

The rest of this chapter is structured as follows. Section 2.2 introduces time series processing using examples and explains the typical processing pipeline. Section 2.3 then introduces the principles of the popular convolutional and recurrent architectures. Section 2.4 provides a concise review of the two important perspectives of metric learning and model interpretability for time series analysis. Finally, Section 2.5 concludes and gives an outlook.

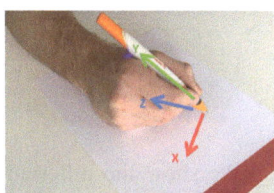

(a) Driver state detection using sequential data from multiple modalities: (1) electrophysiological signals, (2) facial expression from videos and (3) eye tracker data. (from [44], licensed under CC-BY 4.0)

(b) Hand-held tool measuring inertial motion, magnetic field and sound for quality assurances.

(c) Sensor augmented pencil measures 7 degrees of freedom: accelerometer, gyroscope, and pressure.

Fig. 2.1: Applications for time series analysis are diverse. (a) The measurement setup depicted here is used to analyse cognitive load in a road driving secenario. (b) Quality assurance can be supported by sensor-augmented smart tools. (c) Handwriting recognition can be implemented via a sensor-augmented pen.

2.2 Time Series Processing

This section focuses on the signal-processing perspective of time series analysis, which finds applications in a large variety of real-world use-cases [42, 39, 49, 48, 24, 40]. We consider the type of mining of time series data, that analyzes the shape of data sequentially sampled over time [22]. In contrast to sample-wise prediction, time series are captured over time and thus the samples are generally not identically and independently distributed [11]. Models may exploit that values in sequences may be correlated [11].

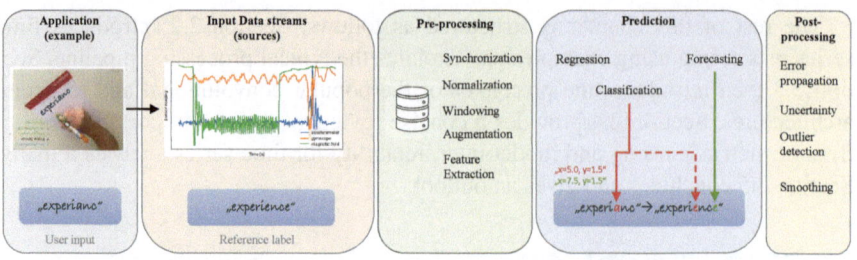

Fig. 2.2: Time series processing pipeline. Data flow from the left to the right.

We show the different characteristics of time series processing via exemplary tasks of domain-specific signal-processing in Figure 2.1. First, Figure 2.1a shows a recording of biosignals, video and eye tracking data used for detecting the cognitive load of drivers [44]. Here, biosignals may show only short-term signal deformations and only a combination of sensors allows the detection of long-term variations [45]. In contrast, the application in Figure 2.1b classifies the activity of hand-held tools from shorter patterns in the sensors' signal amplitude. The final application example in Figure 2.1c shows a combination of both characteristics. It predicts letters, words and whole sentences from spatio-temporal sensor data recorded using a smart pencil. Still, all applications have their ML pipeline in common. Figure 2.2 presents the peculiarities, that distinguish the pipeline for time series from other forms of data. The seasonality of time series data can be identified in the orange curve, outliers in the blue curve, and autocorrelation between the orange and blue curves of the data streams.

2.2.1 Time Series Data Streams

For time series processing, a complete description of the data format is fundamental. The input data may be a continuous stream of several multi-dimensional sensors, for example, a writer's force of pressure and a pen's accelerations, or a hand-held tool's vibrations. Time series data may be multi-variate, recorded at different sampling rates and levels of noise, and have gaps or other issues. Common sensors measure inertial motion, sound or radio frequencies.

Figure 2.3 explains the data type's characteristics using a simplified sample of an inertial measurement unit (IMU) of the activity recognition application from Fig 2.1b. These properties differentiate them from other types of data. Similarly to the augmented pen, the hand-held tool's IMU collects signals from accelerometers, gyroscopes, and a magnetic field sensor. The sensors sample data along three orthogonal axes. We only show one axis of each.

A recorded sensor data stream may be sampled at different sampling rates, see Figure 2.3 for a higher and lower rate. This has practical implications on the ob-

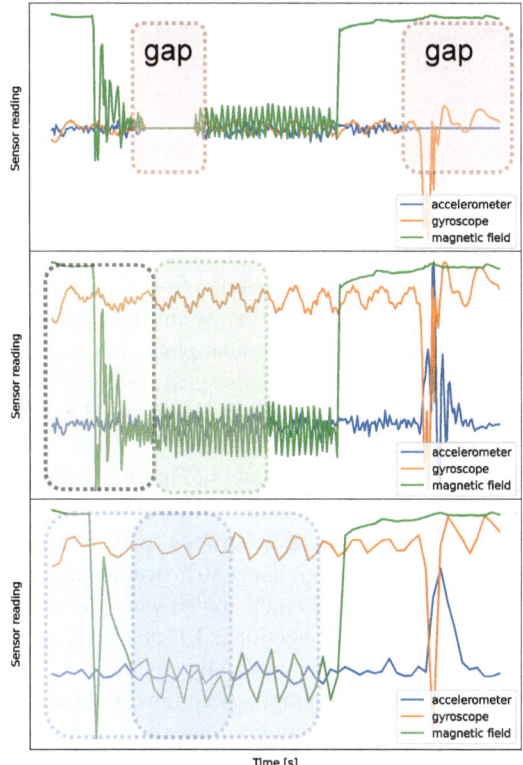

Fig. 2.3: Inertial measurements showing accelerometer and gyroscope data. Windows are shown as dashed rectangles. A *window* represents an application-specific (pre-)selected set of data samples over time, which are processed by a learning algorithm at each inference step. Typically, sliding windows are applied with a specific window length, which overlap at a specific offset and scroll over the data sequence. The top figure presents two possible types of data loss: (1) complete loss and (2) partial loss of sensor data. The middle figure presents two non-overlapping windows and the bottom figure presents two overlapping windows.

servability of phenomena, comparable to the difference in low and a high-resolution photography in computer vision tasks. However, it also entails similar costs and complexity. A higher sampling rate allows handwriting recognition algorithms to detect smaller movements of the writer and thus may lead to enhanced recognition, but also requires more processing power.

A second dimension that influences the density of data per unit of time is the dimensionality of the sensor streams. It can feature more dimensions and is analogous to the difference of monochrome and color photography. We simplify the visualization of the recorded data and show only three of the nine recorded dimensions.

However, for some applications it can be very valuable to add additional views to enable predictions for its use-case, such as the force sensor to detect the contact between a digital pen and the writing surface [49].

2.2.2 Pre-Processing

Next, the pre-processing step for time series data of multiple sensors can require a synchronization of the data that may be streamed from separate sensors. In addition, raw sensory data can require normalization before any operations such as windowing, augmenting or even feature extraction can take place. Popular libraries for feature extraction and other time series specific tasks, such as *seglearn* [12], are able to extract generic features like the "mean" for downstream usage. However, Deep Learning methods may not need such engineering efforts.

We show the effects of generating windows in Figure 2.3. ML models may ingest a continuous stream of samples or subdivide the time series into windows of different sizes and with different fractions of overlap between each other. In contrast to sliding windows, which we show with approximately 50% overlap, tumbling windows have no overlap with each other. It depends on the model what approach is available, e.g., TCNs process fix-sized windows (see Section 2.3.1) and RNNs can process a sample at a time while memorizing the history internally (see Section 2.3.2). Furthermore, it may be crucial that window-based sampling captures all relevant features and does not cut off those that provide the relevant information to perform predictions.

A common problem of time series processing is missing data, e.g., the complete loss of all sensor axes or only a partial loss of one sensor. A complete loss of data may not only occur in case of system defects, but could happen if a sensor edge device with little on-board memory is disconnected from processing in a cloud for a prolonged amount of time or its perception could be obscured. Some partial loss may be recovered from orthogonal sensors such as the gyroscope, and forecasting models may use additional information such as trends or seasonality effects to recover from imputed or missing data.

2.2.3 Predictive Modelling

In the main processing step, the models may perform predictions for different tasks based on extracted features or the data itself. For example, a decision tree classifier may classify the state of the pen for a windowed sub-sequence of input data based on the mean value of the force sensor as "writing". Upon ingesting character-level features, a forecasting model could predict the next character, e.g., after seeing the sequence "experienc_", the model could predict the next character to be an "e". A regression model could process the acceleration and rotation data of an inertial measurement unit in order to predict the pen's motion over the paper and help reconstruct its trajectory.

2.2.4 Post-Processing

A complete ML pipeline requires post-processing of the results for robustness. Post-processing of time series data involves applying additional techniques or transformations to the data after initial analysis or modeling to improve its quality or interpretability. This contains smoothing, filtering, resampling, detrending, differencing, seasonal adjustment, denoising, scaling, and feature engineering. Some models, such as softmax-based classification, output their predictive uncertainty besides the predictions themselves. These likelihoods, together with prediction errors or the detection of outlier samples may be propagated to downstream processing logic or reported for review.

2.3 Methods

Time series analysis includes methods for performing inferences from time series data to predict statistical features and other abstract characteristics such as future developments. The assumption is that the current values of one or more dependent time series influence the current value of another time series [58]. As time series data have a natural temporal order, stochastic regression models assume that observations that are close in time are more closely related than the ones that are further apart. In addition, stochastic regression models assume that observations of a certain period depend on past ones and that future values can be predicted solely from past observations [58]. This section presents TCNs in Section 2.3.1 and RNNs in Section 2.3.2.

2.3.1 Temporal Convolutional Networks

Feed forward neural networks (FFNNs) are nonlinear mappings, where each input is mapped to an output node by matrix multiplication. Although these networks are theoretically capable of learning any function, they lack this in practical terms due to their computational inefficiency, as they require matrix multiplication for every input sample to each output sample. While these FFNNs can process time series data when the time dimension is flattened and a sequence represents the complete time series, this has significant disadvantages. FFNNs are not adaptable for sequences of arbitrary length [21]. Since the architecture only works with complete sequences, the entire relevant history must always be saved. Also, FFNNs are not computationally

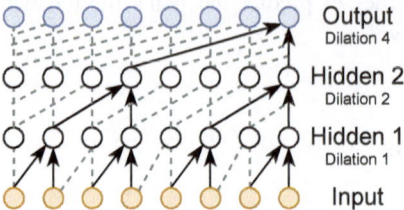

Fig. 2.4: The structure of a temporal convolutional network (TCN) consists of layers of dilated causal convolutions. Each output learns over a large input window of past time steps.

efficient when the sequence is a rolling window or when prediction is required at each time step, since the network then has to process the entire input sequence at each time step [21]. Since FFNNs have no inherent concept of time, time-delayed input sequences deliver completely different results [21]. Recall the example of cognitive load detection from Figure 2.1a. There it is essential to take the subjects' past features to predict their present cognitive load. However, these biological readings are highly individual and the sequences of psycho-physiological data are very long. Therefore, FFNNs become infeasible.

As a simple but effective network architecture, TCNs [3] are based on convolutional operations and learn using a hierarchical representation of the time steps over several levels. Figure 2.4 shows this temporal structure, that uses dilation to create a sparse receptive field for the output units, covering the whole fixed time sequence. With every layer, the receptive field of the output neurons grows. This way, TCNs can process very long sequences. Due to their simple structure, TCNs can be optimized in parallel. They are much more resource-efficient than RNNs and enable efficient training for very large amounts of data if the time series are short and have a fixed, predefined length. However, during inference, the feed-forward structure of the TCNs requires expensive calculations of the entire history of the time series instead of only one sample at a time. As the sequence length increases, the breadth and depth of the architecture increases, and the sequence length cannot be changed. This makes it difficult to perform the calculations on resource-constrained embedded hardware. In contrast, RNNs [21, 27] compress the history into a fixed-size representation that requires minimal computation at each time step.

2.3.2 Recurrent Neural Networks

RNNs were proposed by Elman et al. [21] in 1990 and differ from FFNNs as they employ a special type of neural layer called a recurrent layer that allows the network to maintain state between layers of the network. This recurrent layer is a hidden state vector and memory to store information about the past and current context. This hidden state is computed by the network at each time step and then fed back as input for the next time step. The network uses the hidden state to convert the relevant features from the input history into a more compact representation. This allows efficient computation of the output at each time step t and a representation limited only by the size of the hidden state vector. The assumption is that the model has a state and transition function that computes subsequent states from its predecessor and any model input. Therefore, RNNs are suitable for the processing of time series data [10].

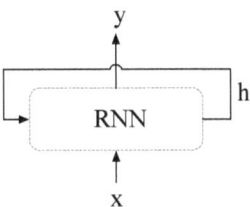

Fig. 2.5: Simplified representation of the basic structure of a RNN. RNNs are an adapted form of FFNNs that adopt a hidden state vector h, a kind of memory, to store information x about the history and the current context. h is computed at each time step and then passed as input for the next time step. The mesh uses h to compress relevant features from the input history. This allows efficient computation of the output y, limited only by the size of h.

In an RNN, all neurons have both incoming connections, which emanate from all neurons of the previous layer, and outgoing connections, which lead to all neurons of the subsequent layer. In contrast to FFNNs, RNNs also have recurring connections in the recurrent layer, which pass on information between the neurons of the same layer, including the same neuron of a layer. A recurrent layer with r neurons has a total of r^2 recurrent connections [10].

With RNNs the lifetime of the network entity can be divided into discrete time steps. At each time step, the model is supplied with the next input sample. The feed-forward connections in an RNN represent the flow of information from one neuron to the next, with the transmitted data representing the computed neuronal activation of the current time step. Figure 2.6 represents the information flow of recurrent connections, where the data shows the stored neural activation from the previous time step, i.e., influence from the left and from below. Thus, the activations of the neurons in an RNN represent the accumulating state of the network entity. The initial activations of the neurons in the recurrent layer are parameters of the model. When training RNNs, the optimal weights w are sought to approximate the network output y to the training target $\hat{y}$, i.e., to minimize the error E. With a fixed lifetime, such as t time steps, an RNN instance can be represented as a *unrolled* irregularly structured FFNN. [10]

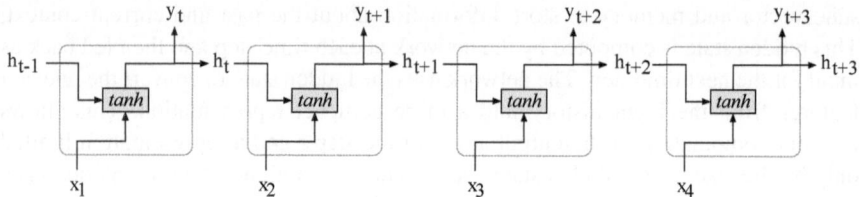

Fig. 2.6: Simplified representation of the flow of information through an unrolled RNN, resembles a FFNN. Forward pass: processing a series of inputs x_1 to x_4. A simplified version of the cells is shown for a better overview.

Vanishing and exploding gradients. To determine the gradient for a longer-term dependency, backpropagation through time (BPTT) is carried out several times. This optimization method "unfolds" the recurrent network's temporal steps into copies of itself and then propagates the gradients back along the network "through time". This may lead to exploding or vanishing gradients for large sequences [5]. Both of these problems prevent learning weight matrices for long-term dependencies as the weight updates are based on this gradient information. Various methods are proposed to overcome these problems [4, 27, 70].

The long-short-term memory (LSTM) proposed by Hochreiter and Schmidhuber [27] in 1997 uses a different cell design, developed to account for long-term dependencies and the problems of vanishing and exploding gradients. The LSTM cell uses structures called *gates* to manipulate and control the flow of the hidden state and is designed in such a way that useful information is effectively preserved in the memory cell over many time steps [10]. In contrast to the RNN, the LSTM cell structure enables a separation between this cell and the initial state. This separation and the fact that the cell state is not directly affected by a nonlinear activation function reduces the vanishing gradients.

2.3.3 Transformer

A drawback of training recurrent networks with BPTT is that its graph cannot be parallelized during training and inference. Due to this, RNNs suffer from a computational complexity that cannot be optimized by parallelizing gradient operations on modern GPUs. Recent techniques such as Transformers use the so-called attention mechanism and combine the strenghts of FFNNs, i.e., extraction of characteristic features, and of RNNs, i.e., memory of causal relationships over long periods of time. Transformers are also used in AI frameworks such as ChatGPT [50].

The transformer neural network introduced by Vaswani et al. [67] is an encoder-decoder architecture. It computes representations of time series solely by making use of attention layers. Hence, it significantly reduces the training time, compared to RNNs, by reducing the need for sequential computation. The ability to resolve relationships between two distant points depends, among others, on the path between the position in the output representation and the position in the input sequence [26].

The attention mechanism used in the transformer has comparatively short paths, resulting in an increased performance for particularly long sequences. However, this comes at the cost of computational complexity, which grows by $O(n^2)$ as the sequence length increases. Recent research focuses on optimizing the attention mechanism towards efficiency by either reducing the number of computations [68] or by optimizing memory access [16]. Furthermore, the architecture requires a large amount of training data due to its tendency to overfit [69]. As the transformer employs neither convolutional nor recurrent layers, it is not capable to detect the order of elements in a sequence. Thus, a positional encoding is added to the time series embedding (see Figure 2.7), which can be fixed or learned [69]. While the transformer architecture was originally introduced in the context of sequence transduction, it has been applied to various time series tasks such as forecasting [74] or classification [73].

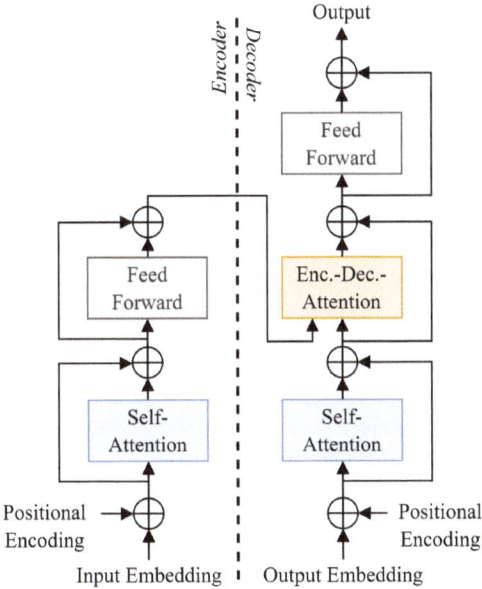

Fig. 2.7: Simplified overview of Transformer architecture. Depicted are a single encoder and a single decoder block. Several operations are neglected to show the overall concept of the architecture.

The type of attention blocks used in the transformer architecture is called multi-head attention. It computes multiple representations of the input data, projected via learned transformations, in parallel. This allows to attend to multiple positions as well as subspaces in the data. The transformer employs two variants of multi-head attention: (1) self-attention and (2) encoder-decoder-attention. (1) self attention is the first layer in the encoder and decoder blocks. It computes the representation of a sequence by modeling dependencies in the sequence itself. (2) the decoder uses encoder-decoder-attention to incorporate the context information provided by the encoder. Figure 2.7 depicts self-attention as a blue box and encoder-decoder-attention as an orange box. The final layers in the encoder and decoder blocks are position-wise feed-forward networks that apply a fully connected layer that shares its weights for all positions but is only applied to the embedding per position in the sequence. All layers are bypassed by a residual connection [67].

2.4 Perspectives

The practical use of DL models like the TCNs, RNNs or LSTMs for time series analysis may quickly run into difficulties, which we explore in this perspectives section. First, many techniques require a similarity (or distance) metric defined on the input data space, e.g., on trajectory data. Hence, Section 2.4.1 describes suitable functions for distance computation. Alternatively, deep metric learning (DML) learns these functions directly from data (see Section 2.4.1.1). The common issue of domain shift that appears between training and test data sources affects machine learning in general. Hence, we summarize the specific domain adaption (DA) techniques for time series applications in Section 2.4.2. Lastly, safety, non-discrimination and fairness, and explanations are important but still open questions [18]. Accordingly, Section 2.4.3 discusses interpretability methods and their limitations for black-box time series models.

2.4.1 Time Series Similarity

In many applications, the distance between two time series is a crucial measure to compare the discrepancy. See for example Figure 2.9 where the goal is to reconstruct a trajectory of the pen tip of the sensor-enhanced pen of Figure 2.2 by minimizing the distance between the ground truth and predicted trajectories. The distance (inverse of similarity) between two time series is measured as the cost of transforming one time series

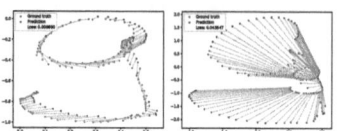

Fig. 2.9: Ground truth (red) and reconstructed (blue) trajectories [47].

into another using a distance measure or function. The existing similarity measures can be classified into two classes (see Figure 2.8): (1) Spatial similarity that focuses on finding time series with similar geometric shapes that ignore the temporal dimension, and (2) spatio-temporal similarity that takes into account both the spatial and the temporal dimensions of time series data [41, 33].

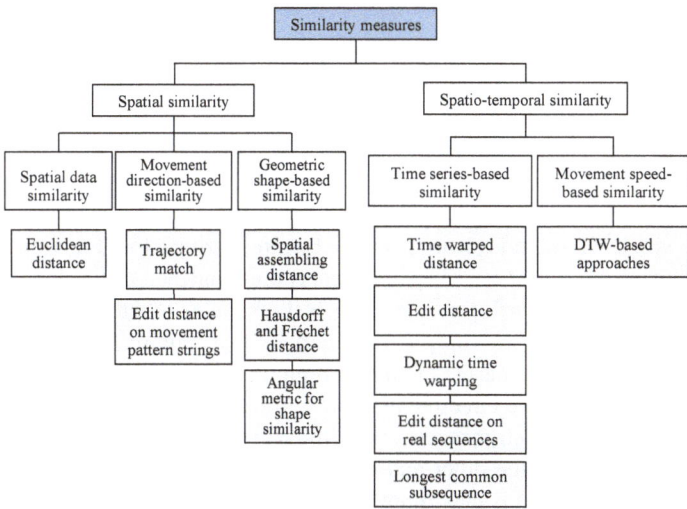

Fig. 2.8: Overview of spatial and spatio-temporal similarity measures [41].

It is well-established to use a sum-of-square-based objective function to measure the average difference between all samples of two time series. A challenge that arises is the possibility of a varying number of data points between two time series. Partial curve mapping (PCM) [72] utilizes a combination of sub time series length and area to measure the similarity between two time series. The performance of PCM is adversely affected by the presence of noise in the data, as the noise leads to an artificial increase in the arc length. The Area method [28] constructs quadrilaterals between two time series and calculates the area for each quadrilateral. It is necessary for the two time series to have the same number of points.

The discrete Fréchet (DF) [19] distance preserves the time series order of data along the series. The DF distance reflects the shortest possible path between two time series. If we consider a series $\mathbf{A}$ with a number of points and a series $\mathbf{B}$ with b number of points, the DF distance has a fixed quadratic run time of $O(ab)$. However, the Fréchet distance is sensitive to outliers. In contrast, dynamic time warping (DTW) [6] measures the similarity between to time series with varying time steps. It aligns the two sequences by warping their respective time axes to find the optimal alignment that minimizes the distance between corresponding points.

The Hausdorff distance (HD) [65] is a popular dissimilarity metric used to compare sets of points. HD is a max-min distance, which offers an advantage in that it accounts for the spatial position of each individual point. HD has nearly-linear complexity. For a comparison of trajectory popular distances for handwriting recognition, see [47], and for certain sports trajectory data, see [56].

2.4.1.1 Deep Metric Learning

To learn an optimal representation between data samples or models, DML techniques are required to compare distances of feature embeddings. DML is a sub-field of ML that aims to learn a function that maps inputs into a feature space, where distances between data points correspond to semantic similarities, e.g., the semantic similarity between two pieces of text or two scenes of football [35] measures how close their meanings are. DML methods learn a feature by minimizing a loss function that takes into account the distances between pairs of data points and their corresponding labels. The main challenge of DML for time series is the variable length of samples, which can make it challenging to apply DML methods that require fixed-length inputs. In the following, we summarize the most common DML functions. The Euclidean loss has been shown to be effective in many tasks. Another commonly used metric is cross-correlation, which measures the similarity between two signals by sliding one of the signals over the other and computing the dot product at each position. Recently well established is maximum mean discrepancy (MMD) [8] that measures the distance between two probability distributions. MMD is a non-parametric metric that is able to capture complex patterns in data distributions. MMD has been used in tasks such as Domain Adaption (DA) and image generation. Correlation alignment (CORAL) [63] and higher-order moment matching (HoMM) [14] align the higher-order moments of the features of two data points. The idea is to match the distributions of the features of two data points. The Sinkhorn [15] distance is based on optimal transport theory. Sinkhorn has the advantage of being able to handle discrete and continuous data, and has been applied to time series classification in [20, 46].

2.4.2 Transfer Learning & Domain Adaptation

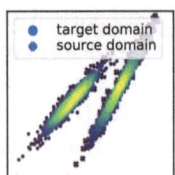 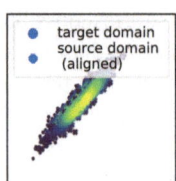 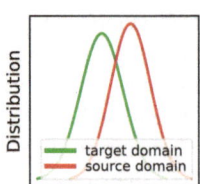 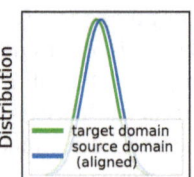

Fig. 2.10: Alignment of the source and target domain distributions, 3rd (before alignment) and 4th (after alignment) plot, to align their representations, 1st (before) and 2nd plot (after) (from [46], licensed under CC-BY 4.0).

For many applications, a domain shift appears between training and test data sources, for instance, between sensor data of right-handed and left-handed writers of the application presented in Figure 2.2 [46]. Domain adaption is a sub-category of transfer learning that adapts a predictive model trained on a source domain to a target domain where the distribution of the data may be different. The characteristics of the time series data may have changed between the source and target domains (training and application data distributions). Additionally, the distribution of the time series data may shift over time due to changes in the underlying system or changes in the measurement process. DA approaches use techniques that focus on aligning the distributions of the source and target domains and aim to learn a shared representation of the data that is invariant to the differences between the domains (see Figure 2.10). The common issue of shifting distributions between training data and real-world application is especially challenging for time series data, because it often has complex temporal dependencies that must be captured by the model.

The following methods to mitigate the domain shift between data sources often make use of DML metrics presented in Section 2.4.1.1. For example, the approach known as minimum discrepancy estimation for deep DA (MMDA) [51] employs the minimization of conditional entropy as a means of integrating MMD and CORAL alignment functions. Deep domain confusion (DDC) [66] is based on the minimization of the MMD distance between domains. The domain-adversarial NN (DANN) [25] utilizes a gradient reversal layer to encourage the emergence of features that are discriminative for the source domain but indiscriminative w.r.t. the shift between the source and target domains. Further methods are the ones proposed by [34, 37, 57, 71].

2.4.3 Model Interpretability

Deep Learning can outperform human experts in certain domains. Still, there are additional criteria besides raw performance [18]: safety, non-discrimination and fairness, and legal requirements such as the right to explain.[1] These criteria are hard to quantify and thus also difficult to optimize for. One way to explain the reasoning of a model and verify its predictions is to make it interpretable [18]. This is especially difficult for time series analysis due to its often unintuitive data compared to, e.g., image data.

The literature differentiates methods into three broad categories [13, 43]. "Pre-model" interpretability inspects the data, "in-model" refers to "intrinsically" interpretable models [52], and "post-model" methods interpret models after training. These post-model (or post hoc) methods [13, 43] are either "model-specific" or "model-agnostic", and may be global, local [18, 13, 43] or in between [43]. A global interpretation may comprehend the entire model at once, whereas a local method explains a single prediction from a small region of interest.

---

[1] General Data Protection Regulation, Regulation (EU) 2016/679 of the European Parliament and of the Council of April 27th 2016, recital 71 https://eur-lex.europa.eu/legal-content/ EN/TXT/HTML/?uri=CELEX:32016R0679&from=EN#d1e40-1-1, see also Chapter 5 for a deeper discussion on the EU AI act.

2.4.3.1 Interpretability for Time Series

Local interpretability methods are popular because widely applicable to many DL architectures such as TCNs or RNNs, even without intrusive changes, and directly visualize the relevance of input features for a model's prediction. We distinguish gradient-based methods, e.g., Gradient [60], Backpropagation [62], Integrated Gradients [64], SmoothGrad [61] or GradCAM [55], and perturbation-based methods (model agnostic) methods like LIME [52], that fits an interpretable surrogate to explain black box models, and Kernel SHAP [38], that uses concepts from Shapley values.

Extending these approaches to time series data builds on insights of Schlegel et al. [54], who propose that saliency methods may explain patterns that underlie the time series data itself, such as temporal correlations. Subsequent work specifies these ideas further. Kusters et al. [31] propose patterns for a conceptual explanation for model-agnostic interpretations, that manipulate input data using transformations of the data, similarly to TCAV [30] that use hypothesis tests of concepts. The time series specific patterns include offset or trend removal, a moving average, low or high pass filters, and additive noise, but experts may additionally design more advanced filters. Similarly, Abanda et al. [1] propose time-dependent perturbations using warp, scale, noise and slice operations for post hoc perturbation-based explanations to visualize the relevance of sub-sequences of the input data. For example, warping the duration of an activity in a dataset can have semantic meaning, e.g., it differentiates the classes of pulling a gun from its holster from pointing a finger in the GunPoint dataset [17]. Abanda et al. [1] show that the construction of semantic perturbations to create interpretations of models shows to be valid for simple time series datasets from the UCR repository [17]. Perturbing important regions changes the prediction of the models, whereas perturbing unimportant regions has a smaller effect.

2.4.3.2 Trusting Interpretations

Recent studies question the effectiveness of interpretations of time series models w.r.t. their reliability. Still, no evaluation of interpretations answers all the possible doubts [36, 64, 32, 7]. On one hand, qualitative metrics are more user-targeted, e.g., towards the general public. On the other hand, quantitative metrics provide deeper insights for experts [53]. Quantitative evaluation metrics forgo a human who performs a qualitative evaluation of an explanation and uses a proxy metric for evaluation [18]. The faithfulness metric is generally seen as a foundational evaluation concept, and it was adapted for time series data [2, 54], e.g., it may perturb temporally connected segments. Faithfulness changes inputs with high relevance w.r.t. saliency, and uses its influence on the model prediction, i.e., its accuracy, as a proxy for an explanation's correctness. A high score in faithfulness means that the saliency is representative of input features that correlate with the model's predictive accuracy. Overall, the choice of visual interpretation methods for time series and the trust in their explanations depends on many factors, e.g., the choice of models and the dataset, and there is no "silver bullet" interpretation method [32].

2.5 Conclusion and Outlook

Time series analysis reaches all aspects of our lives, such as our health, the way we work or learn, or how we spend time with our families. Our contribution focuses on a small set of applications that process time-dependent data, and describes essential properties of commonalities in the data and its differences compared to a generic machine learning processing pipeline. Additionally, we also described the three DL models Temporal Convolutional Network, Recurrent Neural Network, and Transformers that are foundational for many other use-cases as well, and have contrasting benefits and drawbacks. In the remainder, we addressed common difficulties in applied research. The topic of similarity functions informs many learning approaches, and Deep Metric Learning specializes this field further. Finally, we discussed explanation and interpretation methods for predictions of black-box time series models and whether these may be trusted, raising important questions for applied machine learning.

In this chapter, we have presented an overview of learning from time series data, covering a variety of data sources and properties such as streaming and spatio-temporal data. ML models are becoming increasingly adept at capturing complex temporal patterns, accommodating the intricacies of time series data such as seasonality, trends, and irregularities. Here prominent examples are foundation models such as ChatGPT for natural language processing. With the increasing importance of time-dependent data, the discussed methods offer a valuable foundation for further exploration and implementation in various domains. With the advancements of sensor data, IoT devices, and high-frequency trading, the demand for improved time series forecasting, anomaly detection, and pattern recognition solutions is on the rise. Moreover, the integration of DL architectures, i.e., RNNs, CNNs, and Transformers, enables the extraction of hierarchical features from sequential data, further enhancing the accuracy and scalability of time series analysis.

Acknowledgments

We would like to express our sincere gratitude to N. Beck, J. Deuschel, N. Lang-Richter, D. Dzibela, B. M. Eskofier, A. Foltyn, R. Gruber, D. Hartmann, S. Kram, N. Lang, C. Mutschler, A. Porada, N. L. Raichur, L. Reeb, A. Richter, G. Röder, J. Schemm, J. Seitz, B. Sonnleitner, M. Stahlke, U. Wetzker, N. Witt and H. G. Zimmermann, for their valuable insights and expertise in the field of learning from time series data.

References

1. A. Abanda, U. Mori, and J. Lozano. Ad-hoc explanation for time series classification. 252:109366.
2. D. Alvarez-Melis and T. S. Jaakkola. On the robustness of interpretability methods.
3. S. Bai, J. Z. Kolter, and V. Koltun. An empirical evaluation of generic convolutional and recurrent networks for sequence modeling. *arXiv:1803.01271*, 2018.
4. Y. Bengio, N. Boulanger-Lewandowski, and R. Pascanu. Advances in optimizing Recurrent networks. In *Proc. Intl. Conf. Acoustics, Speech and Signal Processing*, pages 8624–8628. Vancouver, Canada, 2013.
5. Y. Bengio, P. Y. Simard, and P. Frasconi. Learning long-term dependencies with gradient descent is difficult. *Trans. on Neural Networks*, 5(2):157–166, 1994.
6. D. J. Berndt and J. Clifford. Using Dynamic Time Warping to Find Patterns in Time Series. In *Intl. Conf. on Knowledge Discovery and Data Mining*, volume 359-370, July 1994.
7. A. Boggust, H. Suresh, H. Strobelt, J. V. Guttag, and A. Satyanarayan. Beyond Faithfulness: A Framework to Characterize and Compare Saliency Methods.
8. K. M. Borgwardt, A. Gretton, M. J. Rasch, H.-P. Kriegel, B. Schölkopf, and A. J. Smola. Integrating Structured Biological Data by Kernel Maximum Mean Discrepancy. In *Bioinformatics*, volume 22(14), pages e49–e57, July 2006.
9. G. E. Box, G. M. Jenkins, G. C. Reinsel, and G. M. Ljung. *Time series analysis: forecasting and control*. John Wiley & Sons, 2015.
10. N. Buduma and N. Locascio. *Fundamentals of deep learning: designing next-generation machine intelligence algorithms*. O'Reilly Media, 1st edition, 2017.
11. D. M. Burns and C. M. Whyne. Seglearn: A python package for learning sequences and time series. (March), 2018. arXiv: 1803.08118.
12. D. M. Burns and C. M. Whyne. Seglearn: A python package for learning sequences and time series. *The Journal of Machine Learning Research*, 19(1):3238–3244, 2018.
13. D. V. Carvalho, E. M. Pereira, and J. S. Cardoso. Machine learning interpretability: A survey on methods and metrics. 8(8):1–34.
14. C. Chen, Z. Fu, Z. Chen, S. Jin, Z. Cheng, X. Jin, and X. sheng Hua. HoMM: Higher-Order Moment Matching for Unsupervised Domain Adaptation. In *Proc. of the AAAI Conf. on Artificial Intelligence (AAAI)*, volume 34(4), pages 3422–3429, Apr. 2020.
15. N. Courty, R. Flamary, D. Tuia, and A. Rakotomamonjy. Optimal Transport for Domain Adaptation. In *IEEE Trans. on Pattern Analysis and Machine Intelligence (TPAMI)*, volume 39(9), pages 1853–1865, Oct. 2016.
16. T. Dao, D. Fu, S. Ermon, A. Rudra, and C. Ré. Flashattention: Fast and memory-efficient exact attention with io-awareness. *Advances in Neural Information Processing Systems*, 35:16344–16359, 2022.
17. H. A. Dau, A. Bagnall, K. Kamgar, C.-C. M. Yeh, Y. Zhu, S. Gharghabi, C. A. Ratanamahatana, and E. Keogh. The UCR Time Series Archive.
18. F. Doshi-Velez and B. Kim. Towards A Rigorous Science of Interpretable Machine Learning.
19. T. Eiter and H. Mannila. Computing Discrete Frechet Distance. May 1994.
20. E. Eldele, M. Ragab, Z. Chen, M. Wu, C. K. Kwoh, and X. Li. Label-efficient Time Series Representation Learning: A Review. In *arXiv preprint arXiv:2302.06433*, Feb. 2023.
21. J. L. Elman. Finding structure in time. *Cognitive Science*, 14(2):179–211, 1990.
22. P. Esling and C. Agon. Time-series data mining. *ACM Computing Surveys*, 45(1):1–34, Nov 2012.
23. T. Feigl, S. Kram, P. Woller, R. H. Siddiqui, M. Philippsen, and C. Mutschler. Rnn-aided human velocity estimation from a single imu. *Sensors*, 20(13), 2020.
24. T. Feigl, D. Roth, S. Gradl, M. Wirth, M. E. Latoschik, B. M. Eskofier, M. Philippsen, and C. Mutschler. Sick moves! motion parameters as indicators of simulator sickness. *IEEE Transactions on Visualization and Computer Graphics*, 25(11):3146–3157, Nov 2019.
25. Y. Ganing, E. Ustinova, H. Ajakan, P. Germain, H. Larochelle, F. Laviolette, M. March, and V. Lempitsky. Domain-Adversarial Training of Neural Networks. In *Journal of Machine Learning Research (JMLR)*, volume 17(59), pages 1–35, 2016.

26. S. Hochreiter, Y. Bengio, P. Frasconi, J. Schmidhuber, et al. Gradient flow in recurrent nets: the difficulty of learning long-term dependencies, 2001.
27. S. Hochreiter and J. Schmidhuber. Long short-term memory. *Neural Computation*, 9(8):1735–1780, 1997.
28. C. F. Jekel, G. Venter, M. P. Venter, N. Stander, and R. T. Haftka. Similarity Measures for Identifying Material Parameters from Hysteresis Loops using Inverse Analysis. In *Intl. Journal of Material Forming*, May 2018.
29. A. Karpathy and L. Fei-Fei. Deep visual-semantic alignments for generating image descriptions. In *Proc. IEEE Conf. on Computer Vision and Pattern Recognition*, pages 3128–3137, 2015.
30. B. Kim, M. Wattenberg, J. Gilmer, C. Cai, J. Wexler, F. Viegas, and R. Sayres. Interpretability beyond feature attribution: Quantitative Testing with Concept Activation Vectors (TCAV). In *35th International Conference on Machine Learning, ICML 2018*, volume 6, pages 4186–4195.
31. F. Kusters, P. Schichtel, S. Ahmed, and A. Dengel. Conceptual Explanations of Neural Network Prediction for Time Series. In *2020 International Joint Conference on Neural Networks (IJCNN)*, pages 1–6. IEEE.
32. X.-H. Li, Y. Shi, H. Li, W. Bai, C. C. Cao, and L. Chen. *An Experimental Study of Quantitative Evaluations on Saliency Methods*, volume 1. Association for Computing Machinery.
33. H. Liu and M. Schneider. Similarity Measurement of Moving Object Trajectories. In *Intl. Workshop on GeoStreaming*, pages 19–22, Nov. 2012.
34. Q. Liu and H. Xe. Adversarial Spectral Kernel Matching for Unsupervised Time Series Domain Adaptation. In *Proc. of the Intl. Joint Conf. on Artificial Intelligence (IJCAI)*, pages 2744–2750, 2021.
35. C. Löffler, K. Fallah, S. Fenu, D. Zanca, B. Eskofier, C. J. Rozell, and C. Mutschler. Active learning of ordinal embeddings: A user study on football data. *Transactions on Machine Learning Research*, 2023.
36. C. Löffler, W.-C. Lai, B. Eskofier, D. Zanca, L. Schmidt, and C. Mutschler. Don't get me wrong: How to apply deep visual interpretations to time series. *arXiv preprint arXiv:2203.07861*, 2022.
37. M. Long, Z. Cao, J. Wang, and M. I. Jordan. Conditional Adversarial Domain Adaptation. In *Advances of Neural Information Processing Systems (NIPS)*, 2018.
38. S. M. Lundberg and S.-I. Lee. A Unified Approach to Interpreting Model Predictions. page 10.
39. C. Löffler, C. Nickel, C. Sobel, D. Dzibela, J. Braat, B. Gruhler, P. Woller, N. Witt, and C. Mutschler. *Automated Quality Assurance for Hand-Held Tools via Embedded Classification and AutoML*, volume 2, page 532–535. 2021.
40. C. Löffler, L. Reeb, D. Dzibela, R. Marzilger, N. Witt, B. M. Eskofier, and C. Mutschler. Deep siamese metric learning: A highly scalable approach to searching unordered sets of trajectories. *ACM Transactions on Intelligent Systems and Technology*, 13(1):1–23, Feb 2022.
41. N. Magdy, M. A. Sakr, T. Mostafa, and K. El-Bahnasy. Review on Trajectory Similarity Measures. In *Intl. Conf. on Intelligent Computing and Information Systems (ICICIS)*, 2015.
42. S. Meyer, T. Windisch, A. Perl, D. Dzibela, R. Marzilger, N. Witt, J. Benzler, G. Kirchner, T. Feigl, and C. Mutschler. Contact tracing with the exposure notification framework in the german corona-warn-app. In *2021 International Conference on Indoor Positioning and Indoor Navigation (IPIN)*, page 1–8, Lloret de Mar, Spain, Nov 2021. IEEE.
43. C. Molnar. *Interpretable Machine Learning: A Guide for Making Black Box Models Explainable*. 2 edition.
44. M. P. Oppelt, A. Foltyn, J. Deuschel, N. R. Lang, N. Holzer, B. M. Eskofier, and S. H. Yang. ADABase: A Multimodal Dataset for Cognitive Load Estimation. 23(1):340.
45. M. P. Oppelt, M. Riehl, F. P. Kemeth, and J. Steffan. Combining Scatter Transform and Deep Neural Networks for Multilabel Electrocardiogram Signal Classification. In *2020 Computing in Cardiology*, page 4. IEEE.

46. F. Ott, D. Rügamer, L. Heublein, B. Bischl, and C. Mutschler. Domain Adaptation for Time-Series Classification to Mitigate Covariate Shift. In *ACM Intl. Conf. on Multimedia (ACMMM)*, pages 5934–5943, Lisboa, Portugal, Oct. 2022.

47. F. Ott, D. Rügamer, L. Heublein, B. Bischl, and C. Mutschler. Joint Classification and Trajectory Regression of Online Handwriting using a Multi-Task Learning Approach. In *Winter Conf. for Applications on Computer Vision (WACV)*, pages 266–276, Waikoloa, HI, Jan. 2022.

48. F. Ott, D. Rügamer, L. Heublein, T. Hamann, J. Barth, B. Bischl, and C. Mutschler. Benchmarking Online Sequence-to-Sequence and Character-based Handwriting Recognition from IMU-Enhanced Pens. In *Intl. Journal on Document Analysis and Recognition (IJDAR)*, volume 25(12), page 385–414, Sept. 2022.

49. F. Ott, M. Wehbi, T. Hamann, J. Barth, B. Eskofier, and C. Mutschler. The OnHW Dataset: Online Handwriting Recognition from IMU-Enhanced Ballpoint Pens with Machine Learning. In *ACM on Interactive, Mobile, Wearable and Ubiquitous Technologies (IMWUT)*, volume 4(3), Cancún, Mexico, Sept. 2020.

50. A. Radford, K. Narasimhan, T. Salimans, I. Sutskever, et al. Improving language understanding by generative pre-training. 2018.

51. M. M. Rahman, C. Fookes, M. Baktashmotlagh, and S. Sridharan. On Minimum Discrepancy Estimation for Deep Domain Adaptation. In *Domain Adaptation for Visual Understanding, Springer, Cham.*, Jan. 2020.

52. M. T. Ribeiro, S. Singh, and C. Guestrin. "Why Should I Trust You?": Explaining the Predictions of Any Classifier. In *Proceedings of the 22nd ACM SIGKDD International Conference on Knowledge Discovery and Data Mining*, pages 1135–1144. ACM.

53. T. Rojat, R. Puget, D. Filliat, J. Del Ser, R. Gelin, and N. Díaz-Rodríguez. Explainable Artificial Intelligence (XAI) on TimeSeries Data: A Survey.

54. U. Schlegel, H. Arnout, M. El-Assady, D. Oelke, and D. A. Keim. Towards a rigorous evaluation of XAI methods on time series. pages 4197–4201.

55. R. R. Selvaraju, M. Cogswell, A. Das, R. Vedantam, D. Parikh, and D. Batra. Grad-CAM: Visual Explanations From Deep Networks via Gradient-Based Localization. pages 618–626.

56. L. Sha, P. Lucey, Y. Yue, P. Carr, C. Rohlf, and I. Matthews. Chalkboarding: A new spatiotemporal query paradigm for sports play retrieval. In *Proceedings of the 21st International Conference on Intelligent User Interfaces*, pages 336–347, 2016.

57. R. Shu, H. H. Bui, H. Narui, and S. Ermon. A DIRT-T Approach to Unsupervised Domain Adaptation. In *Intl. Conf. on Learning Representations (ICLR)*, 2018.

58. R. H. Shumway. *Time series analysis and its applications: with R examples*. Springer Science+Business Media, 2017.

59. J. Siebert, J. Groß, and C. Schroth. A systematic review of python packages for time series analysis. (arXiv:2104.07406), Jun 2021. arXiv:2104.07406 [cs].

60. K. Simonyan, A. Vedaldi, and A. Zisserman. Deep Inside Convolutional Networks: Visualising Image Classification Models and Saliency Maps.

61. D. Smilkov, N. Thorat, B. Kim, F. Viégas, and M. Wattenberg. Smoothgrad: removing noise by adding noise. (arXiv:1706.03825), Jun 2017. arXiv:1706.03825 [cs, stat].

62. J. T. Springenberg, A. Dosovitskiy, T. Brox, and M. Riedmiller. Striving for Simplicity: The All Convolutional Net.

63. B. Sun, J. Feng, and K. Saenko. Correlation Alignment for Unsupervised Domain Adaptation. In *arXiv preprint arXiv:1612.01939*, Dec. 2016.

64. M. Sundararajan, A. Taly, and Q. Yan. Axiomatic Attribution for Deep Networks. In *PMLR*, volume 70, page 10.

65. A. A. Taha and A. Hanbury. An Efficient Algorithm for Calculatingthe Exact Hausdorff Distance. In *Trans. on Pattern Analysis and Machine Intelligence (TPAMI)*, volume 37(11), Nov. 2015.

66. E. Tzeng, J. Hoffman, N. Zhang, K. Saenko, and T. Darrell. Deep Domain Confusion: Maximizing for Domain Invariance. In *arXiv preprint arXiv:1412.3474*, Dec. 2014.

67. A. Vaswani, N. Shazeer, N. Parmar, J. Uszkoreit, L. Jones, A. N. Gomez, Ł. Kaiser, and I. Polosukhin. Attention is all you need. *Advances in neural information processing systems*, 30, 2017.
68. S. Wang, B. Z. Li, M. Khabsa, H. Fang, and H. Ma. Linformer: Self-attention with linear complexity. *arXiv preprint arXiv:2006.04768*, 2020.
69. Q. Wen, T. Zhou, C. Zhang, W. Chen, Z. Ma, J. Yan, and L. Sun. Transformers in time series: A survey. *arXiv preprint arXiv:2202.07125*, 2022.
70. R. J. Williams and J. Peng. An efficient gradient-based algorithm for on-line training of Recurrent network trajectories. *Neural Computation*, 2(4):490–501, 1990.
71. G. Wilson, J. R. Doppa, and D. J. Cook. Multi-Source Deep Domain Adaptation with Weak Supervision for Time-Series Sensor Data. In *Proc. of the ACM Intl. Conf. on Knowledge Discovery & Data Mining (SIGKDD)*, pages 1768–1778, Aug. 2020.
72. K. Witowski and N. Stander. Parameter Identification of Hysteretic Models Using Partial Curve Mapping. In *Intl. Conf. on Multidisciplinary Analysis and Optimization (ISSMO)*, Sept. 2012.
73. C.-H. H. Yang, Y.-Y. Tsai, and P.-Y. Chen. Voice2series: Reprogramming acoustic models for time series classification. In *International conference on machine learning*, pages 11808–11819. PMLR, 2021.
74. T. Zhou, Z. Ma, Q. Wen, X. Wang, L. Sun, and R. Jin. Fedformer: Frequency enhanced decomposed transformer for long-term series forecasting. In *International Conference on Machine Learning*, pages 27268–27286. PMLR, 2022.

Chapter 3
Learning from Experience

Christopher Mutschler[1], Georgios Kontes[1], Sebastian Rietsch[1]

Abstract Reinforcement Learning (RL) is one of the branches of Machine Learning (ML) that aims to learn from the interaction with an environment. In contrast to approaches such as supervised or unsupervised learning, where data samples usually are assigned to a ground truth label (supervised learning) or where they follow some stationary distribution (unsupervised learning), in RL, the agent is learning in direct interaction with the environment. This also defines what data is being collected as a result of which actions are being executed. The agent is hence *learning from experience*. While more traditionally, RL was focused purely on continuously arriving data, lately also approaches that resort to a given data pool of past environment interactions have gained more and more interest. This chapter covers the basics of RL and discusses the latest research in interactive environments, learning with available data or knowledge, and challenges that arise from the actual deployment of agents to the real world.

Key words: Reinforcement Learning, Markov Decision Processes, Exploration-Exploitation, Model-based RL, Offline RL, Safe RL.

3.1 Introduction

Sequential decision-making is a fundamental concept in artificial intelligence (AI) [4], decision theory [41], and operations research (see Chapter 7). It involves making a sequence of decisions over time, where the outcome of each decision depends on previous decisions and affects future decisions. This type of decision-

[1]Fraunhofer Institute for Integrated Circuits IIS, Fraunhofer IIS, Nuremberg, Germany

Corresponding author: Christopher Mutschler
e-mail: `christopher.mutschler@iis.fraunhofer.de`

© The Author(s) 2024   49
C. Mutschler et al. (eds.), *Unlocking Artificial Intelligence*,
https://doi.org/10.1007/978-3-031-64832-8_3

making is pervasive in many real-world applications, such as game playing [59], robotics [98], autonomous driving [82, 80], finance [94], and healthcare [108].

Traditionally, the theory behind sequential decision making is studied across many disciplines, including engineering (optimizing control loops and minimizing costs), mathematics (simulating and optimizing stochastic processes), economics (e.g., in game theory is studied how people make decisions considering social, political and human sciences), psychology (theory of conditioning and behavior) and neuroscience (dopamine system and the study of how the brain makes decision). Computer science usually addresses sequential decision making within the field of machine learning. At least traditionally, supervised and unsupervised ML work on statically available datasets (e.g., classifying objects, events, or classes, and finding representations for hidden structures within the data). In contrast, reinforcement learning works with a trial-and-error scheme: (i) data is acquired using the interaction of a learning agent with its environment (i.e., the learning process itself enables the agent to draw data samples from the environment to continue training), (ii) there are no labels (i.e., feedback to the agent is only provided by means of reward), and (iii) the reward might arrive with a large delay and it is not obvious what the causal relationship between a sequence of actions and a received reward actually is (credit assignment problem).

As we cannot provide a comprehensive view on the area of RL and sequential decision making, we try to limit ourselves to topics that are of utmost importance for the actual deployment and training of such systems in real world applications. Figure 3.1 proposes a motivation for this chapter (we will not cover topics such as model-free RL and digital twins as deeply). In order to make RL usable in practice, we have to consider that we cannot fully leverage the trial-and-error scheme that RL resorts to, as breakage and damages to material is often not a viable option. Hence, we will focus on how to leverage offline data (e.g., available through hard-coded regulators that can be observed), model-based RL (which makes use of available physics-models that can and should be used to accelerate learning or that learn a physical approximation of the real world dynamics), and safe RL (to account for constraints that are given by the actual application). This chapter will not focus on the area of interpretability and explainabliltiy (although this is a fourth important factor), as there is a large overlap with the content presented in Chapter 5.

3.2 Concepts of Reinforcement Learning

3.2.1 Markov Decision Processes (MDPs)

The general sequential decision-making framework can be modeled as a Markov decision process (MDP). An MDP is formally defined as a six-tuple $(S, \mathcal{A}, P, R, \gamma, s_0)$, where S defines a set of states the agent may visit, $\mathcal{A}$ defines a (fixed) set of actions the agent may take during an interaction, P is a transition probability matrix that defines how likely it is for the agent to move between two states given a particular

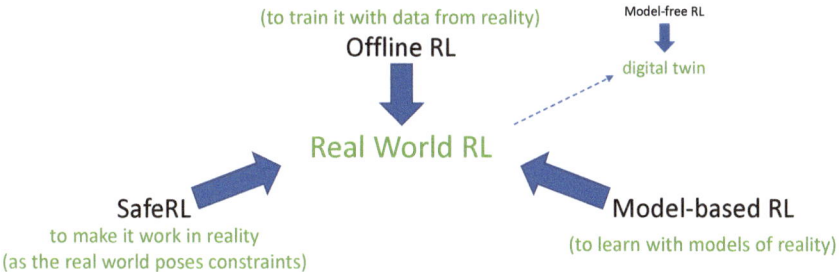

Fig. 3.1: The perspective that we will focus on in this chapter: (1) leveraging offline data, (2) using available information in form of models, and (3) accounting for unpredictable and uninterpretable behaviour through SafeRL.

action has been executed, R is a reward function, γ is a discount factor that allows to under-weigh future rewards over immediate rewards, and s_0 is a distribution over the states that define the starting state of the agent.

Time evolves in discrete time steps. At each time step t, the agent observes the current state s_t (or receives an observation o_t from the environment), receives a reward r_t and then selects an action a_t according to a policy $\pi : S \mapsto \mathcal{A}$, which is a function that maps states to actions. The action modifies the environment and changes its state (given some probability, induced by a random variable $\mathcal{P}$ defined over P). The goal of reinforcement learning is to learn an optimal policy π^* that maximizes the expected cumulative reward over a finite or infinite time horizon:

$$\pi^* = \operatorname*{argmax}_{\pi} \sum_{t=0}^{\infty} \mathop{\mathbb{E}}_{\substack{a_t \sim \pi \\ s_{t+1} \sim \mathcal{P}}} \left[\gamma^t R(s_t, a_t) | s_0 = s \right] \tag{3.1}$$

3.2.2 Dynamic Programming

A common way to solve MDPs is dynamic programming (DP). DP computes the optimal state-value function $V^\pi(s)$ (defining the *value* of being in state s, i.e., the (discounted) reward the agent can expect upon being in state s and following its policy), the optimal state-action-value function $Q^\pi(s,a)$[1] (defining the *value* of being in state s and taking action a, i.e., the (discounted) reward the agent can expect upon being in state s, taking action a, and afterwards following its policy), or an (explicit) policy π. The value function represents the expected cumulative reward starting from a given state and following a given policy. The optimal policy is the

---

[1] While computing the state-action-value function Q requires to estimate $|S| \times |\mathcal{A}|$ values (instead of only $|(S)|$ for the state-value-function $V^\pi(s)$), it is much more convenient to work with Q in practice as it directly estimates the value of each action in a state. At the same time, for V, we also need to account for where an action a brings us in the environment.

one that maximizes the value function for each state. All such approaches recursively evaluate the Bellman equation. The Bellman equation sets up a system of equations that relates the values of all states to each other; for instance, the Bellman optimality equation, i.e., the Bellman equation for the optimal Q-function, is given by:

$$Q^\pi(s, a) = R(s, a) + \gamma \cdot \sum_{s' \in S} \mathcal{P}(s'|s, a) \cdot \max_{a' \in \mathcal{A}} Q^\pi(s', a') \qquad (3.2)$$

However, computing the value function exactly in practice is often infeasible using dynamic programming. First, large and complex MDPs (which we usually are confronted with in practice) require estimating a large amount of state(-action) values recursively (several times). Second, DP is model-based and requires a model of the environment dynamics, the state transition probabilities $\mathcal{P}$, which is usually not available in real-world applications.

3.2.3 Model-free Reinforcement Learning

A different approach to solving MDPs is through a (model-free) trial-and-error interaction with the environment. The basic idea is to directly use the experience samples that the agent receives from the environment and to apply an iterative approximative update of the Q-function (for instance, using Q-learning):

$$Q(s, a) = Q(s, a) + \alpha \cdot \left[R + \gamma \cdot \max_a Q(s', a) - Q(s, a) \right], \qquad (3.3)$$

where $Q(s, a)$ is updated towards a target value using a small learning rate α. The state-action-value $Q(s, a)$ is again recursively described by the sum of the reward R that was recently observed and the maximal Q-value that can be selected from the successive state the agent ends up in. Updating this function as the agent moves through the environment lets us (in the limit) approximate the optimal Q-function.

3.2.4 General Remarks

Sampling complexity is the key performance indicator (KPI) to optimize. Common metrics to evaluate the performance of supervised or unsupervised ML algorithms include, for instance, the number of epochs required to learn a model until convergence or the predictive accuracy of a trained model on the test data set. In RL, there is no explicit concept of a *ground truth label*. However, RL still follows an active learning strategy: the agent is not only tasked to build a model that represents the value function or the policy directly. In essence, the agent is also tasked to select the samples to learn from, i.e., through executing the actions that directly correlate to the samples the agent will observe and use for learning.

The IID assumption. How the agent collects data and learns from it also influences model training. A central assumption for training models with ML is violated, as data is not *independent and identically distributed*. First, data samples are not *independent*. The samples we receive and use to update our model using, e.g., Eq. 3.3 are highly correlated as they constitute and resemble the trajectories that the agent observes (sequentially). Second, training and test distribution are not *identical*.[2] The distribution that underlies the data-generating process is constantly changing. Considering our central idea of the expected return from Eq. 3.1, the data we use within the expectation is based on actions we sample from π – but π is anything else than a stationary distribution as it is the central element we keep on training.

The Deadly Triad. Training an RL agent is always prone to instability and divergence, which arise from a combination of elements that make training different from other ML algorithms, see Sutton et al. [92]:

- Function approximation: usually the number of elements we approximate using V or Q is too large to fit into a computer's main memory, so we approximate those functions using neural networks. Those approximators have an error in their approximation which multiplies within the estimation problem of V and Q itself.

- Bootstrapping: we update targets (e.g., in Eq. 3.3) that include existing estimates rather than relying exclusively on actual rewards and complete returns. In the limits, we *hope* for convergence of all the values.

- Off-policy training: we train on the distribution of transitions that are different from that produced by the target policy. Sweeping through the state space and updating all states uniformly, as in dynamic programming, does not respect the target policy and is an example of off-policy training.

3.3 Learning purely through Interaction

3.3.1 Exploration-Exploitation

The exploration-exploitation-tradeoff [92] is one of the biggest challenges in RL. In its essence, it is a matter of deciding whether to *exploit* the knowledge that the agent has already obtained, i.e., to select actions from which we can expect high rewards/returns, or to *explore* the environment by selecting actions that have not or only rarely been selected to obtain potentially much higher rewards. Modern RL algorithms that optimize for the best returns can efficiently achieve good exploitation, while exploration remains an open research topic.

Classic exploration work builds on top of *Multi-armed Bandits (MABs)*. The idea comes from a casino use-case where you face K different slot machines, and each of them is configured with an unknown probability ξ_k of how likely you

---

[2] Note, that while there are no explicit training and test datasets as in the usual ML setting, in RL, we model $\mathcal{P}$ a bit differently between training and test environments.

can get a reward at one play (i.e., one-time step t). The question is: *What is the best strategy to achieve the highest long-term rewards?* This simple but effective problem formulation allows for a theoretical analysis of the exploration-exploitation-tradeoff. More formally, with $Q(a_t) = \mathbb{E}(r|a) = \xi$ being the expected reward of a_t (i.e., the interaction with slot machine k) and $r_t = \mathcal{R}(a_t)$, returning 1 with probability Q_a and 0 otherwise, we want to maximize the cumulative reward $\sum_{t=0}^{T} r_t$.[3] With the optimal reward probability given as ξ^*:

$$\xi^* = Q(a^*) = \max_{a \in \mathcal{A}} Q(a) = \max_{1 \leq i \leq K} \xi_i = \max_{1 \leq i \leq K} \mathbb{E}\left[r_t | a_t = a\right], \tag{3.4}$$

we can formulate the *regret*:

$$\mathcal{L}_T = \mathbb{E}\left[\sum_{t=1}^{T} (\underbrace{\xi^*}_{1} - \underbrace{Q(a_t)}_{2}))\right] = \sum_{a \in \mathcal{A}} \underbrace{N_T(a)}_{3} \underbrace{\Delta_a}_{4}, \tag{3.5}$$

where **(1)** is what the agent should have selected (i.e, the maximum rewarding action), **(2)** is what the agent actually selected at time t, **(3)** is the action selection counter, and **(4)** is the per-action regret. Good exploration algorithms minimize the total regret and maximize the cumulative rewards.

3.3.1.1 Exploration Strategies

As practical real world problems involve using deep RL (and its means to approximate large or continuous state and action spaces), we want to focus on exploration in deep RL. Hence, we only briefly cover classic work. Well-known exploration strategies are:

ϵ-**greedy (and variants).** This is a simple but often effective method: the agent exploits the knowledge, i.e., $a_t = \arg\max_{a \in \mathcal{A}} Q_t(a)$ with probability 1-ϵ and occasionally selects a random action (with probability ϵ). ϵ-greedy with a constant ϵ has a *linear expected total regret*. In practice, ϵ is often set to a high value at the beginning and then is slowly annealed, resulting in a greedy agent (but the expected total remains linear).

(Bayesian) Upper Confidence Bounds. The idea is to estimate the upper confidence bound $U_t(a)$ such that with a high probability, we satisfy $Q(a) \leq \hat{Q}_t(a) + U_t(a)$ and then select an action that maximizes the upper confidence bound: $a_t^{UCB} = \arg\max_{a \in \mathcal{A}} [Q_t(a) + U_t(a)]$. $\hat{Q}_t(a)$ is where the average rewards associated with action a up to time t and $U_t(a)$ is a function reversely proportional to how many times action a has been taken. Hence, small $U_t(a)$ follow from big $N_t(a)$ and certain/accurate value estimates and vice versa. While originally, UCB1 [92] only counts the selections and rewards, Bayesian UCB [49]

---

[3] Note that this formulation (in contrast to any other work in RL) explicitly considers the rewards obtained over the *course* of training. It quantifies the training performance/progress.

models $U_t(a)$ with a Beta-distribution [48] and hence makes use of confidence intervals. Both UCB1 and Bayesian UCB show *logarithmic expected total regret*.
Probability Matching: Thompson Sampling (TS). TS [78] uses a Beta distribution to keep track of the current belief of probabilities and then directly samples from this distribution. Intuitively, highly rewarding actions become more likely to be sampled (while low rewarding actions still occasionally keep being selected), and the observed data is used to update the prior belief. Thompson sampling also shows *logarithmic expected total regret* while often being more effective than UCB and variants (while also being easier to be implemented).

3.3.1.2 Exploration in Deep RL

Exploration in Deep RL is more difficult than in small MDPs not only because the state space becomes larger. While exploration research on small MDPs as before can be theoretically well-grounded (we can, in principle, *know* the best strategy, define the total regret and benchmark the algorithm, and define upper and lower bounds on its mistakes), this is not possible when working with deep neural networks approximating the state and policy in deep RL. In addition to that, especially for more recent deep RL research, there are two descriptive problems on which exploration methods are being evaluated: (1) the *hard-exploration* problem and (2) the *Noisy-TV*-problem.

The hard exploration problem is presented in environments with very sparse or even *deceptive* rewards. In those environments, a random exploration is likely prone to failure as it will rarely find successful states or obtain meaningful feedback from the environment. One such example is MONTEZUMA'S REVENGE, where an agent traverses several rooms and looks for keys and treasures. The noisy-TV-problem [19] is a hypothetical scenario where a novelty-seeking agent solves a 3D-maze environment (in first-person perspective) and encounters a TV that shows uncontrollable and unpredictable white noise – novelty-seeking agents usually then become a *couch potato* and keep looking at the TV forever.

The key idea is that we count the number of times we visited a state or a state-action-pair (instead of counting the number of arms in the bandit). In other words, we use $N(s)$ (or $N(s, a)$) and add an exploration bonus to the reward that is provided by the environment:

$$r^+(s, a) = r(s, a) + \beta \cdot \mathcal{B}(N(s)), \tag{3.6}$$

where β is a hyperparameter that adjusts the balance between exploration and exploitation. Generally speaking, the first component $r_t^e = r(s, a)$ is often called the *extrinsic* reward (i.e., provided by the environment) and $r_t^i = \mathcal{B}(N(s))$ is called the *intrinsic* reward (i.e., an exploration bonus added by the agent that decreases with larger $N(s)$). The intrinsic component rewards discovering novel states and, hence, possibly gaining knowledge about the environment. Recent research can be grouped into different categories: (1) Count-based Exploration, (2) Prediction-based Exploration, and (3) Memory-based Exploration.

(1) Count-based Exploration deals with the problem of counting states in high-dimensional state spaces. Encountering the same state (e.g., in manipulation tasks) is unlikely, but some states are similar to others, and exactly this relationship should be represented in the bonus that is calculated. Density Models [13] approximate the frequency of visits using a parametric model $p(s; \theta)$ and derive a pseudo count from the model. As the agent observes new data, the density models are incrementally updated. As the density model only needs to correlate high density with large visitation counts (we do not need to sample from the model), we can resort to a large variety of models such as context switching trees [14, 13], PixelCNNs [65, 101], GMMs [112], etc. Another idea is to use *hashing* such as Locality-Sensitive Hashing (LSH) [95, 21]. The idea is to find or learn a hash function that maps the high-dimensional state space into lower-dimensional hash codes that preserve distance information between states, i.e., similar states will have similar hash codes. Simple distance metrics such as the angular distance work reasonably well for low-dimensional state spaces. For larger state spaces, LSH proposes to learn a compression using autoencoders with a regularized loss function that enforces a special representation in the bottleneck (i.e., the hash code).

(2) Prediction-based Exploration follows the concept of *curiosity* and *intrinsic motivation* instead of just counting state visitations [79]. The key idea is to learn about the environment, e.g., its reward structure or dynamics. In the easiest case, the agent uses all the experiences (s_t, a_t, s_{t+1}) seen so far and learns a forward prediction model $f_\theta : (s_t, a_t) \mapsto s_{t+1}$ and derives an extrinsic reward $r_t^e(s_t, a_t) = ||f(s_t, a_t) - s_{t+1}||_2^2$, i.e., large prediction errors result in a high bonus (i.e., a high intrinsic reward) and vice versa. While early approaches such as Intelligent Adaptive Curiosity (IAC) [66] resort to feature-engineering state representations, recent work such as Deep Predictive models [89] learn such forward dynamics models end-to-end using autoencoders. However, in many applications and environments, the observations change without being explicitly affected by the agent's actions. Consider, for instance, a tree with leaves that move due to wind – such factors are not affected by the actions and cannot be controlled by the agent. Hence, those elements should not be encoded in the state space and not targeted to being predicted. The Intrinsic Curiosity Module (ICM) [68, 19] jointly learns an inverse dynamics model g : $(\phi(s_t), (\phi(s_{t+1}))) \mapsto a_t$, where $\phi(s_t)$ is a lower-dimensional embedding of the state observation, i.e., g enforces a bottleneck representation from which an action a_t can be inferred, hence focusing on the crucial information contained in the state space. The forward prediction, the inverse dynamics models, and the policy are then jointly optimized. Self-Supervised Exploration via Disagreement [69] combines a policy network with an ensemble of neural networks (whose disagreement is the bonus) that predict forward predictions and learn all networks end-to-end (as r_t^i is differentiable). However, for all previous approaches, the prediction errors still remain large if (i) the prediction target remains stochastic, or information is missing and (ii) if the model class or capacity of the predictor is too limited to fit the complexity of the target function [34]. Hence, Random Network Distillation (RND) [19, 18] predicts something different from the actual task using two neural networks: (1) a randomly initialized but *fixed* network $f(s_t; \theta)$ that transforms a state into a feature space, and

(2) a network $\hat{f}(s_t;\hat{\theta})$ that is trained to predict the same features as the first network, i.e., $\hat{f}(s_t;\hat{\theta}) = f(s_t;\theta)$; the exploration bonus is $r^i(s_t) = ||\hat{f}(s_t;\hat{\theta}) - f(s_t;\theta||_2^2$. The intuition is that similar states have similar features, and if the agent has already seen them, it should also have a lower error in predicting them. RND makes the prediction target deterministic and ensures that the target is within the class of functions that can be represented.

(3) **Memory-based Exploration** uses external memories to address a problem that remains present in previous approaches: the exploration bonus is non-stationary (i.e., it drops for a particular state when visiting it several times). As a result, intrinsic rewards will vanish and no longer provide a signal for the agent.[4] Never Give Up (NGU) [9] combines an RND (for lifelong learning) with an episodic novelty module for rapid *in-episode* adaptation. Hence, visiting the same states within one episode is rapidly discouraged while revisiting states that have been visited many times across episodes is only slowly discouraged. Agent57 [8] enhances NGU using a population of policies (that all have their exploration parameters, encouraging exploration either more towards the beginning or the end of the training course) and using a separate estimation of Q-values that decompose the influence of the extrinsic and intrinsic rewards. Agent57 has been the first general-purpose algorithm that outperforms the standard human benchmark on all 57 Atari games. While there are much more approaches in that area, methods such as Go-Explore [31] present a complementary solution. The idea is to especially address the detachment and derailment problems, i.e., that intrinsic rewards are consumed and vanish and that interesting states should be revisited, respectively. The idea is to use goal-conditioned policies through self-imitation learning [32] or through a trajectory-conditioned policy based on a memory of demonstrations [37] that enable an agent to revisit a point (through a sequence of low-intrinsic rewards) to keep exploring other areas of the state space (e.g., keep exploring the tree at the end of the maze in the example before).

3.4 Learning with Data or Knowledge

3.4.1 Model-based RL with continuous Actions

In Model-based RL (MBRL – sometimes also referred to as model-assisted reinforcement learning), the idea is to learn an estimator/model $f_\phi : \mathcal{S} \times \mathcal{A} \to \mathcal{S}$ from data for the transition dynamics that would assist the search for the optimal policy. In several cases, a model of the reward can also be learned. Model training follows the supervised learning paradigm and utilizes standard loss functions such as MSE. The field of MBRL is quite diverse [60], including different types of algorithms that

---

[4] Consider, e.g., long sequences within a maze where the agent traverses a tunnel where at the end, the agent might again take several different decisions/ways. However, the intrinsic rewards that lead the agent to enter the tunnel in the first place will vanish, the agent will not enter the tunnel again, and the tree that spans up at the end of the tunnel remains mostly unexplored.

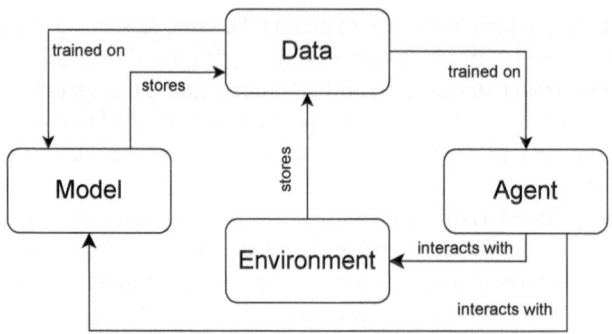

Fig. 3.2: The high-level concept of Model-based Reinforcement Learning

operate in different representations of the state, like features [61], full observations like images [33] or event latent states [105, 39]).

The way the learned model is used is also diverse. For example, algorithms like Dyna [91] or Model-based Policy Optimization (MBPO) [46] augment the real data (that is actually collected using the policy in the actual environment) with "imaginary" data (that is generated by running a model of the environment for several time steps from selected initial states).

Other approaches use the model as a look-ahead approximator. Here, imaginary rollouts are performed from an initial state using the model, and either a policy is trained to select good actions or an open-loop optimization routine (called the *planner*) is utilized to generate a sequence of actions to be applied in the next time-steps. Several types of planners can be used, spanning from well-established planners from control engineering practice like the iterative-linearization and trajectory optimization iLQR planner [96], to population-based approaches that leverage modern parallel computation capabilities to iteratively simulate several trajectories using the model and select the most promising ones to be applied in the real system [106, 28].

One of the downsides of MBRL is that since the model is learned from data, it will always fail to capture the complexity of the environment perfectly. There will always be errors due to the uncertainty of the measurements and the underlying process/dynamics, as well as model extrapolation errors due to a lack of data. Modern MBRL algorithms try to alleviate this problem in two ways (often combined):

- **Online Re-planning:** Following the paradigm of Model Predictive Control (MPC) [20], only the first action of the action sequence generated by an open-loop planner is applied in the environment, and the planning process re-initiates from the new environment state.
- **Uncertainty Estimation:** The idea is that probabilistic models can be utilized to capture both *aleatory uncertainty*, i.e., the uncertainty that is inevitable due to inherent noise in the process or measurements, and *epistemic uncertainty* due to the lack of training data in some regions of the state-action space. Here, as modeling errors can compound with longer rolling horizons, the uncertainty is

propagated since the policy/planner that utilizes the model takes into account the model's confidence of each next state prediction. Various models and policies have been utilized to achieve this, from analytical Gaussian Processes with end-to-end closed-loop policy learning [29] to ensembles of probabilistic neural networks with population-based online planners [24, 61].

MBRL algorithms have been utilized in a variety of tasks, often leading to impressive results [61]. On one hand, one of their downsides is related to the complexity of the (uncertainty-informed) online planner, which could potentially hinder the real-time application. On the other hand, they offer a way of safe exploration and operation [15, 42], as the model can prevent unsafe actions from execution. In contrast, the model's uncertainty level in sampled states indicates that the model can be unreliable for prediction in these instances.

3.4.2 MBRL with Discrete Actions: Monte Carlo Tree Search

The availability of an environment model allows us to view many decision-making problems as a simple search problem over potential future trajectories. Generally speaking, it permits us to simulate the outcome of possible action sequences and, therefore, estimate the optimal value function by building a search tree. Considering a problem with b possible actions (breadth) and a maximum episode length of d (depth), we must scan over b^d possible action sequences, which quickly becomes infeasible to compute in practice. To subdue the computational complexity of the search, Monte Carlo Tree Search (MCTS) has evolved into one of the go-to methods in the field. Instead of an exhaustive search, MCTS uses a combination of Monte Carlo simulation and tree search as a heuristic search algorithm.

MCTS executes four recurring steps: Selection, expansion, simulation, and backpropagation. In the first step, MCTS *selects* a node by traversing the tree until an unexpanded leaf node is reached. It *expands* the selected leaf node by adding one or more child nodes to it, and *simulates* one or more trajectories until a terminal state is reached, also called *rollouts*. This way, an initial value estimate of the state is created. Finally, the results of the simulation phase are *backpropagated*, updating the statistics of all nodes traversed during the iteration (typically the *empirical state return* and *state visit count*).

MCTS must sufficiently balance exploration and exploitation during selection to achieve robust estimates. A widespread method is Upper Confidence bounds applied to Trees (UCT) [52], which treats selection as a multi-armed bandit problem and uses the UCB1 formula for action selection, i.e., $a_t = \mathrm{argmax}_a\, Q(s_t, a) + c\sqrt{\frac{\ln N(s_t)}{N(s_t, a)}}$, where $Q(s_t, a)$ is the empirical reward for taking action a in state s_t, $N(s_t)$ is the total number of trajectories explored from s_t and $N(s_t, a)$ the number of times action a was explored, and c is a hyperparameter. Intuitively, UCB1 favors exploration when estimates are uncertain but leans towards exploitation once more information has been gathered. It can be shown that this strategy achieves regret that grows only logarithmic in the number of iterations [6].

An important scientific breakthrough was achieved when AlphaGo [84] became the first computer program to defeat a professional human Go player. In essence, AlphaGo employs MCTS with a value network to assess the value of a board position and a policy network to bias the action selection, thereby lifting the requirement for accurately simulating rollouts and focusing the search on promising actions. For the latter, a modified version of PUCT [76] is employed, $a_t = \text{argmax}_a\ Q(s_t, a) + c_{puct} P(s_t, a) \frac{\sqrt{N(s_t)}}{1+N(s_t,a)}$, where $P(s_t, a)$ is the output of the policy network. Further, Dirichlet noise is added to the prior estimates to stimulate exploration. Due to the two-player nature of many board games, another key technique is self-play, where recent model versions compete as similarly skilled opponents during training.

From AlphaGo, a whole line of work has evolved, intending to remove more and more prior assumptions from the method. Whereas AlphaGo initializes value and policy networks through pre-training on an expert game dataset, AlphaGo Zero [86] can surpass its ancestor's performance without resorting to expert data (learning "tabula rasa"). At the same time, AlphaGo Zero drops hand-engineered features, uses a simplified neural network structure, and removes MCTS rollout simulations by entirely relying on value network estimates. In AlphaZero [85], the method is generalized to the game of Chess and Shogi. Another vital breakthrough was achieved with MuZero [83], which learns the environment model and employs it for planning with MCTS. This means MuZero does not require access to a resettable simulator, i.e., providing the game's rules to the learning method. They also generalized the method to the single-player Atari domain with intermediate rewards. EfficientZero [107] reduces the sample complexity of MuZero and shows promising performance in the limited training data regiment. The latest improvement is Gumbel MuZero and Gumbel AlphaZero [27], which replaces adding Dirichlet noise by sampling actions without replacement using the Gumbel-Top-k trick and sequential halving, allowing for an effective reduction of the number of required MCTS iterations without degrading performance.

3.4.3 Offline Reinforcement Learning

In many real-world use cases, an extensive trial-and-error procedure as required by traditional (online) RL is not feasible. This might be due to ethical reasons (such as medical treatment or autonomous driving) or simply because it would take too long for an agent to learn some reasonable behavior. The idea of Offline RL is to resort to existing rollouts (from other agents/policies, collected by observing humans, or by traditionally implemented controllers) and to learn a policy based on this offline data.

Figure 3.3 (top left) and (top right) both illustrate the online RL setup, where the agent either follows the policy it is optimizing (on-policy) or where the agent is using data from different (earlier versions of the current) policies to optimize its policy

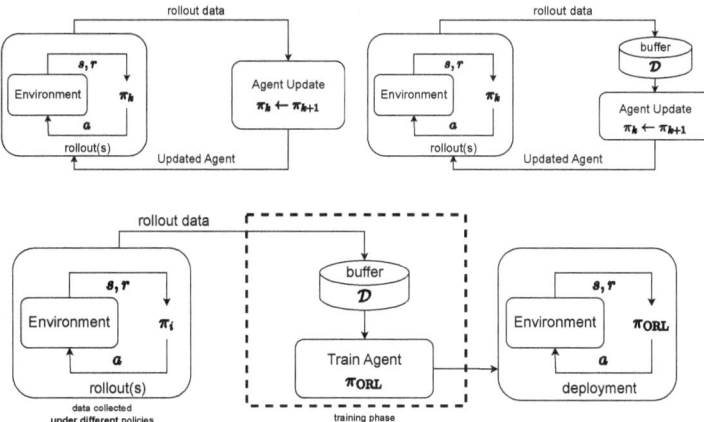

Fig. 3.3: Offline RL in context: On-Policy RL (top left), Off-Policy RL (top right), Offline RL (bottom), based on [56].

(usually realized by maintaining a buffer of transitions that have been collected under different policies at earlier stages of learning). Figure 3.3 (bottom) shows the Offline RL setup: as in off-policy RL, we maintain data from a variety of policies from which we want to learn a unified policy (instead of iterating/updating new policies) that can be deployed into the environment. In contrast to Behavioral Cloning [56], which requires data $\mathcal{D}$ from a single expert policy π_β to get good results and which is prone to failure due to poor generalization in unexpected situations, the expectation of Offline RL is to outperform all the individual policies π_i from which it learned from.

Figure 3.4 motivates where Offline RL should do a better job. Assume that $\mathcal{D}$ contains two trajectories, one from A to B and one from B to C, each trajectory takes several time-steps, and the agent gets penalty of -1 per time-step. Given the data present in the dataset we expect an agent trained on $\mathcal{D}$ to come up with a policy shown on the right, which takes the direct route from A to C (which is supposed to be the optimal solution here).

Interestingly, a simple naive example already fails. Consider a dataset with expert demonstrations and successful task completions. Training in an off-policy setting, iterating on the transitions from the dataset does not converge to a usable policy;

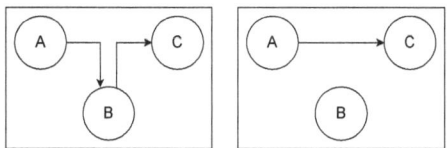

Fig. 3.4: Stitching.

in some settings, even Q-values start to diverge (i.e., take overly large values). The problem arises during training from bootstrapping the value function on Q-values under a slight distribution shift, i.e., in each state, the policy is deemed to bootstrap from the best (= max) action available in a particular state, which is prone to approximation errors (as data might be incomplete). This max-term over-estimates the real Q-value and, while being corrected in an online setting (by actually taking this route and getting grounded), we will get no feedback about this error in the offline setting and hence take routes through the state space that are different from what the data covers.

Existing solutions to Offline RL can be subdivided into methods that either (1) constrain the policy to stay in *a region of trust*, i.e., where we have knowledge from the dataset (which covers that regions), or (2) that estimate the state-value function conservatively based on confidence (from what we can infer from the dataset).

Policy-constrained methods tackle the problem in the policy improvement step, i.e., they try to keep π close to π_β. Batch-constrained Q-Learning (BCQ) [36] restricts the action space to force the agent to behave close to on-policy *concerning the subset of the given data*, hence keeping the extrapolation error of the bootstrapped actions low. While this works reasonably well, it is overly restrictive as BCQ implements a distribution matching strategy (i.e., it also tries to minimize the KL-divergence between π and π_β) and hence, if the behavior policy is e.g. uniform, the learned policy is also required to be uniform. Instead, Bootstrapping Error Accumulation Reduction (BEAR) [53] only requires π to lie in the support of π_β, i.e., $\pi_\beta : \pi(a|s) > 0 \Rightarrow \pi_\beta(a|s) > \epsilon$.

Conservative methods tackle the problem in the policy evaluation step by being conservative in the state(-action)-value estimation in areas of high uncertainty (i.e., for transitions that are not in the dataset), which implicitly then keeps the policy away from out-of-distribution actions. Conservative Q-Learning (CQL) [54] learns a value function that minimizes $Q(s, a)$ on $s \in \mathcal{D}, a \sim (s)$, and additionally maximizes for cases where $a \sim \pi_\beta(s)$ (as in those cases there is no need to be conservative). It is proven that the return estimate of the resulting policy from CQL is lower than the actual policy performance. Model-based Offline Policy Optimization (MOPO) [110] estimates $r(s, a)$ and $p(s'|s, a)$ from $\mathcal{D}$ and applies online RL to the learned model. As still the model can be queried in unknown regions if π diverges from π_β, MOPO learns an uncertainty estimate $u(s, a)$ and provides a pessimistic reward to the agent $\tilde{r}(s, a) = r(s, a) - \lambda u(s, a)$. However, as a correct estimation of $u(s, a)$ is difficult (especially with neural networks), Conservative Offline Model-Based Policy Optimization (COMBO) [109] combines the idea of MOPO (using rollouts from a model) with CQL: it samples transitions either from $\mathcal{D}$ or from the model. but assigns more *trust* to those state-action pairs that are observed in $\mathcal{D}$. As a result, COMBO uses more data than CQL (i.e., added data from the model), and the data is correlated with the policy.

Decision Transformers [22] are an alternative way of solving the Offline RL problem by positing the sequential decision-making problem as a language or sequence modeling task. The idea is to use strong and advanced neural network architectures (such as Transformers [102], due to their ability to capture long-horizon

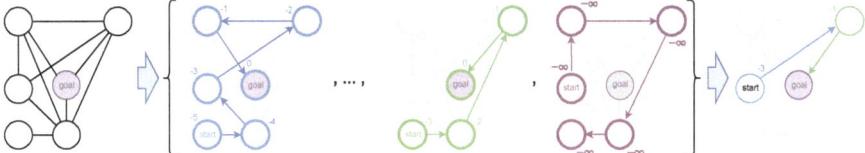

Fig. 3.5: Decision Transformers, based on [22].

dependencies) instead of algorithmic frameworks that make use of Bellman back-ups: the (return, state, action) tuples are processed in a sequence and learned in an autoregressive fashion, where the task is to predict the next tuple from a hidden state that represents the past sequence (i.e., the Markov state). With regard to the principle of Decision Transformers consider Figure 3.5. Working on a fixed graph and observing/training on random rollouts (the edges show the accumulated reward), the decision transformer can be conditioned on a desired performance and produces a sequence of actions that satisfies the condition, i.e., that delivers the requested per-formance. A very related but less noticed approach is Trajectory Transformer [47], which additionally uses discretization and employs beam search for trajectory plan-ning.

3.4.4 Hierarchical RL

As the name suggests, Hierarchical Reinforcement Learning (HRL) has a hierarchy of policies. Lower-level policies are responsible for solving a sub-task of the problem (for example, they might implement specific motor skills), while higher-level poli-cies (also called master policies) implement a "strategy" that combines lower-level policies to solve a given task. The idea is that this decomposition can help with problems like exploration (since the high-level policy can enable data gathering for under-represented sub-tasks), data efficiency (since combining low-level policies can converge faster compared to end-to-end training), and generalization (since a large number of tasks can potentially be solved by learning to combine an available library of sub-policies). Figure 3.6 illustrates the high-level concept of HRL.

More formally, in the simple case where we have only a two-level hierarchy, we assume M low-level policies parameterized with $(\pi_{\omega_1}, \pi_{\omega_2}, \pi_{...}, \pi_{\omega_M})$ and a higher-level, coordinating policy π_θ. There are two important questions: i) how are the low-level policies provided or trained, and ii) how does the high-level policy learn to utilize and combine the low-level ones efficiently.

One of the earliest and most elegant ideas to formalize this problem is the *options framework* [93]. Here, an option $\omega \in \Omega$ is defined by:

- an intra-option policy π_ω, which is the policy to be used when the specific option is selected;

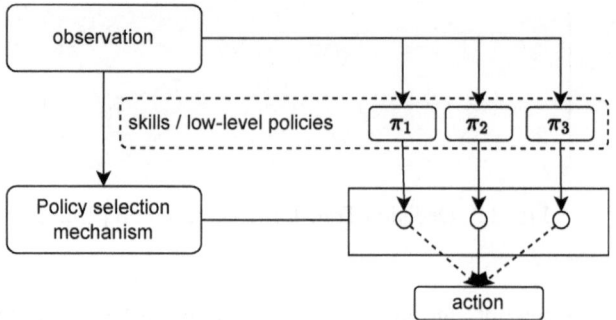

Fig. 3.6: Hierarchical Reinforcement Learning concept.

- a termination function $\beta_\omega : \mathcal{S} \to [0, 1]$ that indicates the probability of terminating (stop using) the option in the current state;
- an initiation set $\mathcal{I} \subseteq \mathcal{S}$, indicating in which states each option can be selected/activated (even though in many practical problems, this is the entire state space $\mathcal{S}$).

It has been shown [93] that augmenting MDPs with options results in Semi-MDPs, which in turn have optimal V and Q functions ($V_\Omega^*(\omega)$ and $Q_\Omega^*(s, \omega)$, respectively).

When an option is activated, the intra-option policy π_ω is utilized to select actions in each environment time-step. Once the action is executed and the next state and the reward are acquired, the option terminates with probability $\beta_\omega(s')$. If the option is terminated at state s', one of all available options is activated according to $Q_\Omega(s', \cdot)$.

The options framework has been extended to methods that learn both the optimal Q-function over options and the intra-option policies in an end-to-end manner [7], as well as to approaches that support an arbitrary number of hierarchy levels [74]. One of their main limitations is defining the number of options *a-priori*.

Of course, there are also successful HRL approaches that do not follow the options framework. In [35], a combination of Meta Learning and hierarchical policy structure is proposed. Here, several low-level policies are selectively activated by a master policy, with the latter being called within fixed time intervals. The low-level and master policies are trained sequentially over several episodes. Still, in low-level training, the reward of a specific episode is considered, while in the master policy, the collective reward from all training episodes is utilized. Apart from the proper tuning of the master policy call time interval, also in this approach, the "correct" number of low-level policies has to be selected beforehand.

In a different line of thought, [103] proposes using FeUdal Networks for HRL. In the simplest case, a "Manager" implements a high-level policy that generates a set of sub-goals (in the form of specific states) in the latent space that a "Worker" must learn how to reach. Through training, the Manager learns to select sub-goals that lead to solving the task. Even though there is no need for pre-selection of the number of low-level policies, selecting proper sub-goals from the Manager policy is an open-ended task on its own that can hinder the scalability of the algorithm.

In several cases, HRL is a design choice to enable safe and/or interpretable solutions to a given task. For example, in [55], a safe and reliable vehicle obstacle avoidance strategy is learned by combining a set of MPC low-level policies with a high-level DQN selection policy. In [81], utilizing low-level driving "primitives" allows distilling the master policy to a binary decision tree, thus allowing a certain level of interpretability.

3.5 Challenges for Agent Deployment

Besides astounding progress and movement in the field of RL, several key challenges remain open for this machine learning technology to be easily and widely applicable in the real world. In this section, we will cover the aspects of safety, policy generalizability, and the issue of defining reward signals.

3.5.1 Safety through Policy Constraints

In many applications, an agent has to behave safely in the sense that it should not put itself or entities of its environment at risk. This is especially true for real-world applications, where RL-controlled robots and machines could cause financial damage or risk human health or lives. At the same time, safety concerns should not disturb reward performance too much, as this would undermine the benefits of RL in contrast to less performant but provably safe traditional methods. This consideration motivates the field of *Safe RL*, which aims at providing methods that bring about optimal reward performance while satisfying safety constraints.

Constrained Markov Decision Processes (CMDPs) are a widespread formalism to model safe RL as a constraint optimization problem. The CMDP problem is an extension of the standard MDP $\mathcal{M}$ with a constraint set $C = \{(c_i, b_i)\}_{i=0}^{m}$, where $c_i(s, a)$ is a cost function that returns a cost value for a state and action pair s, a, and b_i is the safety constraint bound, and we have m such cost constraints. In many cases, constraints are defined as discounted cumulative cost constraints, i.e., the expression $C_i(\pi) = \mathbb{E}_\pi[\sum_{t=0}^{\inf} \gamma^t c_i(s_t, a_t)] \leq b_i$ must hold for all $i \in \{0, \ldots, m\}$, where C_i is the expected cost. However, mean valued, probabilistic, and other constraint formulations are also utilized in the literature [58].

Different theoretical constructs to analyze and solve CMDPs have been proposed in the literature. One such category are Lagrangian approaches, which translate the CMDP formulation into a primal-dual optimization problem,

$$(\pi^*, \lambda^*) = \arg\min_{\lambda \geq 0} \max_{\pi \in \Pi} \{J(\pi_\theta) - \lambda^T(c(\pi) - b)\},$$

with $J(\pi_\theta) = \mathbb{E}_{s \sim p_0}[V_\pi(s)]$ being the original RL objective to maximize. Besides Lagrangian-PPO and Lagrangian-TRPO [2], a highly influential method of this family is Reward Constrained Policy Optimization (RCPO) [97], which incorporates the constraint as a penalty signal into the reward function into a multi-timescale approach. Further, it has recently been shown that the primal-dual formulation has zero duality gap [67].

Another line of work are trust-region methods, which aim to improve the policy through a local policy search while enforcing the constraints in every update. Among them are Constrained Policy Optimization (CPO) [3], Projected-based Constrained Policy Optimization (PCPO) [62] and other derivative works.

3.5.2 Generalizability of Policies

A central reason for the growing interest in ML lies in the ability to train models that adapt properly to new, previously unseen data. Even though this is not necessarily true for all RL applications, e.g., learning a policy that behaves optimally inside the training environment could be all that we want, it is of vital importance for the applicability of RL in the real world, where agents will need to be robust to variations in their environments or capable of handling even harder forms of generalization [51]. This starkly contrasts with typical RL benchmarks like Atari and MuJoCo [99], where policies are evaluated directly inside the training environment, making it an often-overlooked algorithmic characteristic in the RL literature.

More technically, generalization in RL requires an agent to deploy successfully across various tasks or environment instances. This encompasses variations in the state space, for example, due to changes in the initial state distribution, the environment dynamics, visual aspects for image-based observation spaces, the reward function, etc. Contextual MDPs [40] is one construct to formalize this. They add a context distribution that, loosely speaking, allows for modeling distributions over MDPs. Analogous to supervised learning, the need for generalization emerges for differing train and test context sets. Depending on the distribution of the varying environmental factors, policies must learn to interpolate and/or extrapolate from the training experience to close the train-test performance gap.

An important distinction can be made based on the occurring type of distribution shift. Training and testing are often assumed to be IID, i.e., the MDP contexts are samples of one mutual distribution. The opposite case of out-of-distribution (OOD) test environments, termed *domain generalization* [57], is especially hard to accomplish. Because today's state-of-the-art RL methods typically have low sample efficiency, agents are trained in simulation even though their deployment domain is a real-world environment. Transferring agents to a physical environment, or *sim-to-real transfer*, requires OOD generalization if no additional measures are employed because camera inputs, actuator feedback, etc. can typically be simulated accurately only to some degree [113]. The issue is often alleviated by highly randomizing the simulation dynamics [70] or visuals [100] (*domain randomization*), or closing the visual domain gap through image-to-image GAN models [73]. Further, the field of *Robust RL* aims to tackle specific forms of environment model misspecification through worst-case optimization [111].

We can also characterize approaches by how much target domain data they can access. This paints a spectrum from *zero-shot generalization*, where no explicit target domain data is available [51], to *online adaptation*, which often assumes access to at least a few training episodes in the test environment (often addressed through methods of *Meta RL* [12])), to *Continual RL*, where MDP components are assumed non-stationary and the agent must continually learn to adapt to this property [50]. In the following, we will give a more detailed view of zero-shot transfer, which can be seen as the most general and straightforward form of generalization.

Zero-Shot Transfer. Similar to methods stated for sim-to-real transfer, a common technique is to increase the similarity between source and target domain data. This can be achieved through data augmentation, as done by UCB-DrAC [72], or domain randomization techniques, like ADR [63]. A line of work aims to optimally sample from the set of randomized MDPs with approaches like POET [104] and PAIRED [30]. The latter trains an adversarial agent that designs the environment levels to guarantee solvability and maximize generalization.

Analogous to data alignment, another field aims to learn more robust features not specific to the training domains. One method is encoding inductive biases, for example, by encouraging internal features that are not predictive of time within an episode, as has been done in IDAAC [71], when we know that optimal policies should not have time-dependence. Also, techniques like $L2$ weight decay and policy entropy regularization have improved generalization [26], as simple models can be expected to generalize the best. Finally, methods that aim to learn invariant, context-independent feature representations like IPO [88] have proved successful.

More fundamental work is done to optimize the RL objective without overfitting. Iterated Relearning (ITER) [43] introduces repeated knowledge distillation of the policy to counteract the memory effects of neural-network caused by the non-stationarity of RL training. Phasic Policy Gradients (PPG) [25] splits the policy and value heads training regiment into two separate phases. Finally, initial investigations have recently shown that MBRL might inherently facilitate better generalizability [5] and might be an interesting future research direction in this area.

3.5.3 Lack of a Reward Function

Unlike in RL applications in games, designing a reward function for most real-world problems is a complex task, requiring extensive domain knowledge and experience, as well as a significant amount of fine-tuning. There are voices in the community [87] that firmly state that the reward function should be representative of the problem, even if the reward signal is sparse (meaning the reward is a rare event and a significant amount of exploration is required), to avoid inducing bias in the way the policy is designed. Nevertheless, designing a more rich reward signal can lead to high-quality solutions with less training data/iterations [38].

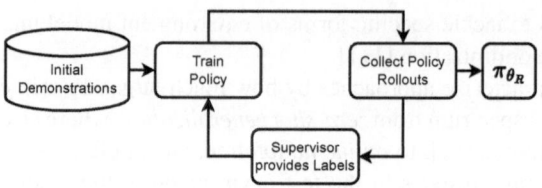

Fig. 3.7: The Dagger algorithm.

There are several ways to make a reward signal mode dense, but the two main variants differ on the explicit knowledge of a goal/terminal state, that is assumed to hold positive reward for the agent. If we know the goal state, we can easily create an inverse curriculum and progressively learn to reach the goal from further initial states [44]. If we don't know the terminal state, we could augment the reward function with auxiliary rewards that encourage exploration [45].

In several real-world problem cases, we are not required to define a reward function. Here, demonstrations of successful interactions with the system (e.g., from human operators) are available in the form of state-action trajectories $\xi \in \Xi$ and the task is to learn to replicate the actions of the expert in similar states. This problem can be formulated as a typical supervised learning problem:

$$\pi_\theta = \underset{\theta}{\text{argmin}} \sum_{\xi \in \Xi} \sum_{s \in \xi} \mathcal{L}(\pi_\theta(s), \pi^*(s)), \qquad (3.7)$$

where π_θ is the policy to be learned, $\xi \in \Xi$ are the available expert trajectories, $\mathcal{L}$ is a loss function (e.g., MSE for continuous actions and cross-entropy for discrete) and π^* is the expert policy that generated the logged actions.

This setup, even though conceptually and implementation-wise simple, rarely works in practice without some form of data augmentation [16, 10], because of compounding errors due to the different data distribution between training and application settings/environments. Here, Imitation Learning (IL) [64] can address the problem. In its most simplest form, the Dataset Aggregation (DAgger) algorithm [77], it consists of the following steps repeating until convergence (illustrated also in Figure 3.7):

1. Train a policy using supervised learning, to predict the selected action for all state-action pairs in the available expert trajectories;
2. Use the policy to interact with the system and collect new data;
3. Query the expert (a more complex algorithm or a human labeler) for the optimal actions in the newly visited states;
4. Aggregate the dataset with the new state / optimal action pairs and re-train the policy using supervised learning;

Note here that IL can be a way to distill an optimal policy of type A (e.g., a complex neural network) to a policy type B (e.g, an interpretable Binary Decision Tree as in the case of [11]). This gives the possibility to learn a complex policy to

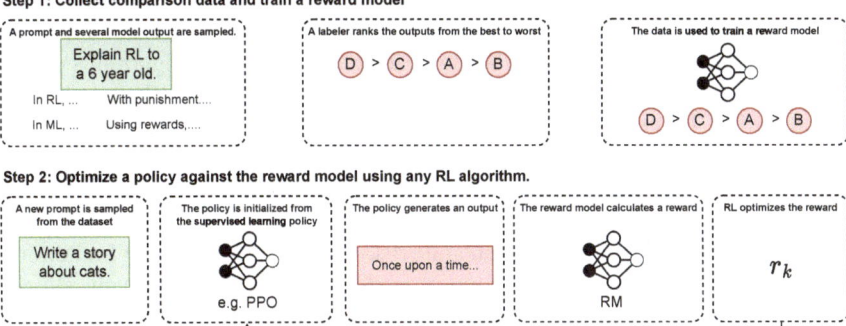

Fig. 3.8: Reinforcement Learning from Human Feedback.

solve a given problem, which can be in turn distilled to a simpler, interpretable (and even verifiable in some cases [11]) policy that could be deployed in devices with stringent memory or real-time execution requirements.

Finally, in some cases, approximating the policy that generated the provided demonstration trajectories might not be enough, as this expert policy might not be the optimal one. Here, a line of work called Inverse Reinforcement Learning (IRL) attempts to learn the reward function from the available data.

Classical algorithms [1, 114, 75] use only the available data from the demonstration trajectories ξ. Recently, through the success stories of ChatGPT and GPT-4 [17] a reinforcement learning approach, called Reinforcement Learning from Human Feedback (RLHF) [23, 90], that incorporates an interactive reward learning method has been popularized.

The high-level concept is shown in Figure 3.8. The current policy generates different outputs/action sequences for the same input and a human labeler is tasked to sort these in a specific order, depending on their quality. This creates a number of pair-wise reward combinations (i.e., action/output A is better than action/output B), based on which a reward model is learned. This reward model is in turn used as the reward function in a "classical" RL algorithm, aiming to improve the policy.

3.6 Conclusion and Outlook

In this chapter we outlined some challenges and avenues in the application of RL to the real world. Hence, we focused on particular aspects. After a general introduction to the topic, we first focus don methods that employ available models, called *model-based RL*. Next, we focused on a more recent trend, which uses available offline data to learn a policy. Finally, we also presented current research and trends in the area of safe reinforcement learning.

Contemporary research into reinforcement learning (RL) in the real world poses a variety of challenges, ranging from theoretical considerations to practical implementations. In the field of offline RL, the search for effective learning algorithms without continuous interaction with the environment remains an important challenge. Methods that utilize historical data while ensuring stability and convergence are promising to overcome this challenge. Furthermore, safe RL introduces a critical dimension that requires the development of algorithms that balance exploration and exploitation while ensuring safety compliance. Finding safe strategies in unsafe environments requires innovative approaches that integrate risk-aware decision-making frameworks. In addition, model-based RL is a cornerstone for improving the efficiency and generalization of sampling. The integration of learned models into decision-making processes provides opportunities for robust and adaptive behavior in complex domains. The research landscape in real-world RL faces significant advances driven by interdisciplinary collaboration and the convergence of theoretical insights and practical applications.

References

1. P. Abbeel and A. Y. Ng. Apprenticeship learning via inverse reinforcement learning. In *Proceedings of the twenty-first international conference on Machine learning*, page 1, 2004.
2. J. Achiam and D. Amodei. Benchmarking safe exploration in deep reinforcement learning. In *preprint*, 2019.
3. J. Achiam, D. Held, A. Tamar, and P. Abbeel. Constrained policy optimization. In *International Conference on Machine Learning*, 2017.
4. E. Amir. *Reasoning and decision making*, page 191–212. Cambridge University Press, 2014.
5. A. Anand, J. Walker, Y. Li, E. V'ertes, J. Schrittwieser, S. Ozair, T. Weber, and J. B. Hamrick. Procedural generalization by planning with self-supervised world models. *ArXiv*, abs/2111.01587, 2021.
6. P. Auer, N. Cesa-Bianchi, and P. Fischer. Finite-time analysis of the multiarmed bandit problem. *Machine learning*, 47(2):235–256, 2002.
7. P.-L. Bacon, J. Harb, and D. Precup. The option-critic architecture. In *Proceedings of the AAAI conference on artificial intelligence*, 2017.
8. A. P. Badia, B. Piot, S. Kapturowski, P. Sprechmann, A. Vitvitskyi, Z. D. Guo, and C. Blundell. Agent57: Outperforming the Atari human benchmark. In H. D. III and A. Singh, editors, *Proceedings of the 37th International Conference on Machine Learning*, volume 119 of *Proceedings of Machine Learning Research*, pages 507–517. PMLR, 13–18 Jul 2020.
9. A. P. Badia, P. Sprechmann, A. Vitvitskyi, Z. D. Guo, B. Piot, S. Kapturowski, O. Tieleman, M. Arjovsky, A. Pritzel, A. Bolt, and C. Blundell. Never give up: Learning directed exploration strategies. In *8th International Conference on Learning Representations, ICLR 2020, Addis Ababa, Ethiopia, April 26-30, 2020*. OpenReview.net, 2020.
10. M. Bansal, A. Krizhevsky, and A. Ogale. Chauffeurnet: Learning to drive by imitating the best and synthesizing the worst. *arXiv preprint arXiv:1812.03079*, 2018.
11. O. Bastani, Y. Pu, and A. Solar-Lezama. Verifiable reinforcement learning via policy extraction. *Advances in neural information processing systems*, 31, 2018.
12. J. Beck, R. Vuorio, E. Z. Liu, Z. Xiong, L. M. Zintgraf, C. Finn, and S. Whiteson. A survey of meta-reinforcement learning. *ArXiv*, abs/2301.08028, 2023.
13. M. G. Bellemare, S. Srinivasan, G. Ostrovski, T. Schaul, D. Saxton, and R. Munos. Unifying count-based exploration and intrinsic motivation. In D. D. Lee, M. Sugiyama, U. von Luxburg, I. Guyon, and R. Garnett, editors, *Advances in Neural Information Processing Systems 29:*

Annual Conference on Neural Information Processing Systems 2016, December 5-10, 2016, Barcelona, Spain, pages 1471–1479, 2016.

14. M. G. Bellemare, J. Veness, and E. Talvitie. Skip context tree switching. In *Proceedings of the 31th International Conference on Machine Learning, ICML 2014, Beijing, China, 21-26 June 2014*, volume 32 of *JMLR Workshop and Conference Proceedings*, pages 1458–1466. JMLR.org, 2014.

15. F. Berkenkamp, M. Turchetta, A. Schoellig, and A. Krause. Safe model-based reinforcement learning with stability guarantees. *Advances in neural information processing systems*, 30, 2017.

16. M. Bojarski, D. Del Testa, D. Dworakowski, B. Firner, B. Flepp, P. Goyal, L. D. Jackel, M. Monfort, U. Muller, J. Zhang, et al. End to end learning for self-driving cars. *arXiv preprint arXiv:1604.07316*, 2016.

17. S. Bubeck, V. Chandrasekaran, R. Eldan, J. Gehrke, E. Horvitz, E. Kamar, P. Lee, Y. T. Lee, Y. Li, S. Lundberg, et al. Sparks of artificial general intelligence: Early experiments with gpt-4. *arXiv preprint arXiv:2303.12712*, 2023.

18. Y. Burda and H. Edwards. Reinforcement learning with prediction-based rewards. https://openai.com/research/reinforcement-learning-with-prediction-based-rewards. Accessed: 2013-06-06.

19. Y. Burda, H. Edwards, A. J. Storkey, and O. Klimov. Exploration by random network distillation. In *7th International Conference on Learning Representations, ICLR 2019, New Orleans, LA, USA, May 6-9, 2019*. OpenReview.net, 2019.

20. E. F. Camacho and C. B. Alba. *Model predictive control*. Springer science & business media, 2013.

21. M. Charikar. Similarity estimation techniques from rounding algorithms. In J. H. Reif, editor, *Proceedings on 34th Annual ACM Symposium on Theory of Computing, May 19-21, 2002, Montréal, Québec, Canada*, pages 380–388. ACM, 2002.

22. L. Chen, K. Lu, A. Rajeswaran, K. Lee, A. Grover, M. Laskin, P. Abbeel, A. Srinivas, and I. Mordatch. Decision transformer: Reinforcement learning via sequence modeling. In M. Ranzato, A. Beygelzimer, Y. Dauphin, P. Liang, and J. W. Vaughan, editors, *Advances in Neural Information Processing Systems*, volume 34, pages 15084–15097. Curran Associates, Inc., 2021.

23. P. F. Christiano, J. Leike, T. Brown, M. Martic, S. Legg, and D. Amodei. Deep reinforcement learning from human preferences. *Advances in neural information processing systems*, 30, 2017.

24. K. Chua, R. Calandra, R. McAllister, and S. Levine. Deep reinforcement learning in a handful of trials using probabilistic dynamics models. *Advances in neural information processing systems*, 31, 2018.

25. K. Cobbe, J. Hilton, O. Klimov, and J. Schulman. Phasic policy gradient. In *International Conference on Machine Learning*, 2020.

26. K. Cobbe, O. Klimov, C. Hesse, T. Kim, and J. Schulman. Quantifying generalization in reinforcement learning. *ArXiv*, abs/1812.02341, 2018.

27. I. Danihelka, A. Guez, J. Schrittwieser, and D. Silver. Policy improvement by planning with gumbel. In *International Conference on Learning Representations*, 2022.

28. P.-T. De Boer, D. P. Kroese, S. Mannor, and R. Y. Rubinstein. A tutorial on the cross-entropy method. *Annals of operations research*, 134:19–67, 2005.

29. M. Deisenroth and C. E. Rasmussen. Pilco: A model-based and data-efficient approach to policy search. In *Proceedings of the 28th International Conference on machine learning (ICML-11)*, pages 465–472, 2011.

30. M. Dennis, N. Jaques, E. Vinitsky, A. M. Bayen, S. J. Russell, A. Critch, and S. Levine. Emergent complexity and zero-shot transfer via unsupervised environment design. *ArXiv*, abs/2012.02096, 2020.

31. A. Ecoffet, J. Huizinga, J. Lehman, K. O. Stanley, and J. Clune. Go-explore: a new approach for hard-exploration problems. *CoRR*, abs/1901.10995, 2019.

32. A. Ecoffet, J. Huizinga, J. Lehman, K. O. Stanley, and J. Clune. First return, then explore. *Nat.*, 590(7847):580–586, 2021.
33. C. Finn and S. Levine. Deep visual foresight for planning robot motion. In *2017 IEEE International Conference on Robotics and Automation (ICRA)*, pages 2786–2793. IEEE, 2017.
34. L. Fox, L. Choshen, and Y. Loewenstein. DORA the explorer: Directed outreaching reinforcement action-selection. In *6th International Conference on Learning Representations, ICLR 2018, Vancouver, BC, Canada, April 30 - May 3, 2018, Conference Track Proceedings*. OpenReview.net, 2018.
35. K. Frans, J. Ho, X. Chen, P. Abbeel, and J. Schulman. Meta learning shared hierarchies. *arXiv preprint arXiv:1710.09767*, 2017.
36. S. Fujimoto, D. Meger, and D. Precup. Off-policy deep reinforcement learning without exploration. In K. Chaudhuri and R. Salakhutdinov, editors, *Proceedings of the 36th International Conference on Machine Learning, ICML 2019, 9-15 June 2019, Long Beach, California, USA*, volume 97 of *Proceedings of Machine Learning Research*, pages 2052–2062. PMLR, 2019.
37. Y. Guo, J. Choi, M. Moczulski, S. Feng, S. Bengio, M. Norouzi, and H. Lee. Memory based trajectory-conditioned policies for learning from sparse rewards. In H. Larochelle, M. Ranzato, R. Hadsell, M. Balcan, and H. Lin, editors, *Advances in Neural Information Processing Systems 33: Annual Conference on Neural Information Processing Systems 2020, NeurIPS 2020, December 6-12, 2020, virtual*, 2020.
38. A. Gupta, A. Pacchiano, Y. Zhai, S. Kakade, and S. Levine. Unpacking reward shaping: Understanding the benefits of reward engineering on sample complexity. *Advances in Neural Information Processing Systems*, 35:15281–15295, 2022.
39. D. Ha and J. Schmidhuber. World models. *arXiv preprint arXiv:1803.10122*, 2018.
40. A. Hallak, D. D. Castro, and S. Mannor. Contextual markov decision processes. *ArXiv*, abs/1502.02259, 2015.
41. S. O. Hansson. *Decision Theory: A Brief Introduction*, pages 1–. Royal Institute of Technology, 01 2005.
42. L. Hewing, K. P. Wabersich, M. Menner, and M. N. Zeilinger. Learning-based model predictive control: Toward safe learning in control. *Annual Review of Control, Robotics, and Autonomous Systems*, 3:269–296, 2020.
43. M. Igl, G. Farquhar, J. Luketina, W. Boehmer, and S. Whiteson. Transient non-stationarity and generalisation in deep reinforcement learning. In *International Conference on Learning Representations*, 2020.
44. B. Ivanovic, J. Harrison, A. Sharma, M. Chen, and M. Pavone. Barc: Backward reachability curriculum for robotic reinforcement learning. In *2019 International Conference on Robotics and Automation (ICRA)*, pages 15–21. IEEE, 2019.
45. M. Jaderberg, V. Mnih, W. M. Czarnecki, T. Schaul, J. Z. Leibo, D. Silver, and K. Kavukcuoglu. Reinforcement learning with unsupervised auxiliary tasks. *arXiv preprint arXiv:1611.05397*, 2016.
46. M. Janner, J. Fu, M. Zhang, and S. Levine. When to trust your model: Model-based policy optimization. *Advances in neural information processing systems*, 32, 2019.
47. M. Janner, Q. Li, and S. Levine. Offline reinforcement learning as one big sequence modeling problem. *Advances in neural information processing systems*, 34:1273–1286, 2021.
48. N. L. Johnson, S. Kotz, and N. Balakrishnan. *Chapter 25: Beta Distributions*, pages 210 – 276. Wiley, 1995.
49. E. Kaufmann, O. Cappe, and A. Garivier. On bayesian upper confidence bounds for bandit problems. In N. D. Lawrence and M. Girolami, editors, *Proceedings of the Fifteenth International Conference on Artificial Intelligence and Statistics*, volume 22 of *Proceedings of Machine Learning Research*, pages 592–600, La Palma, Canary Islands, 21–23 Apr 2012. PMLR.
50. K. Khetarpal, M. Riemer, I. Rish, and D. Precup. Towards continual reinforcement learning: A review and perspectives. *J. Artif. Intell. Res.*, 75:1401–1476, 2020.
51. R. Kirk, A. Zhang, E. Grefenstette, and T. Rocktäschel. A survey of zero-shot generalisation in deep reinforcement learning. *J. Artif. Intell. Res.*, 76:201–264, 2021.

52. L. Kocsis and C. Szepesvári. Bandit based monte-carlo planning. In *Proc. ECML*, volume 2006, pages 282–293, 09 2006.
53. A. Kumar, J. Fu, M. Soh, G. Tucker, and S. Levine. Stabilizing off-policy q-learning via bootstrapping error reduction. In H. M. Wallach, H. Larochelle, A. Beygelzimer, F. d'Alché-Buc, E. B. Fox, and R. Garnett, editors, *Advances in Neural Information Processing Systems 32: Annual Conference on Neural Information Processing Systems 2019, NeurIPS 2019, December 8-14, 2019, Vancouver, BC, Canada*, pages 11761–11771, 2019.
54. A. Kumar, A. Zhou, G. Tucker, and S. Levine. Conservative q-learning for offline reinforcement learning. In H. Larochelle, M. Ranzato, R. Hadsell, M. Balcan, and H. Lin, editors, *Advances in Neural Information Processing Systems 33: Annual Conference on Neural Information Processing Systems 2020, NeurIPS 2020, December 6-12, 2020, virtual*, 2020.
55. D. Landgraf, A. Völz, G. Kontes, C. Mutschler, and K. Graichen. Hierarchical learning for model predictive collision avoidance. *IFAC-PapersOnLine*, 55(20):355–360, 2022.
56. S. Levine, A. Kumar, G. Tucker, and J. Fu. Offline reinforcement learning: Tutorial, review, and perspectives on open problems. *CoRR*, abs/2005.01643, 2020.
57. B. Li, V. Franccois-Lavet, T. V. Doan, and J. Pineau. Domain adversarial reinforcement learning. *ArXiv*, abs/2102.07097, 2021.
58. Y. Liu, A. Halev, and X. Liu. Policy learning with constraints in model-free reinforcement learning: A survey. In *International Joint Conference on Artificial Intelligence*, 2021.
59. V. Mnih, K. Kavukcuoglu, D. Silver, A. Graves, I. Antonoglou, D. Wierstra, and M. Riedmiller. Playing atari with deep reinforcement learning. In *NIPS Deep Learning Workshop 2013*, 2013.
60. T. M. Moerland, J. Broekens, A. Plaat, C. M. Jonker, et al. Model-based reinforcement learning: A survey. *Foundations and Trends® in Machine Learning*, 16(1):1–118, 2023.
61. A. Nagabandi, K. Konolige, S. Levine, and V. Kumar. Deep dynamics models for learning dexterous manipulation. In *Conference on Robot Learning*, pages 1101–1112. PMLR, 2020.
62. K. Narasimhan. Projection-based constrained policy optimization. *ArXiv*, abs/2010.03152, 2020.
63. OpenAI, I. Akkaya, M. Andrychowicz, M. Chociej, M. Litwin, B. McGrew, A. Petron, A. Paino, M. Plappert, G. Powell, R. Ribas, J. Schneider, N. A. Tezak, J. Tworek, P. Welinder, L. Weng, Q. Yuan, W. Zaremba, and L. M. Zhang. Solving rubik's cube with a robot hand. *ArXiv*, abs/1910.07113, 2019.
64. T. Osa, J. Pajarinen, G. Neumann, J. A. Bagnell, P. Abbeel, J. Peters, et al. An algorithmic perspective on imitation learning. *Foundations and Trends® in Robotics*, 7(1-2):1–179, 2018.
65. G. Ostrovski, M. G. Bellemare, A. van den Oord, and R. Munos. Count-based exploration with neural density models. In D. Precup and Y. W. Teh, editors, *Proceedings of the 34th International Conference on Machine Learning, ICML 2017, Sydney, NSW, Australia, 6-11 August 2017*, volume 70 of *Proceedings of Machine Learning Research*, pages 2721–2730. PMLR, 2017.
66. P. Oudeyer, F. Kaplan, and V. V. Hafner. Intrinsic motivation systems for autonomous mental development. *IEEE Trans. Evol. Comput.*, 11(2):265–286, 2007.
67. S. Paternain, L. F. O. Chamon, M. Calvo-Fullana, and A. Ribeiro. Constrained reinforcement learning has zero duality gap. *ArXiv*, abs/1910.13393, 2019.
68. D. Pathak, P. Agrawal, A. A. Efros, and T. Darrell. Curiosity-driven exploration by self-supervised prediction. In D. Precup and Y. W. Teh, editors, *Proceedings of the 34th International Conference on Machine Learning, ICML 2017, Sydney, NSW, Australia, 6-11 August 2017*, volume 70 of *Proceedings of Machine Learning Research*, pages 2778–2787. PMLR, 2017.
69. D. Pathak, D. Gandhi, and A. Gupta. Self-supervised exploration via disagreement. In K. Chaudhuri and R. Salakhutdinov, editors, *Proceedings of the 36th International Conference on Machine Learning, ICML 2019, 9-15 June 2019, Long Beach, California, USA*, volume 97 of *Proceedings of Machine Learning Research*, pages 5062–5071. PMLR, 2019.
70. X. B. Peng, M. Andrychowicz, W. Zaremba, and P. Abbeel. Sim-to-real transfer of robotic control with dynamics randomization. *2018 IEEE International Conference on Robotics and Automation (ICRA)*, pages 1–8, 2017.

71. R. Raileanu and R. Fergus. Decoupling value and policy for generalization in reinforcement learning. *ArXiv*, abs/2102.10330, 2021.
72. R. Raileanu, M. Goldstein, D. Yarats, I. Kostrikov, and R. Fergus. Automatic data augmentation for generalization in deep reinforcement learning. *ArXiv*, abs/2006.12862, 2020.
73. K. Rao, C. Harris, A. Irpan, S. Levine, J. Ibarz, and M. Khansari. Rl-cyclegan: Reinforcement learning aware simulation-to-real. *2020 IEEE/CVF Conference on Computer Vision and Pattern Recognition (CVPR)*, pages 11154–11163, 2020.
74. M. Riemer, M. Liu, and G. Tesauro. Learning abstract options. *Advances in neural information processing systems*, 31, 2018.
75. S. Rosbach, V. James, S. Großjohann, S. Homoceanu, and S. Roth. Driving with style: Inverse reinforcement learning in general-purpose planning for automated driving. In *2019 IEEE/RSJ International Conference on Intelligent Robots and Systems (IROS)*, pages 2658–2665. IEEE, 2019.
76. C. D. Rosin. Multi-armed bandits with episode context. *Annals of Mathematics and Artificial Intelligence*, 61(3):203–230, 2011.
77. S. Ross, G. Gordon, and D. Bagnell. A reduction of imitation learning and structured prediction to no-regret online learning. In *Proceedings of the fourteenth international conference on artificial intelligence and statistics*, pages 627–635. JMLR Workshop and Conference Proceedings, 2011.
78. D. Russo, B. V. Roy, A. Kazerouni, I. Osband, and Z. Wen. A tutorial on thompson sampling. *Found. Trends Mach. Learn.*, 11(1):1–96, 2018.
79. J. Schmidhuber. A possibility for implementing curiosity and boredom in model-building neural controllers. In *Proceedings of the First International Conference on Simulation of Adaptive Behavior on From Animals to Animats*, page 222–227, Cambridge, MA, USA, 1991. MIT Press.
80. L. M. Schmidt, J. Brosig, A. Plinge, B. M. Eskofier, and C. Mutschler. An introduction to multi-agent reinforcement learning and review of its application to autonomous mobility. In *25th IEEE International Conference on Intelligent Transportation Systems, ITSC 2022, Macau, China, October 8-12, 2022*, pages 1342–1349. IEEE, 2022.
81. L. M. Schmidt, G. Kontes, A. Plinge, and C. Mutschler. Can you trust your autonomous car? interpretable and verifiably safe reinforcement learning. In *2021 IEEE Intelligent Vehicles Symposium (IV)*, pages 171–178. IEEE, 2021.
82. L. M. Schmidt, S. Rietsch, A. Plinge, B. M. Eskofier, and C. Mutschler. How to learn from risk: Explicit risk-utility reinforcement learning for efficient and safe driving strategies. In *25th IEEE International Conference on Intelligent Transportation Systems, ITSC 2022, Macau, China, October 8-12, 2022*, pages 1913–1920. IEEE, 2022.
83. J. Schrittwieser, I. Antonoglou, T. Hubert, K. Simonyan, L. Sifre, S. Schmitt, A. Guez, E. Lockhart, D. Hassabis, T. Graepel, T. P. Lillicrap, and D. Silver. Mastering atari, go, chess and shogi by planning with a learned model. *Nature*, 588:604 – 609, 2019.
84. D. Silver, A. Huang, C. J. Maddison, A. Guez, L. Sifre, G. van den Driessche, J. Schrittwieser, I. Antonoglou, V. Panneershelvam, M. Lanctot, S. Dieleman, D. Grewe, J. Nham, N. Kalchbrenner, I. Sutskever, T. Lillicrap, M. Leach, K. Kavukcuoglu, T. Graepel, and D. Hassabis. Mastering the game of Go with deep neural networks and tree search. *Nature*, 529(7587):484–489, jan 2016.
85. D. Silver, T. Hubert, J. Schrittwieser, I. Antonoglou, M. Lai, A. Guez, M. Lanctot, L. Sifre, D. Kumaran, T. Graepel, T. P. Lillicrap, K. Simonyan, and D. Hassabis. Mastering chess and shogi by self-play with a general reinforcement learning algorithm. *ArXiv*, abs/1712.01815, 2017.
86. D. Silver, J. Schrittwieser, K. Simonyan, I. Antonoglou, A. Huang, A. Guez, T. Hubert, L. Baker, M. Lai, A. Bolton, Y. Chen, T. Lillicrap, F. Hui, L. Sifre, G. van den Driessche, T. Graepel, and D. Hassabis. Mastering the game of go without human knowledge. *Nature*, 550:354–, Oct. 2017.
87. D. Silver, S. Singh, D. Precup, and R. S. Sutton. Reward is enough. *Artificial Intelligence*, 299:103535, 2021.

88. A. Sonar, V. Pacelli, and A. Majumdar. Invariant policy optimization: Towards stronger generalization in reinforcement learning. In *Conference on Learning for Dynamics & Control*, 2020.

89. B. C. Stadie, S. Levine, and P. Abbeel. Incentivizing exploration in reinforcement learning with deep predictive models. *CoRR*, abs/1507.00814, 2015.

90. N. Stiennon, L. Ouyang, J. Wu, D. Ziegler, R. Lowe, C. Voss, A. Radford, D. Amodei, and P. F. Christiano. Learning to summarize with human feedback. *Advances in Neural Information Processing Systems*, 33:3008–3021, 2020.

91. R. S. Sutton. Dyna, an integrated architecture for learning, planning, and reacting. *ACM Sigart Bulletin*, 2(4):160–163, 1991.

92. R. S. Sutton and A. G. Barto. *Reinforcement learning - an introduction*. Adaptive computation and machine learning. MIT Press, 1998.

93. R. S. Sutton, D. Precup, and S. Singh. Between mdps and semi-mdps: A framework for temporal abstraction in reinforcement learning. *Artificial intelligence*, 112(1-2):181–211, 1999.

94. A. Tamar, Y. Chow, M. Ghavamzadeh, and S. Mannor. Sequential decision making with coherent risk. *IEEE Transactions on Automatic Control*, 62(7):3323–3338, 2017.

95. H. Tang, R. Houthooft, D. Foote, A. Stooke, X. Chen, Y. Duan, J. Schulman, F. D. Turck, and P. Abbeel. #exploration: A study of count-based exploration for deep reinforcement learning. In I. Guyon, U. von Luxburg, S. Bengio, H. M. Wallach, R. Fergus, S. V. N. Vishwanathan, and R. Garnett, editors, *Advances in Neural Information Processing Systems 30: Annual Conference on Neural Information Processing Systems 2017, December 4-9, 2017, Long Beach, CA, USA*, pages 2753–2762, 2017.

96. Y. Tassa, T. Erez, and E. Todorov. Synthesis and stabilization of complex behaviors through online trajectory optimization. In *2012 IEEE/RSJ International Conference on Intelligent Robots and Systems*, pages 4906–4913. IEEE, 2012.

97. C. Tessler, D. J. Mankowitz, and S. Mannor. Reward constrained policy optimization. *ArXiv*, abs/1805.11074, 2018.

98. S. Thrun, W. Burgard, and D. Fox. *Probabilistic robotics*. Intelligent robotics and autonomous agents. MIT Press, 2005.

99. E. Todorov, T. Erez, and Y. Tassa. Mujoco: A physics engine for model-based control. In *IROS*, pages 5026–5033. IEEE, 2012.

100. J. Tremblay, A. Prakash, D. Acuna, M. Brophy, V. Jampani, C. Anil, T. To, E. Cameracci, S. Boochoon, and S. Birchfield. Training deep networks with synthetic data: Bridging the reality gap by domain randomization. *2018 IEEE/CVF Conference on Computer Vision and Pattern Recognition Workshops (CVPRW)*, pages 1082–10828, 2018.

101. A. van den Oord, N. Kalchbrenner, L. Espeholt, K. Kavukcuoglu, O. Vinyals, and A. Graves. Conditional image generation with pixelcnn decoders. In D. D. Lee, M. Sugiyama, U. von Luxburg, I. Guyon, and R. Garnett, editors, *Advances in Neural Information Processing Systems 29: Annual Conference on Neural Information Processing Systems 2016, December 5-10, 2016, Barcelona, Spain*, pages 4790–4798, 2016.

102. A. Vaswani, N. M. Shazeer, N. Parmar, J. Uszkoreit, L. Jones, A. N. Gomez, L. Kaiser, and I. Polosukhin. Attention is all you need. In *NIPS*, 2017.

103. A. S. Vezhnevets, S. Osindero, T. Schaul, N. Heess, M. Jaderberg, D. Silver, and K. Kavukcuoglu. Feudal networks for hierarchical reinforcement learning. In *International Conference on Machine Learning*, pages 3540–3549. PMLR, 2017.

104. R. Wang, J. Lehman, J. Clune, and K. O. Stanley. Paired open-ended trailblazer (poet): Endlessly generating increasingly complex and diverse learning environments and their solutions. *ArXiv*, abs/1901.01753, 2019.

105. M. Watter, J. Springenberg, J. Boedecker, and M. Riedmiller. Embed to control: A locally linear latent dynamics model for control from raw images. *Advances in neural information processing systems*, 28, 2015.

106. G. Williams, N. Wagener, B. Goldfain, P. Drews, J. M. Rehg, B. Boots, and E. A. Theodorou. Information theoretic mpc for model-based reinforcement learning. In *2017 IEEE International Conference on Robotics and Automation (ICRA)*, pages 1714–1721. IEEE, 2017.

107. W. Ye, S.-W. Liu, T. Kurutach, P. Abbeel, and Y. Gao. Mastering atari games with limited data. *ArXiv*, abs/2111.00210, 2021.

108. C. Yu, J. Liu, S. Nemati, and G. Yin. Reinforcement learning in healthcare: A survey. *ACM Comput. Surv.*, 55(1), nov 2021.

109. T. Yu, A. Kumar, R. Rafailov, A. Rajeswaran, S. Levine, and C. Finn. COMBO: conservative offline model-based policy optimization. In M. Ranzato, A. Beygelzimer, Y. N. Dauphin, P. Liang, and J. W. Vaughan, editors, *Advances in Neural Information Processing Systems 34: Annual Conference on Neural Information Processing Systems 2021, NeurIPS 2021, December 6-14, 2021, virtual*, pages 28954–28967, 2021.

110. T. Yu, G. Thomas, L. Yu, S. Ermon, J. Y. Zou, S. Levine, C. Finn, and T. Ma. MOPO: model-based offline policy optimization. In H. Larochelle, M. Ranzato, R. Hadsell, M. Balcan, and H. Lin, editors, *Advances in Neural Information Processing Systems 33: Annual Conference on Neural Information Processing Systems 2020, NeurIPS 2020, December 6-12, 2020, virtual*, 2020.

111. H. Zhang, H. Chen, D. S. Boning, and C.-J. Hsieh. Robust reinforcement learning on state observations with learned optimal adversary. *ArXiv*, abs/2101.08452, 2021.

112. R. Zhao and V. Tresp. Curiosity-driven experience prioritization via density estimation. *CoRR*, abs/1902.08039, 2019.

113. W. Zhao, J. P. Queralta, and T. Westerlund. Sim-to-real transfer in deep reinforcement learning for robotics: a survey. *2020 IEEE Symposium Series on Computational Intelligence (SSCI)*, pages 737–744, 2020.

114. B. D. Ziebart, A. L. Maas, J. A. Bagnell, A. K. Dey, et al. Maximum entropy inverse reinforcement learning. In *Aaai*, volume 8, pages 1433–1438. Chicago, IL, USA, 2008.

Chapter 4
Learning with Limited Labelled Data

Christoffer Loeffler[1,2], Rasmus Hvingelby[2], Jann Goschenhofer[2]

Abstract Modern machine and deep learning require large amounts of training data. Yet, even if the data itself is abundantly available, the fraction of annotated data may still be proportionally small or missing. Hence, learning with limited labeled data is an important research field. Two streams of research attack this problem from opposite directions [64]. On the one hand, semi-supervised learning aims to leverage all information by directly incorporating unlabeled data. On the other hand, active learning finds unlabeled data for that annotations would be most beneficial for learning, and queries humans-in-the-loop of model training. This chapter discusses both concepts and their essential principles, methodological overlaps, and strengths and weaknesses. Furthermore, we elaborate on possible combinations and their advantages ands disadvantages. Finally, the conclusion refers to recent state-of-the-art and provides an outlook into the future of learning with few labeled data.

Key words: semi-supervised learning, active learning

4.1 Introduction

One main hurdle in the design and training of machine learning models is their need for large amounts of labeled training data. This labeling process also referred to as the annotation process, can be very time consuming as it requires the knowledge and involvement of domain experts that add annotated input data X with their respective labels Y. Despite this abundance of labeled training data, there often exists a large amount of unlabeled data that was (not yet) annotated by domain experts. Due to

---

[1]Escuela de Ingeniería Informática, Pontificia Universidad Católica de Valparaíso, Valparaíso, Chile
[2]Fraunhofer Institute for Integrated Circuits IIS, Fraunhofer IIS, Nuremberg, Germany

Corresponding author: Christoffer Löffler
e-mail: christoffer.loffler@pucv.cl

© The Author(s) 2024   77
C. Mutschler et al. (eds.), *Unlocking Artificial Intelligence*,
https://doi.org/10.1007/978-3-031-64832-8_4

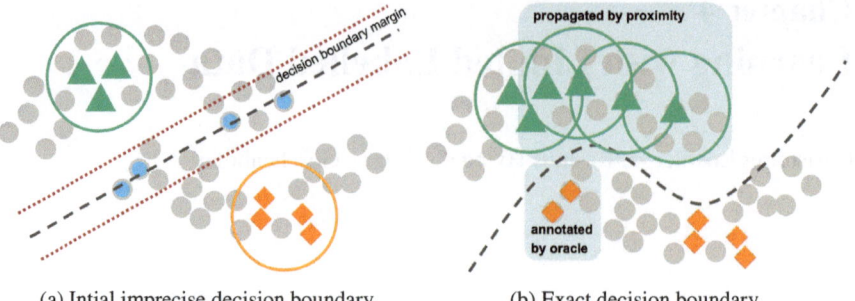

(a) Intial imprecise decision boundary.   (b) Exact decision boundary.

Fig. 4.1: In limited labeled data scenarios, only a subset of samples is annotated (orange and green data points) while the majority is unlabeled (grey data points). (a) shows the resulting imprecise and wrong decision boundary of an exemplary linear binary classifier trained on the labeled data only. As shown in (b), semi-supervised learning assumes that unlabeled samples are of the same class if they are close in proximity in the feature space. Semi-supervised methods such as label propagation, see upper part of (b), exploit this proximity to propagate the class information of the labeled samples. On the other hand, active learning aims to improve the learned model by selecting the most informative samples to annotate. Here, the annotator is queried with the most uncertain samples, i.e., those that are close to the decision boundary as depicted in the lower part of (b).

this, machine learning experts often face the situation of "learning with limited labeled data" where a small dataset of labeled data exists next to a large dataset of unlabeled data. Both semi-supervised and active learning try to leverage the information given in the unlabeled dataset next to the labeled dataset to train strong-performing machine learning models. Thereby, semi-supervised learning focuses on the direct incorporation of unlabeled data in the training process. Active learning on the other hand aims at finding those unlabeled samples that would support model training the most and presents them to a (human) oracle that iteratively annotates subsets of the unlabeled dataset in a model-driven way. Both approaches thereby "attack the same problem from opposite directions" [64] as illustrated in Figure 4.1. While semi-supervised methods exploit what the model thinks it knows about the unlabeled data, active methods attempt to explore the unknown aspects.

In the following, we provide an overview of both approaches and discuss their methodological overlaps, strengths, and weaknesses to give the reader a comprehensive understanding of both fields.

Throughout the chapter, we make use of the following notation. We define an input space $\mathcal{X}$ and use $y^{(i)} \in \mathcal{Y}$ to denote a categorical variable in the target space $\mathcal{Y}$ with a cardinality of $K = |\mathcal{Y}|$. Further, we define a labeled dataset $\mathcal{D}^l$ consisting of n_l tuples of samples and their respective labels $(x_i, y_i), ..., (x_l, y_l) \in \mathcal{D}^l$ as well as an unlabeled dataset $\mathcal{D}^u$ which consists of n_u samples $x_{l+1}, ..., x_u \in \mathcal{D}^u$. The goal of semi-supervised learning is to train a prediction model $f : \mathcal{X} \mapsto \mathcal{Y}$ on a dataset $\mathcal{D} = (\mathcal{D}^l, \mathcal{D}^u)$ which consists of an labeled dataset $\mathcal{D}^l = \{(x^{(i)}, y^{(i)})\}_{i=1}^{n_l}$ and an unlabeled dataset $\mathcal{D}^u = \{x^{(i)}\}_{i=n_l+1}^{n}$ where $n = n_l + n_u$. Model predictions are denoted as $\hat{y}^{(i)} = f(x^{(i)}|\theta)$ where $\hat{y}^{(i)}$ is a class probability vector of dimension K, $\hat{y}^{(ik)}$ denotes the predicted probability for class $k \in 0, ..., K$ for input sample $x^{(i)}$ and θ refers to the model parameters. We consider the case where $n_l \ll n_u$, as usual in SSL. Further, we define one batch of data as $\mathcal{B} \subset \mathcal{D}$, where $\mathcal{B}^l \subseteq \mathcal{D}^l$ contains the labeled samples and $\mathcal{B}^u \subseteq \mathcal{D}^u$ the unlabeled samples in that batch such that $\mathcal{B} = (\mathcal{B}^l, \mathcal{B}^u)$.

4.2 Semi-Supervised Learning

The goal and promise of semi-supervised learning, at the intersection of unsupervised and supervised learning, is to leverage both labeled and unlabeled data for machine learning tasks. The expanding research in this field is mainly driven by the sometimes prohibitively high effort involved in annotating large labeled datasets on the one side and the abundance of unlabeled data on the other. Hence, semi-supervised methods mainly focus on settings with few labeled and many unlabeled training data. While there exists research on semi-supervised learning for a broad variety of learning tasks, we focus on semi-supervised classification for which most research exists.

Semi-supervised learning relies on three interconnected assumptions [72].

1. **Smoothness assumption**: two samples x_i, x_j that are close to each other in a high-density region of the input space should have similar labels y_i, y_j.
2. **Low-density assumption**: the decision boundary of model f should go through low-density areas where $p(x)$ is low, so-called low-density regions. This adds another perspective to assumption 1) as placing the decision boundary in a high-density region would violate this smoothness assumption.
3. **Manifold assumption**: high-dimensional data should lie on lower-dimensional manifolds, subspaces that are locally Euclidean. Hence, two samples x_i, x_j that lie on the same manifold are assumed to have similar labels y_i, y_j. This assumption mainly targets the curse of dimensionality and allows for the translation of the previous assumptions to high-dimensional settings.

Furthermore, semi-supervised algorithms can be distinguished in **inductive** and **transductive** methods. Inductive learning algorithms aim at learning a mapping $f : X \rightarrow Y$ from the data to the input space. After the learning phase at inference time, these models along their estimated model parameters can be used to assign predicted labels from Y to newly, unseen data X. Contrary to this, transductive methods merely aim at annotating the unlabeled D^u using the D^l such that $f : X^u \rightarrow Y$ without the learning of a general decision rule. In that sense, induction is more general as it aims at learning general decision rules while transduction tries to reason from the labeled cases to the specific unlabeled cases.

Following this introduction, we next provide an overview of classical semi-supervised learning methods and then focus on recent developments in deep semi-supervised learning and the different concepts applied therein.

4.2.1 Classical Semi-Supervised Learning

This section gives a rough overview of classical machine learning approaches developed for semi-supervised classification. Following the taxonomy developed in the standard textbook [13], we distinguish these models into four model classes: 1) **Generative models** such as the EM-algorithm for incomplete data [18] that aim at learning the class-conditional density $p(x|y)$ and use the unlabeled data D^u to improve its estimation. 2) Approaches that follow the **Low-Density Separation** rationale try to direct the decision boundary through low-density areas following the low-density assumption using the latent information in D^u to identify these areas. This mainly involves max-margin estimators such as the transductive SVM [17]. 3) **Graph-based methods** that exploit the neighborhood of labeled and unlabeled samples defined via a metric (e.g. defined via a kernel function following the manifold assumption). These neighborhood relationships are then used to propagate class labels from the labeled to their neighboring unlabeled samples. Most of these methods are transductive and Label Propagation [78] is one prominent method in this model class. 4) **Change of Representation**: two-step approaches that e.g. use D^u in the first step to learn a meaningful data representation which is then tailored towards the learning task using D^l in the second step.

4.2.2 Deep Semi-Supervised Learning

In a more recent overview, [72] extended this taxonomy further towards the use of neural networks along the dimensions of transduction and induction. Under transduction, they collect mainly Graph-based models that leverage joint neighborhood structures in $D = (D^l, D^u)$. With that, they follow the structure of Chapelle et al. [13] but extend it towards deep graph-based methods such as Deep Label Propagation [32].

They further differentiate different learning paradigms that mainly aim at extending existing supervised inductive methods, toward using additional unlabeled data D^u next to the labeled data D^l.

1) **Self-training** methods, also referred to as "Wrapper methods" or "Pseudo-Labeling", use a supervised model f trained on D^l to iterative pseudo-label additional unlabeled samples from D^u to augment the training data set and then re-train on this expanded D^{l*}.

2) **Unsupervised preprocessing** methods that use D^u to aid the generation of a meaningful representation of the data in an unsupervised manner. This includes

the extraction of meaningful features from the raw data to find an embedding that is favorable for the initial learning task. Such approaches contain but are not limited to dimensionality reduction techniques such as PCA or autoencoders, again related to the manifold assumption. Further, cluster-then-label approaches use clustering techniques over D or D^u only to facilitate the initial supervised learning task. The final sub-branch of methods mainly targeted at neural-network-based methods summarizes pre-training algorithms that use D^u to initialize the model architecture which is then fine-tuned on D^l.

3) **Inherently semi-supervised approaches** that extend supervised loss functions defined over D^l with tailored loss functions that allow the inclusion of D^u in the training process to enable a semi-supervised model training.

Recent strong-performing semi-supervised learning methods follow at least one of these paradigms or are combinations of them. In the remainder of this chapter, we will focus on 1) Self-training and 3) intrinsically semi-supervised learning as these are the most active research areas at the time of writing.

4.2.2.1 Self-training

Self-training, also referred to as Pseudo-labeling or Self-learning, is one of the oldest approaches to semi-supervised learning [62, 22, 2]. It follows the idea that the model trains itself by iteratively annotating parts of the unlabeled data. The procedure usually alternates between a training and a *pseudo-labeling* step. After the training step, the model selects unlabeled samples via a selection criterion such as model confidence. These selected samples are then assigned the predicted label and added from D^u to the now updated labeled dataset D^{l*}. The model is then trained on this (pseudo-) labeled dataset and this self-training cycle continues until a stopping criterion, such as the fact that no unlabeled data is left, is reached. This concept relates to Active Learning replacing the there often-used (human) oracle with the model f.

Self-training was transferred to deep learning by [40] and since then has sparked the creation of numerous variants. For instance, [3] investigate the confirmation bias that can occur when the model is overconfident but wrong on unlabeled samples D^u. This then leads to wrong pseudo-labels which confuses model training. They use Mixup [77] and the injection of label noise to overcome this issue. In a similar realm, [60] successfully use a combination of prediction confidence and model uncertainty with two distinct thresholds as a pseudo-label selection criterion to overcome this issue. [11] take in another perspective and combine *Curriculum Learning* with Pseudo-Labeling. This enables the model to use adaptive thresholds in the selection criterion and leads to on-par performance with more advanced and complex semi-supervised learning techniques. Next to these extensions, pseudo-labeling remains a crucial component in recent semi-supervised models.

4.2.2.2 Unsupervised Regularization

Alternative inherently semi-supervised methods create additional loss functions L^u defined over D^u or D which are combined with the initial, supervised loss function L^l to allow joint model training over both datasets via the combined loss $L = L^l + \lambda L^u$, where hyperparameter λ controls the impact of D^u. This has a regularizing effect and has the benefit that samples from D^u can be inherently integrated into model training. One early approach in this context is **Minimum Entropy Regularization** (MER) [27] where the prediction entropy serves as unsupervised regularization term such that $L^u(f, x_i) = H(f(x_i))$ for $x_i \in D^u$ leading to the final loss function

$$L(f, B^l, B^u) = - \sum_{(x^{(i)}, y^{(i)}) \in \mathcal{B}^l} \sum_{k=1}^{K} y^{(ik)} \log(\hat{y}^{(ik)}) - \lambda \sum_{(x^{(i)}) \in \mathcal{B}^u} \sum_{k=1}^{K} \hat{y}^{(ik)} \log(\hat{y}^{(ik)})$$

(4.1)

This forces the model to create low entropy predictions, i.e., sharp predictions, over the entire dataset. MER was developed following the observation that unlabeled data does not contribute to the maximum-likelihood estimation of discriminative, supervised models. Thus, it introduces the regularization term as a prior adding an inductive bias to the model driven by the unlabeled data. The penalization of the model for high-entropy predictions over the unlabeled data potentially pushes the model's decision boundary towards low-density regions, following the *low-density assumption*. Originally developed for logistic regression, MER can also be used for neural-network-based classifiers.

The rationale of unsupervised regularization was further extended within models that use **Consistency Regularization**, also termed *perturbation-based methods*. These build up on the *smoothness assumption* such that a slightly perturbed version $\tilde{x}^{(i)} = x^{(i)}$ of the input sample $x^{(i)}$ is expected to have the similar class the clean, non-perturbed version $x^{(i)}$, assuming $x^{(i)}$ lies in a high-density region. This expected *consistency* in model predictions lends this branch of research its name. In recent years, different perturbation methods have been developed from the simple addition of random noise to inputs to the use of more elaborate methods which we will cover in the following.

Noise Perturbation. With the Ladder-Net, [55] introduced an Autoencoder-based approach that injects additive gaussian noise at different intermediate representations of the input samples and calculates a regularization term over changes in these representations. This allows them to a) robustify the model representations and b) train the model on the joint D using both the reconstruction loss of the autoencoder as well as the noise-regularization term. The encoder part of the architecture is used at inference time. Instead of random noise, [51] propose to add directed *adversarial noise* to the unlabeled input samples as a regularization mechanism. In contrast to the addition of noise to the input sample, the Π-Model adds noise in the form of dropout layers to the model architecture. The regularization term is then calculated over different model prediction samples via the MCDropout algorithm [23] which simulates an ensemble of models and enforces consistent model predictions across those.

Temporal Consistency. Another branch of research leverages predictions from different training stages as a perturbation mechanism following the rationale that the model should produce *temporally consistent* model predictions during training. Within the Temporal Ensembling Model, [38] maintain an exponential moving average of model predictions over stochastically augmented, unlabeled input samples from past training epochs. In the current training epoch, these serve as an auxiliary target and the squared distance between those past model predictions is used as an unsupervised loss function L^u. [70] follow this rationale as well in their Mean Teacher architecture. Instead of storing past model predictions of D^u, they maintain a teacher version of the initial student model whose weight parameters are updated via exponential moving averaging of the student model's weights that are directly optimized via gradient descent. Model predictions over D^u from the teacher model here serve as auxiliary targets in the unsupervised loss part L^u. This concept remains an important training paradigm for semi-supervised learning and was used in the Unbiased Teacher architecture for semi-supervised object detection [44].

4.2.3 Self-Training and Consistency Regularization

The use of elaborate data augmentation strategies as perturbation methods in consistency regularization sparked a more recent line of research in this area. Within MixMatch [9], the authors combine a) data augmentation with the different established semi-supervised techniques b) Pseudo-Labeling and Entropy Regularization via a Sharpening function, and c) Mixup [77] in one holistic approach to semi-supervised learning. Model prediction vectors over differently augmented versions of an unlabeled sample $x_u^{(i)}$) are averaged, sharpened via a temperature scaling mechanism, and then used as pseudo-labels. Subsequently, a batch of labeled and pseudo-labeled data are combined via MixUp to create synthetic training samples with are then fed into a Brier-Score as an unsupervised loss function L^u. This combination of different semi-supervised learning paradigms allows MixMatch to yield impressive predictive performance given low levels of supervision. With FixMatch, [68] improve upon these results by introducing the strong- and weak- augmentation scheme: pseudo-labels from weakly augmented samples $x_u^{(i)}$ are selected based on a prediction confidence criterion and serve as training targets in the auxiliary loss L^u. Model predictions over exaggeratedly strong augmented versions of these samples are then used as input to this loss function, allowing model training on both D^u and D^l. This idea has sparked a lot of further research such as FeatMatch [37] which uses data augmentation in the manifold space or FlexMatch [76] which combines this concept with Curriculum Learning.

4.3 Active Learning

Active Learning (AL) algorithms select the most valuable samples and query an annotator with them [64]. This means that models can learn more quickly from a subset of annotated samples and that the intelligent choice of such samples can be better than randomly subsampling a data stream or dataset. This choice can be based on, e.g., insights about the model, the dataset, or on a heuristic.

Cost. The reduction of cost is one of the primary reasons to use Active Learning. The costs arise from different sources, e.g., the annotation task's difficulty and the associated expensive expertise of the annotators. Similarly, creating well-curated, representative datasets may become a financial roadblock for ML projects. A typical example of these issues is the medical field [7], where Active Learning may decrease the time (and money) spent on generating labels.

Active Learning Loop. The human-in-the-loop, that is also called the "oracle", is at the center of the AL loop [64]. Figure 4.2 shows the loops components: the ML model, a pool of labeled training data, and a pool of unlabeled data from which the Active Learner constructs queries to the annotator. A common assumption is that the oracle initially annotates a small subset for a first model training. Next, an AL strategy selects one or more unlabeled samples using an acquisition function. The expectation is that the ML model would learn faster from these than from randomly sampled data. Finally, the expert is queried and the labeled data is added to the labeled pool. The stop condition may be the depletion of some budget or an accuracy threshold.

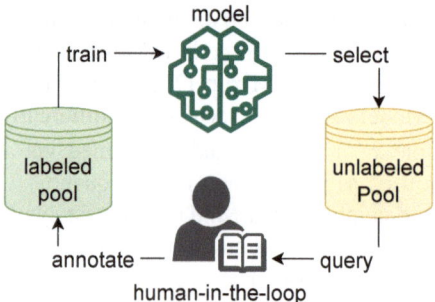

Fig. 4.2: The Active Learning loop has a human-in-the-loop at its center, that serves as the label oracle. The AL method acquires samples to query the oracle with, and the subsequent labels are added to a training pool.

Scenarios. The literature distinguishes how the unlabeled data is made available [64]. In pool-based Active Learning, all unlabeled data is available in a database, and any of these can be selected for the construction of a query. Alternatively, in stream-based scenarios samples may stream into the AL loop over time, and the method incrementally selects queries. Besides the delivery mode of the dataset, the number of simultaneously selected samples further differentiates what acquisition functions can achieve and how queries can be selected. Querying the oracle one sample at a time may choose a best sample but requires frequent re-fitting of the model, whereas batch-mode AL samples whole batches at once and thus speeds up the AL loop considerably. This is especially important when using DNNs as models [4].

Acquisition Functions. The method of how the AL algorithm chooses which unlabeled samples an oracle is queried with is a crucial part of the AL loop. A typical example of an acquisition functions for classification tasks is querying the least confident prediction [41]. These are $\hat{x} = \text{argmax}_x(1 - P(\hat{y}|x))$, where $\hat{y} = \text{argmax}_y(P(y|x))$ is the most likely label $\hat{y}$ for an unlabeled sample x. This uncertainty-based acquisition function selects samples that the model is least certain about.

Before the recent success of DL, a multitude of acquisition functions was proposed [64], such as uncertainty-based sampling, queries by committees of models, based on the expected model change, the expected error reduction, on variance reduction or based on density. The following sections explain the fundamental concepts of Deep Active Learning, such as uncertainty and diversity sampling, and their combination.

4.3.1 Deep Active Learning (DAL)

Deep Neural Networks require large amounts of data to generalize well [58]. This becomes an issue in supervised learning settings that, unlike self-, semi- or unsupervised learning, need annotations to fit models. Using AL may seem the natural choice to reduce the costs for generating training datasets. However, DAL faces challenges that arise from their use of DNNs. Traditional heuristics like discussed in [64] and one-by-one querying showed to be ineffective when used with DNNs [63]. Hence, this section introduces the fundamentals of AL with a focus on models from Deep Learning. The research on DAL [58] developed families of methods that broadly parallel traditional AL strategies [64], but adapted them to DL. Hence, the following sections explain the more traditional AL strategies. See Section 4.5 for an outlook on more recent literature that extends the fundamentals.

4.3.2 Uncertainty Sampling

Selecting those samples for queries, that a model is least certain about [41], intuitively should provide most information on the dataset. For probabilistic models this is a feasible approach [64]. However, depending on the task, obtaining predictive uncertainty for DNNs is unavailable or of lower quality. In classification tasks, the softmax activation tends to quantify uncertainty with overconfidence, and regression usually is not accompanied by an uncertainty measure at all. Obtaining a more reliable uncertainty measure is important to select those samples, that are really informative.

DNNs predictive uncertainty can be interpreted as having two components: on the one hand of aleatoric or data uncertainty, and on the other of epistemic or model uncertainty [31]. The model uncertainty can be reduced via Active Learning by selecting samples for that the model is least certain. While approaches like Bayesian neural networks provide a well-calibrated uncertainty estimation that may be used for AL, they are intractable for larger problem instances.

Recently, Gal et al. [24] proposed a first tractable and efficient approach for estimating a DNN's uncertainty that is implemented as a Bayesian Convolutional Network. This approach provides better calibrated uncertainty by implementing an ensemble- or vote-agreement scheme [39] based on a Monte-Carlo simulation of a model ensemble using the Dropout connections of the DNN. This trick allows to interpret each MC pass as a separate model and thus the epistemic uncertainty is measured more efficiently.

Beluch et al. [7] proposed to use *the power of ensembles* for AL, and show that this source of uncertainty is better calibrated than relying on Dropout connections within a single network. However, this observation was only valid in the few data domain. Interestingly, they show that AL with an ensemble still leads to increases in accuracy in larger problem sizes.

These two uncertainty-based Active Learners provide measures of uncertainty that different acquisition functions use to select queries. We present the three most common functions. The first is based on Shannon's information theory [65]. The **Max Entropy** function selects those points, that maximize the predictive entropy as follows

$$H[y|xD^l] = - \sum (p(y = k)|x, D^l)\log(p(y = k|x, D^l)) \qquad (4.2)$$

This is then adapted for ensembles and MC Dropout, in that the probabilities $p(\cdot|\cdot)$ are summed and averaged over the number of networks in the ensemble [7] or over the number of forward passes [24].

The **Variation Ratio** acquisition function selects those samples, whose predicted classes have the lowest agreement in an ensemble, see Eq. 4.3, or in its Bayesian formulation, whose probability is more dispersed to others, see Eq. 4.4:

$$\text{variation-ratio}(x) = 1 - \frac{m}{N}, \qquad (4.3)$$

$$= 1 - max_y p(y|x, D) \qquad (4.4)$$

4.3.3 Diversity Sampling

Another parallel between classical and Deep Active Learning is the notion of querying the oracle with a diverse set of examples, so that the model learns from a representative training dataset. The selection of a diverse batch seems especially promising for batch-mode learning, because it can help avoiding biased training. Compared to a simple random down-sampling of the training data pool, AL strategies, such as the core-set selection proposed by Sener and Sevarese [63] aim to find an optimal (unbiased) subset. Uncertainty-based sampling generally is also more affected by outliers [64]. Additionally, Sener and Savarese empirically show the limitations of uncertainty-based AL when used with larger datasets.

In traditional optimization, algorithms for selecting a core-set were already used for k-center clustering and other applications [1], and also already for AL, e.g., with Support Vector Machines [71]. However, the extension of this idea to a deep model, such as CNNs, was only recently pioneered [63]. The authors use the DNN to generate an embedding of the pool, and then solve the k-center problem to select a batch query. In addition, Yehuda et al. [74] recently proposed a diversity sampling approach that maximizes Probability Coverage, and that is designed specifically for the low-budget regime.

4.3.4 Balanced Criteria

The combination of multiple selection criteria lends itself especially well to DAL, because model training is most often performed via mini-batches and stochastic gradient descent, and the selection of samples within such a batch enables it. This section explains BALD [29] and BatchBALD [35] as examples for the variety of AL strategies that combine uncertainty-sampling with selecting more diverse batches.

Houlsby et al. [29] propose BALD, that measures the mutual information between the model's parameters and its predictions, which points out whether learning about the true label would provide new information on the parameters as well. BALD uses the following equation at its core:

$$I(y; w|x, D^l) = H(y|x, D^l) - E_{p(w|D^l)}[H(y|x, w, D^l)].$$ (4.5)

The first term measures the prediction's entropy and is high for uncertain predictions. The second term is the expectation of the prediction, given the model and its parameters, and is low if the model is certain. Maximizing the information I leads to choosing samples with a high uncertainty in the prediction, but a low uncertainty in the learned model. However, this does not select for diverse samples and performs bad with larger batch sizes [35].

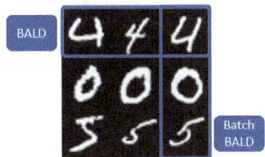

Fig. 4.3: The idealized acquisition of BatchBALD [35] selects a more diverse query compared to BALD [29], that selects the most informative samples, even if they repeat.

Kirsch et al. [35] propose BatchBALD as an extension to BALD, that finds batches of informative data. They extend Eq. 4.7 by estimating the joint of multiple data points $x_1, ..., x_b$:

$$I(y_1, ..., y_b; w | x_1, ..., x_b, D^l) = H(y | x_{1:b}, D^l)$$
$$- E_{p(w|D^l)} [H(y | x_{1:b}, w, D^l)]. \tag{4.6}$$

Kirsch et al. argue that BALD overestimates the joint mutual information of pairs of prediction y_i and sample x_i, whereas the formulation of I for BatchBALD measures the overlap between multiple variables 1 to b from a batch, and thus tends to acquire more diverse queries. Figure 4.3 visualizes this with BALD ignoring the repetitiveness of similar x_i within a batch, whereas BatchBALD considers $x_{1:b}$ in calculating Entropy H.

4.4 Active Semi-Supervised Learning

Both AL and SSL aim to improve learning with limited labeled data. They tackle the problem from two different perspectives, where SSL assumes a static setting where the set of labeled data is fixed, AL assumes a dynamic setting where expanding the labeled data pool is possible. In the AL setting the model will be trained every time new labeled data is available, i.e., in between each of the query iterations. Most AL approaches only use the labeled data for training the model. However, in a pool-based AL setting, both labeled data and unlabeled data are available to the model. Thus it would be a natural idea to use SSL techniques that do not only use labeled data but also unlabeled data for training.

As AL and SSL are compatible, several works have combined them and we will look at how this can be done. First, we will give an example of how AL and SSL can be combined using concepts introduced in previous sections. Second, we will discuss the mutual benefits of SSL and AL based on recent advancements in SSL.

4.4.1 How can SSL and AL Work Together?

Following the AL loop depicted in Figure 4.2 a pool of unlabeled data and a pool of labeled data is available. This data will be used for training a model by minimizing an objective loss function $\mathcal{L}$. Commonly in AL the loss for training the model is only the supervised loss $\mathcal{L} = \mathcal{L}_{sup}$, which could be standard cross-entropy loss for classification. When integrating SSL into the AL loop we utilize the unlabeled data as well by combining the supervised loss with an unsupervised loss $\mathcal{L} = \mathcal{L}_{sup} + \mathcal{L}_{unsup}$. As seen previously, there exists a variety of different loss functions for the SSL loss which can be combined with any AL acquisition function.

Gao et al. [25] suggested to combine SSL and AL and used consistency regularization as basis for their SSL loss. They experimented with multiple acquisition functions such as random sampling, the uncertainty-based max-entropy and diversity-based k-center clustering. However, they note that choosing the right acquisition function is crucial when combining SSL and AL. As they are using consistency regularization as their unsupervised loss $\mathcal{L}_{unsup}$ that enforces consistency of predictions over augmented versions of the sample, they hypothesize that labeling the samples with the most inconsistent predictions should be the most beneficial to the model. This is based on the intuition that these samples must be hard to classify for the model. They propose the following simple metric for acquisition

$$\varepsilon(x^{(i)}) = \sum_{k}^{K} \text{Var}\left[p(y^{(ik)} = k|x^{(i)}), p(y^{(ik)} = k|\tilde{x}_1^{(i)}), \ldots, p(y^{ik} = k|\tilde{x}_N^{(i)})\right],$$

(4.7)

where N is the number of different data augmentations applied.

Combining consistency regularization and a maximum inconsistency acquisition function [25] show that SSL benefits AL and outperforms other AL methods from the literature. They also show that their specific choice of SSL method and AL acquisition function performs better than other combinations.

4.4.2 Are SSL and AL Always Mutually Beneficial?

The recent advances in Deep Semi-Supervised Learning have shown impressive performance in utilizing the unlabeled data together with a small amount of labels [69]. This progress has raised the question of whether the human annotations are beneficial when SSL is able to utilize the unlabeled data so efficiently and therefore questioning the relevance of AL [50, 12, 8].

Utilizing consistency regularization-based SSL in the AL loop, [50] showed that for image classification the combination of SSL and random sampling works better than using AL for sampling. Similarly, [12] do not observe any additional benefit of using more advanced AL algorithms when combined with both SSL and self-supervised methods. [8] experimented with self-supervised models and active learning and demonstrated that self-supervised learning in itself is more efficient than AL at reducing the labeling effort. They also observe that the combination of self-supervised learning and AL is only beneficial only when the labeling budget is high which goes against the purpose of using AL.

Although these recent critiques of AL show the impressive performance of SSL, more research is needed to understand if this is also the case in real-world scenarios. The comparisons between AL and SSL [50, 12, 8] are based on experimental results on well-established benchmark datasets such as CIFAR10 and CIFAR100 where it is well-known what data augmentations and hyperparameters work well. This is not the case in real-world settings where it is hard to find optimal hyperparameters as well as design data augmentations that are label preserving and beneficial to the model.

As semi-supervised methods rely on assumptions it cannot be guaranteed that they will perform well in case these assumptions are broken. Therefore it is important to analyze a real-world scenario and conduct experiments to see if SSL is actually beneficial [72].

In this context, [53] formulated a critique on the evaluation of semi-supervised learning techniques. Among other issues, they observed that the multitude of hyperparameters such as weighting factors, thresholds, or perturbation ratios, require heavy hyperparameter tuning. This in turn requires the presence of reasonably large labeled validation data sets whose size is often magnitudes higher than that of the labeled training dataset – which increases the required amount of labeled training data for practical scenarios. While semi-supervised learning promises the effective use of unlabeled data for model training, its final benefit in practical scenarios heavily depends on the final setting – semi-supervised learning can help alleviate the problem of limited labeled data but is no silver bullet to it.

4.5 Conclusion and Outlook

Semi-supervised learning tries to tackle the limited labeled data problem by using the latent information provided in a large, unlabeled dataset D^u next to a smaller labeled dataset D^l. The field has been around from the early days of Machine Learning research and spans various approaches and related research fields. While recent deep semi-supervised learning approaches yield impressive gains on benchmark datasets, their applicability to practical real-world scenarios depends on the respective task and cannot be taken for granted. For instance, the heavy use of tailored data augmentation strategies in modern, strong-performing semi-supervised learning methods requires strong domain knowledge of the underlying task which could also be used to annotate more training data. These approaches also mainly target scenarios with a balanced class distribution and assume that both D^l and D^u follow the same data distribution, i.e. the absence of any distribution shifts. Further, the design and training of these partially highly elaborate algorithms require intense engineering and modeling efforts which could alternatively be used to annotate more high-quality training data.

Despite this critique, semi-supervised learning offers a high potential for low-label scenarios which fuels evermore increasing research activity in this field. Recent algorithmic development merges the concept of semi-supervision with connected research fields such as contrastive learning. With S4L, semi-supervised self-supervised learning, [75] introduced a self-supervised training scheme building up on contrastive learning for model training in a semi-supervised fashion. Furthermore, [14] successfully combined self-supervised pretraining on unlabeled data via the SimCLR architecture with subsequent semi-supervised fine-tuning and showed impressive classification performance on the ImageNet benchmark with a small 1% fraction of labeled examples. Similarly, [46] leverage self-supervised learning to extend Fix-Match towards barely supervised learning scenarios, where as little as 4 labeled samples per class are provided. With the advent of multi-purpose and multi-modal models [54], we can expect the use of these large pretrained models also for the generation of pseudo-labels in semi-supervised image classification tasks, similar to their use in Natural Language Processing.

In summary, Active Learning is a method to increase the number of annotated samples in the most cost-efficient manner [64]. Two of the fundamental strategies are sampling according to an uncertainty measure of the model [7, 24] or according to a representative measure of the underlying data distribution [63]. A combination [35] of such concepts balances the classical explore/exploit dilemma.

Most recent research on modern Deep Learning [59] can be broadly categorized as learning to active learning. For example, the underlying data distributions can be learned with generative models and this representation exploited [67, 48, 66, 34]. AL can be understood as an optimization problem and solved using Reinforcement Learning [73, 5, 36, 10, 19, 57, 45, 28], by imitating experts [61, 43, 47], or by simply selecting the most suitable strategy from a diverse set of heuristics [6, 30, 16] In closely related fields, AL can be cast as Meta-Learning of learning quickly with few samples [56, 15, 20, 52, 33, 21, 49, 42] or even as a Neural Architecture Search problem [26].

The future of Active Learning and Semi-Supervised Learning could be a combination that leverages both methods' strengths (active sampling) to reduce their weaknesses (cost, uncertainty). When combining the two paradigms, it is important that the chosen acquisition functions are designed to benefit an SSL method where the SSL method is unable to utilize the unlabeled data.

References

1. P. K. Agarwal, S. Har-Peled, and K. R. Varadarajan. Geometric Approximation via Coresets. *Combinatorial and Computational Geometry*, MSRI Publications(52):21, 2005.
2. A. Agrawala. Learning with a probabilistic teacher. *IEEE Transactions on Information Theory*, 16(4):373–379, 1970.
3. E. Arazo, D. Ortego, P. Albert, N. E. O. Connor, and K. Mcguinness. Pseudo-Labeling and Confirmation Bias in Deep Semi-Supervised Learning. In *IJCNN*, 2020.
4. J. T. Ash, C. Zhang, A. Krishnamurthy, J. Langford, and A. Agarwal. Deep Batch Active Learning by Diverse, Uncertain Gradient Lower Bounds. 2019.
5. P. Bachman, A. Sordoni, and A. Trischler. Learning Algorithms for Active Learning. *Proceedings of the 34th International Conference on Machine Learning*, 2017.
6. Y. Baram, R. El-Yaniv, and K. Luz. Online choice of active learning algorithms. *Journal of Machine Learning Research*, 5:255–291, 2004.
7. W. H. Beluch, T. Genewein, A. Nürnberger, and J. M. Köhler. The Power of Ensembles for Active Learning in Image Classification. *Proceedings of the IEEE Computer Society Conference on Computer Vision and Pattern Recognition*, pages 9368–9377, 2018.
8. J. Z. Bengar, J. van de Weijer, B. Twardowski, and B. Raducanu. Reducing label effort: Self-supervised meets active learning. *CoRR*, abs/2108.11458, 2021.
9. D. Berthelot, N. Carlini, I. Goodfellow, N. Papernot, A. Oliver, and C. A. Raffel. Mixmatch: A holistic approach to semi-supervised learning. *Advances in Neural Information Processing Systems*, 32, 2019.

10. A. Casanova, P. O. Pinheiro, N. Rostamzadeh, and C. J. Pal. REINFORCED ACTIVE LEARNING FOR IMAGE SEGMENTATION. page 17, 2020.

11. P. Cascante-Bonilla, F. Tan, Y. Qi, and V. Ordonez. Curriculum labeling: Revisiting pseudo-labeling for semi-supervised learning. *AAAI*, 2021.

12. Y.-C. Chan, M. Li, and S. Oymak. On the Marginal Benefit of Active Learning: Does Self-Supervision Eat Its Cake? *arXiv:2011.08121 [cs]*, Nov. 2020.

13. O. Chapelle, B. Scholkopf, and A. Zien. Semi-supervised learning. *Cambridge, Massachusettes: The MIT Press View Article*, 20(3):542–542, 2009.

14. T. Chen, S. Kornblith, K. Swersky, M. Norouzi, and G. Hinton. Big self-supervised models are strong semi-supervised learners. *NeurIPS*, 2020.

15. Y. Chen, M. W. Hoffman, S. G. Colmenarejo, M. Denil, T. P. Lillicrap, M. Botvinick, and N. De Freitas. Learning to learn without gradient descent by gradient descent. *34th International Conference on Machine Learning, ICML 2017*, 2:1252–1260, 2017.

16. H. M. Chu and H. T. Lin. Can active learning experience be transferred? *Proceedings - IEEE International Conference on Data Mining, ICDM*, pages 841–846, 2017.

17. R. Collobert, F. Sinz, J. Weston, L. Bottou, and T. Joachims. Large scale transductive svms. *Journal of Machine Learning Research*, 7(8), 2006.

18. A. P. Dempster, N. M. Laird, and D. B. Rubin. Maximum likelihood from incomplete data via the em algorithm. *J. Royal Statistical Society: Series B*, 39(1):1–22, 1977.

19. Y. Fan, F. Tian, T. Qin, X.-Y. Li, and T.-Y. Liu. Learning to Teach. pages 1–16, 2018.

20. C. Finn, P. Abbeel, and S. Levine. Model-Agnostic Meta-Learning for Fast Adaptation of Deep Networks. 2017.

21. C. Finn, K. Xu, and S. Levine. Probabilistic Model-Agnostic Meta-Learning. In *NIPS*, number NeurIPS, 2018.

22. S. Fralick. Learning to recognize patterns without a teacher. *IEEE Transactions on Information Theory*, 13(1):57–64, 1967.

23. Y. Gal and Z. Ghahramani. Dropout as a bayesian approximation: Representing model uncertainty in deep learning. In *international conference on machine learning*, pages 1050–1059. PMLR, 2016.

24. Y. Gal, R. Islam, and Z. Ghahramani. Deep Bayesian Active Learning with Image Data. Technical report, 2017.

25. M. Gao, Z. Zhang, G. Yu, S. O. Arik, L. S. Davis, and T. Pfister. Consistency-based Semi-supervised Active Learning: Towards Minimizing Labeling Cost. In *ECCV*, 2020.

26. Y. Geifman and R. El-Yaniv. Deep Active Learning with a Neural Architecture Search. In *Conference on Neural Information Processing Systems*, page 11, Vancouver, Canada, 2019.

27. Y. Grandvalet, Y. Bengio, et al. Semi-supervised learning by entropy minimization. *NeurIPS*, 367:281–296, 2005.

28. M. Haussmann, F. A. Hamprecht, and M. Kandemir. Deep Active Learning with Adaptive Acquisition. In *Intl. Joint Conf. on Artificial Intelligence*, pages 2470–2476, 2019.

29. N. Houlsby, F. Huszár, Z. Ghahramani, and M. Lengyel. Bayesian Active Learning for Classification and Preference Learning, Dec. 2011.

30. W. N. Hsu and H. T. Lin. Active learning by learning. *Proceedings of the National Conference on Artificial Intelligence*, 4:2659–2665, 2015.

31. E. Hüllermeier and W. Waegeman. Aleatoric and epistemic uncertainty in machine learning: An introduction to concepts and methods. 110(3):457–506.

32. A. Iscen, G. Tolias, Y. Avrithis, and O. Chum. Label propagation for deep semi-supervised learning. In *Proceedings of the IEEE/CVF Conference on Computer Vision and Pattern Recognition*, pages 5070–5079, 2019.

33. M. A. Jamal and H. Cloud. Task Agnostic Meta-Learning for Few-Shot Learning. 2018.

34. K. Kim, D. Park, K. I. Kim, and S. Y. Chun. Task-Aware Variational Adversarial Active Learning. In *2021 IEEE/CVF Conference on Computer Vision and Pattern Recognition (CVPR)*, pages 8162–8171, Nashville, TN, USA, June 2021. IEEE.

35. A. Kirsch, J. van Amersfoort, and Y. Gal. BatchBALD: Efficient and Diverse Batch Acquisition for Deep Bayesian Active Learning. (NeurIPS), 2019.

36. K. Konyushkova, R. Sznitman, and P. Fua. Learning Active Learning from Data. In *Proceedings of the 31st Conference n Neural Information Processing Systems*, 2017.
37. C.-W. Kuo, C.-Y. Ma, J.-B. Huang, and Z. Kira. Featmatch: Feature-based augmentationfor semi-supervised learning. In *European Conference on Computer Vision*. Springer, 2020.
38. S. Laine and T. Aila. Temporal ensembling for semi-supervised learning. *ICLR*, 2017.
39. B. Lakshminarayanan, A. Prinzel, and C. Blundell. Simple and Scalable Predictive Uncertainty Estimation using Deep Ensembles. In *NeurIPS*, 2017.
40. D.-H. Lee. Pseudo-label: The simple and efficient semi-supervised learning method for deep neural networks. In *Workshop on challenges in representation learning*, volume 3, page 896, 2013.
41. D. D. Lewis, T. B. Laboratories, and M. Hill. A Sequential Algorithm for Training Text Classifiers. page 10, 1994.
42. M. Li, X. Liu, J. van de Weijer, and B. Raducanu. Learning to Rank for Active Learning: A Listwise Approach. *arXiv*, (i), 2020.
43. M. Liu, W. Buntine, and G. Haffari. Learning how to actively learn: A deep imitation learning approach. In *ACL 2018 - 56th Annual Meeting of the Association for Computational Linguistics, Proceedings of the Conference (Long Papers)*, volume 1, pages 1874–1883. Association for Computational Linguistics, 2018.
44. Y.-C. Liu, C.-Y. Ma, Z. He, C.-W. Kuo, K. Chen, P. Zhang, B. Wu, Z. Kira, and P. Vajda. Unbiased teacher for semi-supervised object detection. *ICLR*, 2021.
45. Z. Liu, J. Wang, S. Gong, D. Tao, and H. Lu. Deep Reinforcement Active Learning for Human-in-the-Loop Person Re-Identification. In *2019 IEEE/CVF International Conference on Computer Vision (ICCV)*, pages 6121–6130, Seoul, Korea (South), Oct. 2019. IEEE.
46. T. Lucas, P. Weinzaepfel, and G. Rogez. Barely-supervised learning: semi-supervised learning with very few labeled images. In *Proceedings of the AAAI Conference on Artificial Intelligence*, volume 36, pages 1881–1889, 2022.
47. C. Löffler and C. Mutschler. Iale: Imitating active learner ensembles. *Journal of Machine Learning Research*, 23(107):1–29, 2022.
48. D. Mahapatra, B. Bozorgtabar, J.-P. Thiran, and M. Reyes. Efficient Active Learning for Image Classification and Segmentation Using a Sample Selection and Conditional Generative Adversarial Network. In A. F. Frangi, J. A. Schnabel, C. Davatzikos, C. Alberola-López, and G. Fichtinger, editors, *Medical Image Computing and Computer Assisted Intervention – MICCAI 2018*, volume 11071, pages 580–588. Cham, 2018.
49. N. Mishra, M. Rohaninejad, X. Chen, and P. Abbeel. A Simple Neural Attentive Meta-Learner. In *ICLR*, pages 1–17, 2018.
50. S. Mittal, M. Tatarchenko, Ö. Çiçek, and T. Brox. Parting with illusions about deep active learning. *ArXiv*, abs/1912.05361, 2019.
51. T. Miyato, S.-i. Maeda, M. Koyama, and S. Ishii. Virtual adversarial training: a regularization method for supervised and semi-supervised learning. *IEEE transactions on pattern analysis and machine intelligence*, 41(8):1979–1993, 2018.
52. A. Nichol, J. Achiam, and J. Schulman. On First-Order Meta-Learning Algorithms. pages 1–15, 2018.
53. A. Oliver, A. Odena, C. A. Raffel, E. D. Cubuk, and I. Goodfellow. Realistic evaluation of deep semi-supervised learning algorithms. *Advances in neural information processing systems*, 31, 2018.
54. A. Radford, J. W. Kim, C. Hallacy, A. Ramesh, G. Goh, S. Agarwal, G. Sastry, A. Askell, P. Mishkin, J. Clark, et al. Learning transferable visual models from natural language supervision. In *International Conference on Machine Learning*, pages 8748–8763, 2021.
55. A. Rasmus, M. Berglund, M. Honkala, H. Valpola, and T. Raiko. Semi-supervised learning with ladder networks. *Advances in neural information processing systems*, 28, 2015.
56. S. Ravi and H. Larochelle. Optimization as a model for few-shot learning. *ICLR*, 2017.
57. S. Ravi and H. Larochelle. Meta-Learning for Batch mode Active Learning. *ICLR workshop*, pages 1–6, 2018.
58. P. Ren, Y. Xiao, X. Chang, P. Y. Huang, Z. Li, X. Chen, and X. Wang. A survey of deep active learning. *arXiv*, 37(4), 2020.

59. P. Ren, Y. Xiao, X. Chang, P.-Y. Huang, Z. Li, B. B. Gupta, X. Chen, and X. Wang. A Survey of Deep Active Learning. *ACM Comput. Surv.*, 54(9):1–40, Dec. 2022.

60. M. N. Rizve, K. Duarte, Y. S. Rawat, and M. Shah. In defense of pseudo-labeling: An uncertainty-aware pseudo-label selection framework for semi-supervised learning. *ICLR*, 2021.

61. S. Ross, G. J. Gordon, and J. A. Bagnell. A Reduction of Imitation Learning and Structured Prediction to No-Regret Online Learning. page 9, 2011.

62. H. Scudder. Probability of error of some adaptive pattern-recognition machines. *IEEE Transactions on Information Theory*, 11(3):363–371, 1965.

63. O. Sener and S. Savarese. Active learning for convolutional neural networks: A core-set approach. In *Intl. Conf. Learning Representations*, Vancouver, CA, 2018.

64. B. Settles. Active Learning Literature Survey. Technical Report 1, Morgan & Claypool Publishers, 2012.

65. C. E. Shannon. A Mathematical Theory of Communication. *The Bell System Technical Journal*, (27):379–423, 1948.

66. C. Shui, F. Zhou, C. Gagne, and B. Wang. Deep Active Learning: Unified and Principled Method for Query and Training. In *Deep Active Learning*, page 10, Palermo, Italy, 2020.

67. S. Sinha, U. C. Berkeley, and T. Darrell. Variational Adversarial Active Learning. 2019.

68. K. Sohn, D. Berthelot, C. Li, Z. Zhang, N. Carlini, E. D. Cubuk, A. Kurakin, H. Zhang, and C. Raffel. Fixmatch: Simplifying semi-supervised learning with consistency and confidence. *CoRR*, abs/2001.07685, 2020.

69. K. Sohn, D. Berthelot, C.-l. L. Zizhao, Z. Nicholas, E. D. Cubuk, A. Kurakin, H. Zhang, and C. Raffel. FixMatch : Simplifying Semi-Supervised Learning with Consistency and Confidence. In *NeurIPS*, 2020.

70. A. Tarvainen and H. Valpola. Mean teachers are better role models: Weight-averaged consistency targets improve semi-supervised deep learning results. *Advances in neural information processing systems*, 30, 2017.

71. I. W. Tsang, J. T. Kwok, P.-M. Cheung, C. U. Hk, C. U. Hk, and C. U. Hk. Core Vector Machines: Fast SVM Training on Very Large Data Sets. *Journal of Machine Learning Research*, 6(13):29, 2005.

72. J. E. Van Engelen and H. H. Hoos. A survey on semi-supervised learning. *Machine Learning*, 109(2):373–440, 2020.

73. M. Woodward and C. Finn. Active One-shot Learning. In *NIPS*, Barcelona, Spain, 2016.

74. O. Yehuda, A. Dekel, G. Hacohen, and D. Weinshall. Active learning through a covering lens. In S. Koyejo, S. Mohamed, A. Agarwal, D. Belgrave, K. Cho, and A. Oh, editors, *Advances in Neural Information Processing Systems*, volume 35, pages 22354–22367, 2022.

75. X. Zhai, A. Oliver, A. Kolesnikov, and L. Beyer. S4l: Self-supervised semi-supervised learning. In *Proceedings of the IEEE/CVF International Conference on Computer Vision*, pages 1476–1485, 2019.

76. B. Zhang, Y. Wang, W. Hou, H. WU, J. Wang, M. Okumura, and T. Shinozaki. Flexmatch: Boosting semi-supervised learning with curriculum pseudo labeling. In M. Ranzato, A. Beygelzimer, Y. Dauphin, P. Liang, and J. W. Vaughan, editors, *Advances in Neural Information Processing Systems*, volume 34, pages 18408–18419, 2021.

77. H. Zhang, M. Cisse, Y. N. Dauphin, and D. Lopez-Paz. Mixup: Beyond empirical risk minimization. *ICLR*, 2018.

78. X. Zhul and Z. Ghahramanih. Learning from labeled and unlabeled data with label propagation. 2002.

Chapter 5
The Role of Uncertainty Quantification for Trustworthy AI

Jessica Deuschel[1], Andreas Foltyn[1], Karsten Roscher[2], Stephan Scheele[1,†]

Abstract The development of AI systems involves a series of steps, including data acquisition and preprocessing, model selection, training, evaluation, and deployment. However, each of these steps involves certain assumptions that introduce inherent uncertainty, which can result in inaccurate outcomes and reduced confidence in the system. To enhance confidence and comply with the EU AI Act, we recommend using Uncertainty Quantification methods to estimate the belief in the correctness of a model's output. To make these methods more accessible, we provide insights into the possible sources of uncertainty and offer an overview of the different available methods. We categorize these methods based on when they are used in the process, accounting for various application requirements. We distinguish between three types: data-based, architecture-modifying and post-hoc methods, and share our personal experiences with each.

Key words: Trustworthy AI, Uncertainty Quantification, Machine Learning, Safety Critical Systems

5.1 Introduction

Artificial Intelligence (AI) systems are increasingly ubiquitous in modern life, easily accessible to commercial and non-technical users, and capable of outperforming humans in various tasks. In particular, data-intensive machine learning approaches such as deep learning [45, 21] have achieved a remarkable success in a wide range of applications, from image and speech recognition to natural language processing

---

[1] Fraunhofer Institute for Integrated Circuits IIS, Erlangen, Germany
[2] Fraunhofer Institute for Cognitive Systems IKS, Munich, Germany

Corresponding author: Stephan Scheele
e-mail: stephan.scheele@uni-bamberg.de

© The Author(s) 2024
C. Mutschler et al. (eds.), *Unlocking Artificial Intelligence*,
https://doi.org/10.1007/978-3-031-64832-8_5

and recently large language transformer models such as OpenAI's ChatGPT [57], generative imaging technology based on stable diffusion [65] or biomedical applications such as AlphaFold [39]. Their success is mainly based on their ability to automatically learn complex features and patterns from large amounts of data that are difficult or impossible to investigate and implement manually.

However, a major challenge with deep learning models is their lack of transparency and explainability [1, 46, 51], which makes it difficult to assess their trustworthiness. Understanding the reasoning behind an AI model's decisions and assessing the reliability and certainty of its predictions can be challenging in complex and uncertain situations. This lack of transparency makes it challenging to identify a model's limitations, evaluate its reliability across various application domains, assess potential risks, and ultimately trust its results. This is especially striking when we adopt Machine Learning (ML) methods for safety-critical application scenarios to enable higher autonomy by solving complex tasks related to perception and planning [8, 3, 77]. Particularly, ensuring the safety and trustworthiness [71, 73, 47, 40, 40, 36] of ML is considered one of the key obstacles to their extensive adoption in high-stakes applications where systems may infringe on our rights or human lives may be at stake. Moreover, deep learning models are limited in their ability to reliably handle data not covered in the training phase, which is particularly problematic in mission-critical systems, where unknown situations can pose significant risks and lead to serious consequences [87].

To get a good introduction to this topic, the structure of this chapter is as follows. In Section 5.2, we will examine the current challenges related to trust in AI systems, particularly in the context of the political landscape. We especially emphasize the potential of uncertainty quantification to increase trust in deep learning. Section 5.3 outlines the various sources of uncertainty in AI systems and provides an overview of the latest uncertainty quantification methods. Finally, in Section 5.4, we discuss the limitations of existing uncertainty quantification methods and suggest potential directions for future research to build more trustworthy AI.

5.2 Towards Trustworthy AI

Ensuring trust in AI systems is essential to promote their reliability and acceptance, both from a practical and legal standpoint. The EU AI Act [11], for instance, mandates that AI systems used in high-risk applications comply with strict rules to ensure their trustworthiness and transparency.

While Explainable AI (XAI) has made significant strides in providing some level of reasoning for AI decisions, the topic of uncertainty in its various forms is not adequately represented in the context of enhancing trust.

We argue that qualitative analysis and understanding of various sources of uncertainty, along with their quantification, hold the potential to offer valuable insights on the confidence and limitations of AI predictions. Ultimately, this understanding can contribute to enhancing the reliability and transparency of AI systems.

5.2.1 The EU AI Act

Numerous countries and governments are advocating for regulatory guidelines to advance the creation of reliable and trustworthy AI. Notable examples include the "Blueprint for an AI Bill of Rights" by the US government [34], the ongoing development of a "National AI Strategy" in the UK [68], and the Artificial Intelligence Act (the "AI Act"), which passed the European Parliament on June 14, 2023 [11].

The EU AI Act as proposed in April 2021 aims to lay the foundation for the development and use of AI systems in the European Union. One of its main objectives is to increase transparency and trust in AI systems. The AI Act is a layered risk-based approach to assess and enforce product safety, divided into the risk classes *unacceptable, high,* and *limited & low* risk, i.e., stricter rules are enforced as the risk increases. These range from voluntary and self-regulatory impact assessments combined with codes of conduct to strict and externally audited compliance obligations covering the entire life-cycle of AI applications.

AI systems falling into the first category of *unacceptable risk* are prohibited because they are considered a threat to safety, health, human rights, ethics or livelihood. Systems belonging to the second category of *high risk* concern the area of safety-critical systems. Such AI systems are subject to stringent requirements before they can be placed on the market as a product, such as risk assessment, quality assurance of the AI pipeline in terms of data and prediction quality, compliance with technical documentation requirements, a high level of robustness, safety and accuracy, and human oversight and transparency measures to minimise risk. The categories of *limited & low risk* refer to AI systems that must comply to minor transparency obligations.

In particular, Chapter 2 [11] of the AI Act introduces compliance requirements such as risk-management, data-governance, book-keeping, transparency and human-oversight on AI providers. The transparency obligations require to identify and quantify known and unknown risks, and accordingly to implement risk mitigation procedures and controls. Transparency requirements demand explanations for the cause-effect relationship of complex machine learning models that are beyond intuitive human understanding, so that "[. . .] their operation is sufficiently transparent to enable users to interpret the system's output and use it appropriately". Particularly, this includes the identification and communication of limitations of a machine learning model to its users. Furthermore, the act imposes the identification and assessment of uncertainty factors of an AI system, as well as explainability and human supervision [11, 25], incorporating "human-machine interface tools". Moreover, the EU parliament identified the need to address uncertainty [20] in AI systems, particularly in the healthcare domain, putting forward that "Future AI solutions for healthcare should be implemented by integrating uncertainty estimation [. . .] to provide clinicians with clinically useful indications on the degree of confidence in AI predictions". By addressing these fields, the EU regulation scheme aims to ensure that AI systems are developed and deployed in a way that is trustworthy, ethical and in line with EU legislation. Besides the AI Act, there exist further legal acts and regulations like the General Data Protection Regulation (GDPR) that imposes further transparency requirements to mitigate the risk potential of automated decision-making systems and governs the processing and movement of personal data within the EU.

While the AI Act places a strong emphasis on transparency and explainability requirements, the topic of uncertainty quantification is not explicitly mentioned but only implicitly subsumed under the key requirement of robustness. In our view, the relevancy of uncertainty quantification for trustworthy AI should be represented more prominently and explicitly in the AI Act.

5.2.2 From Uncertainty to Trustworthy AI

AI-based systems follow a typical pipeline that involves the steps of data collection and pre-processing, selection of a model and a learning algorithm, model training and tuning and drawing inferences in the final step. However, each step inserts inherent uncertainties, where for instance one possible source of uncertainty is related to the collection and representation of real-world data. Furthermore, uncertainty may also concern the pre-processing of data, including cleaning and labeling. Additionally, machine learning models are abstractions of the real world and its learning algorithms may introduce modeling and inferential uncertainties by an oversimplification of their assumptions. Such uncertainties can affect the accuracy and reliability of an AI system's predictions.

We argue that uncertainty quantification is essential to increase trust in AI systems, as outlined in the EU AI Act. Trust is an essential prerequisite for high-stakes and safety-critical applications where the reliability and acceptance of an AI system is critical. Lack of trust in AI systems can lead to users either refusing to use them or abusing them in ways compromising their effectiveness. Trust is particularly important in applications where the stakes are high, such as healthcare, finance or autonomous systems, where errors or biases can have serious consequences. For instance, in healthcare, medical misdiagnosis can lead to incorrect treatment and potentially harmful consequences for a patient. In finance, there is a risk of financial losses when using erroneous AI systems in domains like fraud detection or algorithmic trading. AI systems deployed in autonomous systems such as transportation or robotics pose a safety risk, which may result in severe accidents or even loss of lives. Further examples include privacy risks, when AI systems process personal data, where errors can cause exposure of personal data due to data breaches, or infrastructure failures that can jeopardise the safe operation of critical infrastructure systems such as power grids or water supply systems. Thus, it is important to develop and deploy AI systems carefully to minimize these risks.

Uncertainty quantification can identify and quantify uncertainties in data, models and algorithms, and can also be used as a tool to audit the trustworthiness of an AI system. AI predictions are affected by such uncertainties, which highlights the need for reliable uncertainty estimates, to provide diagnostic assistance for both developers and end-users of an AI system, and to better understand its limitations. These estimates can help developers to identify areas for improvement, such as the data representation process, the need to extend data collection or necessary model tuning and refinement. By revealing the system's internal workings, uncertainty quantification methods can identify potential sources of bias, errors, or inconsistencies in the system's decision-making process. Furthermore, in high-risk applications where explanation support with a high confidence becomes mandatory, quantifying uncertainties of explanations is a prerequisite as well [10].

While uncertainty quantification is crucial in tracking performance insufficiencies of an ML-based system during operation, additional sources of uncertainty within the life cycle of such systems have been recognized [9]. On one end of this spectrum, Machine learning is usually employed when complex tasks cannot be described by a precise algorithmic specification due to their inherent complexity, variability, or lack of deterministic rules. The very nature of such tasks defines a straightforward, rule-based solution, making it challenging to explicitly follow a traditional algorithmic approach. Instead, to address such tasks, utilizing machine learning is necessary, with data essentially serving as the specification and giving rise to what we term 'specification uncertainty'. On the other end of the spectrum, even with machine learning models in place, the evaluation of their performance is not straightforward. The input space is often high-dimensional, and the behavior of continuously growing neural networks is inherently unpredictable, requiring advanced techniques for comprehensive validation and testing. It is therefore important for developers of trustworthy AI components to qualitatively and quantitatively argue certain model properties, such as robustness, limited bias, and absence of known failure modes. However, those arguments are seldom a complete formal proof and include some leap of faith in the process. Therefore, a lack of trust in the collected evidence and derived arguments gives rise to assurance uncertainty. All of those types of uncertainty have to be taken into account by standardization activities and practitioners bringing AI into high-stake applications. Uncertainty quantification can be an effective tool to identify ambiguous inputs providing a direct link to specification uncertainty, i.e., in cases where the boundary between two classes is neither specified nor very clear. Its role towards assurance is less clear. While it is definitely one building block, it is no silver bullet towards reliable and safe systems either [32, 67].

For end-users, accurately specified uncertainties, combined with effective communication and explainability, are essential for better decision-making and hereby can increase the transparency and trust in an AI system. Hence, uncertainty quantification can provide users not only with insights into the decision-making process, but also with an understanding of the uncertainties that can affect the accuracy of an AI system's predictions.

5.3 Uncertainty Quantification

While it's important to acknowledge that deep learning models may not achieve perfect predictive accuray, integrating a reliable measure of uncertainty into the deep learning system can significantly enhance its transparency and acceptance. This is particularly crucial in safety-critical domains, such as credit card fraud detection [26], predicting molecular properties [66], or classifying tissues in histopathology. Information on a method's uncertainty can be useful even in non-critical situations, saving time and costs. For example, in an automatic brake disc check at a production plant, the algorithm can confidently detect flawless (cf. Figure 5.1-A) or broken (C) discs. Uncertain cases (B) require human inspection to avoid accidentally discarding functional brake discs.

In classification tasks, deep learning models typically generate a pseudo-probability for each class using the softmax function, where the maximum softmax output is commonly considered as a measure of certainty. However, this output tends to be overly confident and does not accurately reflect the true probability of each class. To obtain a credible measure of certainty, the predicted class probability should align with the true likelihood, meaning that samples with an 80% certainty measure should yield an accuracy of 80%.

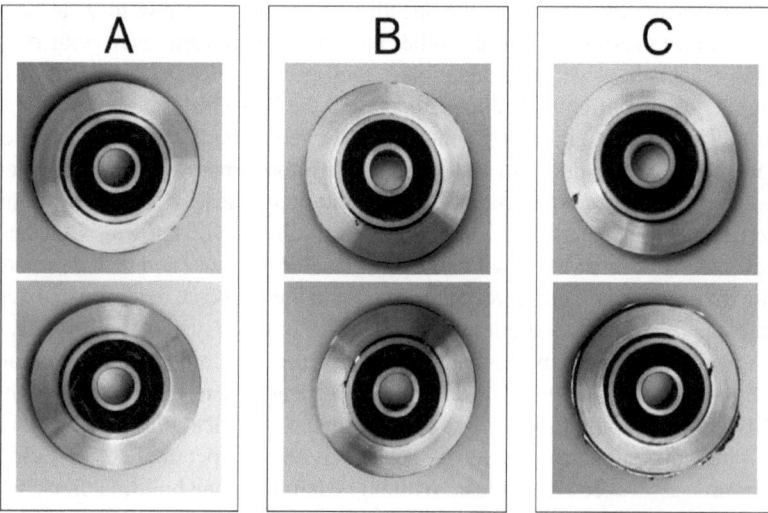

Fig. 5.1: Necessity of uncertainty measures in distinguishing flawless vs. damaged brake discs. A and C show examples of intact or clearly broken brake discs, respectively. The examples in B cannot be clearly assigned to one of the two classes. The supposed artifacts in B may possibly only be a hair or a shadow (taken from [12], licensed under CC-BY 4.0).

5.3.1 Sources of Uncertainty

In the literature, sources of uncertainty are generally divided into two types of uncertainty: aleatoric and epistemic uncertainty.

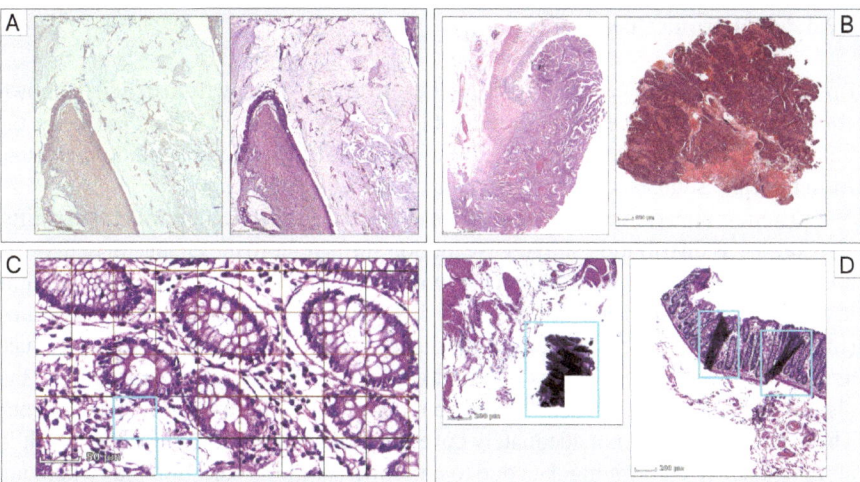

Fig. 5.2: Sources of uncertainty in histopathology. A: Distribution shift when using a different scanner for the same slide, B: Stronger distribution shift by different clinics (different staining process and different scanner of the slides which results in different appearances), C: Label noise: The blue marked patches do not contain enough class specific information to be classified correctly, D: Typical artefacts (within the blue marked regions) that might be present in a histological tissue section and should be recognized as out-of-distribution data.

5.3.1.1 Aleatoric Uncertainty

Aleatoric uncertainty, also called data uncertainty, refers to the statistical uncertainty. It is related to randomness, comparable with flipping a coin where a larger amount of data cannot reduce the randomness of the outcome [35, 24]. This type of uncertainty can stem from various origins, including measurement errors like random noise, as well as from limited information, such as low resolution or imperfect measurements [19]. Another possible source of aleatoric uncertainty is the labeling process, where label noise may arise due to factors such as fuzziness or the inability to achieve a clear separation between classes. For instance, in the context of tissue classification in histopathology (as illustrated in Figure 5.2-C), the classification of a large whole slide image is typically based on smaller image patches. Although the blue patches in image C belong to the same tissue class as the surrounding patches, they offer barely any information about the tissue.

Aleatoric uncertainty in the labeling process can also occur when there is no specific true label as in the case of a soft property. For instance, in cognitive load detection, the question arises as to which aspect serves as the most accurate ground truth: the task itself, the person's performance on the task, or their self-assessment of their cognitive load [58].

5.3.1.2 Epistemic Uncertainty

Epistemic or model uncertainty refers to the uncertainty arising from deficient knowledge. This type of uncertainty is reducible by more evidence, which means in the field of machine and deep learning by more data. Epistemic uncertainty can stem from multiple sources, either from a data or modeling perspective.

Epistemic uncertainty from the data perspective can arise due to dataset shifts, also known as distribution shifts. Distribution shifts refer to the change in real-world data compared to the training data, violating the default assumption in machine learning that the train and test data are independently and identically distributed (i.i.d.) [24, 62]. This can be caused by changes in the environment over time, such as changing weather conditions or variations in the appearance of spam emails but also from the high variability and inherent complexity of real-world environments. The training data may not adequately cover this total variability, which can result in an imbalance in the training data due to unknown sampling selection bias where not all populations are equally represented. Two possible kinds of distribution shifts are depicted in Figure 5.2: the appearance of the slides differs when using a different scanner (A) or when moving to another clinic (B), where the slides are prepared and scanned in a different way. Both kinds of shifts would need to be taken into consideration already during the data collection process, which would mean that all scanners and clinics have to be included for minimal uncertainty. Although it is possible to account for a limited amount of the changes using data augmentation methods to simulate different lighting conditions or enhancing robustness by adding noise, it is impossible to account for all possible variations in advance. As a result, the training data distribution may not exactly match the test distribution, leading to epistemic uncertainty [19]. More extremely, a model might encounter out-of-distribution (OOD) data during testing. OOD data refers to valid samples that a model can potentially process but that do not belong to the task the model addresses, for example new classes that were not present during training [19], but also adversarial examples, that are designed to trick a model into making wrong predictions. Figure 5.2 shows an example of OOD samples. The blue boxes mark artefacts in the slides that do not represent a tissue class.

From a modeling and learning perspective, uncertainty can also arise from the design choices, also called inductive bias, of the deep learning model, such as the size and structure of the model architecture and the hyperparameters of the learning process. Large models commonly lead to overconfident softmax outputs, which can result in poor uncertainty estimates. On the other hand, small models can lead to underfitting and high epistemic uncertainty [23, 19].

5.3.2 Methods for Quantification of Uncertainty and Calibration

In the following we want to give an overview of methods that are designed to give a reliable uncertainty estimation. These methods either provide an uncertainty estimate directly or re-calibrate the softmax output.

We begin by presenting a mathematical setup. We only consider supervised learning, specifically the task of classification. However, most approaches are also applicable for regression problems. For all of the following models let $(\mathbf{x}_i, y_i) \in (X, Y)$ be a data-label pair from the set of all training data X and labels Y. Let $X = \{\mathbf{x}_i\}_i^N$ be the set of N training data points with corresponding class labels $Y = \{\mathbf{y}_i\}_i^N, y_i \in \{1, \ldots, K\}$ for K classes. Further let $f_\theta : X \to \mathbb{R}^K$ be a (deep) learning function with learnable weights θ mapping to a K-dimensional logit space. Thus, for a sample $\mathbf{x}_i \in X$ we obtain logits $\mathbf{z}_i = f_\theta(\mathbf{x}_i)$ from which we can derive the probability for each class $k \in K$: The application of the softmax function σ returns probability values $\hat{\mathbf{p}}_i^{(k)}$ for each class k.

$$\hat{\mathbf{p}}_i^{(k)} = \sigma(\mathbf{z}_i)^{(k)} = \frac{\exp(\mathbf{z}_i^{(k)})}{\sum_{j=1}^K \exp(\mathbf{z}_i^{(j)})} \tag{5.1}$$

The maximum class probability

$$\hat{p}_i = \max_{k \in \{1 \ldots K\}} \hat{\mathbf{p}}_i^{(k)} \tag{5.2}$$

can be interpreted as the confidence of the sample. We refer to it as vanilla or softmax confidence. The prediction is simply $\hat{y}_i = \arg\max_k \mathbf{p}_i^{(k)}$. Generally, the softmax confidence does not necessarily represent the true certainty. Most deep learning networks tend to produce overconfident predictions.

The following methods either aim to improve the calibration of the softmax confidence or they propose other means to estimate the confidence of a sample. We categorize these methods based on the stage of the process at which they approach the task, in order to account for different application requirements.

1. *Data-based methods* operate on the data perspective and aim to smooth the transition between classes, thereby improving calibration.
2. *Architecture modification methods* involve adjusting the architecture or learning process of a model to provide uncertainty estimates or improve calibration.
3. In cases where the model cannot be modified, *post-hoc methods* can adjust the model's calibration. To achieve enhanced calibration, the methods can also be combined.

5.3.2.1 Data-based Methods

On a data level we can exploit regularization techniques such as data augmentation methods to smooth the input space which can provide a better calibration. One such approach is Mixup [84]. Mixup augments the training data by creating convex combinations between random samples and their labels. For two random data-label pairs $(\mathbf{x}_i, \mathbf{y}_i), (\mathbf{x}_j, \mathbf{y}_j) \in (X, Y)$, where $\mathbf{y}_i, \mathbf{y}_j$ are the one-hot encoded labels we calculate

$$
\begin{aligned}
\tilde{\mathbf{x}} &= \lambda \mathbf{x}_i + (1 - \lambda)\mathbf{x}_j \\
\tilde{\mathbf{y}} &= \lambda \mathbf{y}_i + (1 - \lambda)\mathbf{y}_j
\end{aligned}
\tag{5.3}
$$

with interpolation weight $\lambda \sim Beta(\alpha, \alpha)$ and $\alpha \geq 1$ to generate the augmented training samples and soft labels. Thulasidasan et. al. [72] have shown that the regularization and label smoothing of Mixup significantly improve calibration on image data and the model is less prone to making overconfident predictions on OOD and noisy data [72]. Follow-up work investigated the method and proved that Mixup training is approximately a regularized loss minimization [85]. There are multiple variants of Mixup such as CutMix [82], Puzzle Mix [42], Adversarial Mixup Resynthesis [4], PixMix [31] and [86] to improve the calibration of imbalanced data. Each of these variants uses a different approach of mixing the data. Cutmix, for instance, motivated from the perspective of dropout regularization, works on image data and cuts and pastes image patches among different image samples and generates soft labels accordingly. PixMix aims to improve robustness to safety measures like calibration and anomalies by adding structures to images and increasing their complexity.

The application of Mixup methods can be extended beyond the input data and onto the feature space: Manifold Mixup, as suggested by [76], creates a linear interpolation on the feature vectors before the classification layer which can provide a smoother decision boundary. From [15, 14] we conclude that Mixup alone might not be sufficient as an uncertainty method but it provides small improvements in terms of calibration over the baseline and might be useful in combination with other methods as it is easy to implement and cheap to apply.

5.3.2.2 Architecture-Modifying Methods

On the modelling perspective there are multiple methods that either aim to improve the overall calibration or provide a separate uncertainty estimation.

One solution with well-established theoretical foundations is to estimate uncertainty in neural networks from a Bayesian perspective [55, 33]: instead of learning the neural network's weights directly, Bayesian deep learning places a (Gaussian) distribution on the weights and learns its parameters. Sampling from those weights provides different perspectives on the solution space; in other words we can sample infinitely many valid networks.

More formally the theory is built on Bayes' theorem

$$
p(\theta | \mathbf{x}, y) = \frac{p(y | \mathbf{x}, \theta) p(\theta)}{p(y | \mathbf{x})} \propto p(y | \mathbf{x}, \theta) p(\theta)
\tag{5.4}
$$

where $(\mathbf{x}, y)$ is an input-label pair and θ represents the learning weights of the model. The theorem states that the posterior distribution $p(\theta|\mathbf{x}, y)$, that we are looking for, is calculated from the likelihood $p(y|\mathbf{x}, \theta)$, the prior $p(\theta)$ and a constant marginal distribution or evidence $p(y|\mathbf{x})$. Assuming that the posterior is known, we can obtain a prediction for a new sample $\mathbf{x}^*$ by

$$p(y^*|\mathbf{x}^*, \mathbf{x}, y) = \int p(y^*|\mathbf{x}^*, \theta)p(\theta|\mathbf{x}, y)d\theta. \qquad (5.5)$$

The calculation of the posterior is usually intractable. Therefore the inference of Bayesian neural networks usually requires approximations such as sampling approaches like Markov Chain Monte Carlo sampling [55], a Laplace approximation of the log-posterior [13, 64], or (stochastic) variational inference [33, 22]. Techniques such as Bayes by Backprop [5] or VOGN [41], which are specifically proposed for deep learning, leverage variational inference approximation to approximate the true posterior with a simpler distribution. However, the complexity of inference in Bayesian deep learning approaches usually becomes computationally expensive. It is primarily due to high-dimensional and non-convex weight spaces, necessitating intricate and computationally demanding methods to approximate or sample from the posterior distribution over the weights of the model, not only during training but also at test time. These techniques are therefore rarely put into practice [59]. We experienced that Bayes by Backprop and VOGN do not scale well with large networks and require too much computational time. On top, they do not necessarily improve upon simple softmax baselines in our experience.

A more recent approach, SWAG [50], proposes to use the weights of multiple stages of the learning process as samples from the posterior distribution and approximates the posterior by learning a multivariate Gaussian distribution with a low-rank approximation of the covariance matrix over the weights from those samples. As another approximation, Gal and Ghahramani proposed Monte Carlo Dropout (MC Dropout) where a more robust network is learned by randomly dropping neurons from the network both during training and during testing [18]. While SWAG and MC Dropout may successfully capture the details of one mode within the posterior distribution, they are incapable of accommodating multiple modes simultaneously [37].

Ensemble methods make this possible. Ensembles, which refer to the combination of multiple machine learning models, have been widely used to make more accurate and robust predictions. By mitigating the impact of individual model weaknesses and relying on the diversity of the predictors, ensembles can improve overall accuracy and uncertainty estimation. This simple technique has become a popular choice in various applications and has evolved over time to take on different forms. Classical machine learning often employs ensembles using methods like bagging and boosting [6, 17]. In the context of neural networks, deep ensembles are often composed of multiple individually trained models, whose probabilistic outputs are averaged for inference [44]. They have become a gold standard for accurate and well-calibrated predictive distributions, outperforming other popular approaches like Bayesian neural networks or MC Dropout [60]. A possible reason for this is that ensembles explore more diverse modes in the function space and thus create more diverse hypotheses for the data than Bayesian approaches, which oftentimes sample hypotheses around a single mode [16, 81].

While naive ensembling offers substantial benefits to single models, further improvements can be achieved in several ways. One important line of research focuses on improving the diversity of ensembles, which can be achieved by diversifying hyperparameters [79], focusing on different feature subsets [63], or using different architectures for each ensemble member [83]. Another challenge is the parameter- and time-complexity of ensembles. Multiple independent networks must be trained and applied for inference, which can result in high computational demands. Although the inference time can be reduced by parallelization, the training effort can be significant in some applications. One common approach to address these challenges involves weight sharing between ensemble members, as proposed in several studies [78, 27]. Ensembles have also been combined with Bayesian approaches to combine the mode exploration of ensembles with the within-mode exploration of Bayesian approaches. For example, MultiSWAG is an ensemble of multiple independently trained Stochastic Weight Averaging (SWAG) approximations [81].

While all of the above methods require several forward passes, other methods only need one forward pass at test time. Evidential deep learning [69] approaches uncertainty estimation from a Dempster–Shafer Theory of Evidence point of view. They utilize a Dirichlet distribution on the class probabilities as different opinions, which allows them to derive belief masses and an overall uncertainty.

Other methods express uncertainty, specifically in order to reject OOD samples, by distances to class prototypes [75, 49]. DUQ [75], for instance, uses an RBF kernel to measure the distance or uncertainty of an embedded sample to a class centroid or prototype. Samples located too far from the prototype are rejected as OOD. Despite being a sound concept, DUQ resulted in stability issues during our experiments which was also noted in the author's follow-up paper [74].

A new and highly effective method, called SNGP (spectral normalized Gaussian process) [48] also builds on the idea of distance awareness within the network for a meaningful feature representation. They ensure this distance awareness by weight normalizations and by modelling the last layer of the neural network as a Gaussian process. By utilizing a Laplace approximation technique, a more cost-effective sampling can be performed during test time with a single forward pass. This is possible because the samples are drawn from a closed-form posterior. Conveniently, Tensorflow provides an implementation and tutorial for SNGP but also for MC Dropout and Deep Ensembles [70, 53].

5.3.2.3 Post-Hoc Methods

Up to this point, we have only examined techniques that require substantial changes in the architecture and training procedure. However, in some cases, making these substantial modifications during the training process of a machine learning model may not be feasible due to practical considerations such as time or cost, for example due to the requirement of sampling at inference. Post-hoc approaches can enhance the calibration performance of a model after training has already taken place. A simple but effective approach to improve the post-hoc calibration of a model is temperature scaling [61, 23]. The objective of temperature scaling is to determine a value $T > 0$ that scales the predicted logits before the softmax calculation, resulting in a more faithful representation of the true confidence. The confidence is estimated as

$$\hat{p}_i = \max_k \sigma(\mathbf{z}_i/T)^{(k)}. \tag{5.6}$$

for logits $\mathbf{z}_i \in \mathbb{R}^K$ and softmax function σ. The temperature parameter T is typically obtained through the minimization of the negative log likelihood with respect to T on a validation set:

$$\min_T \sum_{i=1}^{N^{\text{val}}} \log \sigma(\mathbf{z}_i/T)^{(y_i)}, \tag{5.7}$$

where N^{val} denotes the size of the validation set. The predicted confidence decreases if $T > 1$, and increases for $T < 1$. As temperature scaling only conducts a re-calibration and does not rearrange the order of classes, the prediction remains unchanged.

The effectiveness of temperature scaling calibration has been assessed across a range of scenarios. The findings of Ovadia et al. [60] indicate that temperature scaling can offer accurate calibration of uncertainty estimates in cases involving i.i.d. data or data that is only slightly shifted from the training distribution. However, as the shift increases, its effectiveness decreases [30] and ensembling methods outperform temperature scaling in calibration performance [60]. The combination of both has shown promising results in a facial analysis scenario with artificially induced discriminatory biases [14], where the optimal value of T was determined on a balanced validation set.

Several variations of this method adopt a recalibration model to learn the temperature in a post-hoc manner following the training phase, e.g. [43, 38].

5.3.3 Evaluation Metrics for Uncertainty Estimation

Researchers employ a variety of metrics for evaluating uncertainty estimates. In this section, we will focus on the most commonly used metrics in current publications. Understanding the behaviour and the limitations of these metrics is crucial for making better decisions. It is important to note that no single metric measures all the desirable properties of uncertainty estimates obtained by a model of interest. Moreover, the appropriate metric depends on the use cases and specifications of a given task. For instance, in a use case where we have the opportunity to set a confidence threshold, we might accept predictions above this threshold while the remainder is classified by a human expert, e.g. in the case of the brake-disc example shown in Figure 5.1. In such cases, it may not be necessary to have perfect calibration, but rather a strong correlation between uncertainty and misclassification rate may suffice. In this case, we are mainly interested in rejection-based metrics. This can also be the case for possible OOD samples, as shown in Figure 5.2 D, where we could use a threshold on the uncertainty estimates to filter out possible artifacts. In the following, the metrics are presented for multiclass classification.

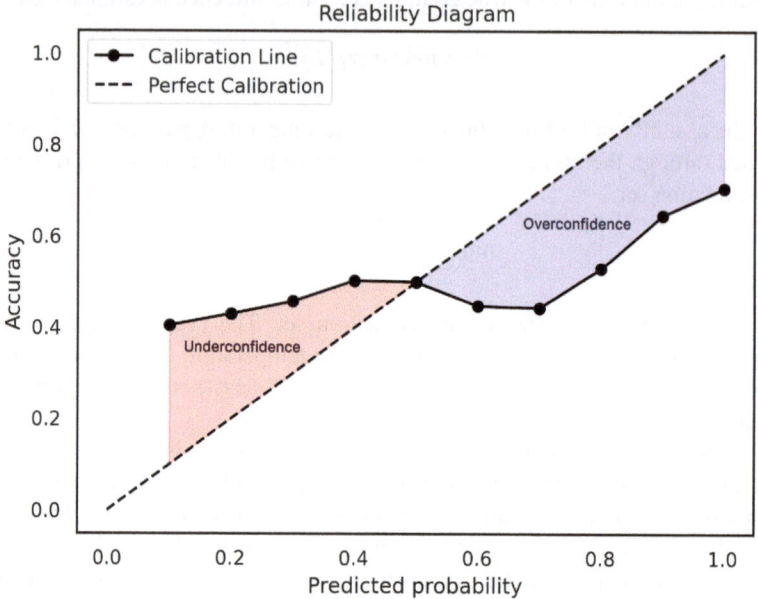

Fig. 5.3: Reliability diagram for evaluating the calibration of a classification model. Points lying on the diagonal line indicate perfect calibration, while points above or below the line indicate areas of under and overconfidence, respectively.

5.3.3.1 Negative Log-Likelihood

Negative Log-Likelihood (NLL) is a widely used evaluation metric for probabilistic models, as it also serves as a loss function for classification tasks. It measures the likelihood of the observed data given the predicted probabilities. A lower NLL indicates better uncertainty estimates. Mathematically, the negative log-likelihood is defined as:

$$NLL = \frac{1}{N} \sum_{i=1}^{N} (-\log p(y = y_i | \mathbf{x}_i)), \qquad (5.8)$$

for input-label pairs $(\mathbf{x}_i, y_i), i \in \{1, \ldots, N\}$. However, maximizing the confidence of a prediction to minimize the NLL can lead to overconfident models that exhibit higher accuracy. Consequently, comparing the quality of uncertainty between models with varying levels of accuracy using NLL can be misleading, as it may favor models with higher accuracy [2].

5.3.3.2 Expected Calibration Error

The Expected Calibration Error (ECE) [54] is a metric used to estimate the calibration error of a model's uncertainty estimates based on a reliability diagram (Figure 5.3). This diagram shows the deviation of estimated probabilities from the observed accuracy. It does this by dividing the probability interval into fixed bins and assigning each predicted probability to its corresponding bin and calculating the accuracy per bin. A model with good calibration will have the calibration curve close to the diagonal line. The ECE is calculated by taking the difference between the accuracy and the confidence of each bin, weighted by the number of predictions in each bin, and averaged across all bins. In a multiclass setting, the maximum probability score is used for this calculation. Formally it is defined as

$$ECE = \sum_{m=1}^{M} \frac{|B_m|}{N} |acc(B_m) - conf(B_m)|, \qquad (5.9)$$

where M is the number of bins, B_m is the m-th bin and $|B_m|$ is the number of samples in bin B_m. The total number of samples is described by N, $acc(B_m)$ is the accuracy of bin B_m, and $conf(B_m)$ is the confidence of bin B_m. The ECE only measures calibration in contrast to metrics like NLL, which is influenced by the accuracy. However, it is sensitive to the number of bins and the thresholding. Therefore, other approaches such as Adaptive Calibration Error [56] were introduced, which spaces the bins such that an equal number of predictions contribute to each bin.

5.3.3.3 Rejection-based Measures

An alternative perspective for utilizing uncertainty estimates involves using them to reject certain data points. This can take various forms, such as setting a confidence threshold and accepting only predictions above it while manually classifying the remaining ones or differentiating between out-of-distribution and in-distribution data. As this task involves binary separation, binary classification measures like the area under the receiver operating characteristic curve (AUROC) can be used or other commonly used metrics for out-of-distribution detection. Another approach is to use accuracy-rejection curves [52], which are functions that represent the accuracy of a classifier as a function of its rejection rate.

5.4 Conclusion and Outlook

Trustworthy AI is a prerequisite for any responsible deployment of AI-based systems in almost all domains. While AI's recent successes and the speed of its development are impressive, our understanding of the inner workings of such systems is falling behind. A reliable estimation of the uncertainty associated with a model's output is necessary to understand when to trust an AI system and when not. This is a first step towards transparency and a tool to avoid certain kinds of – potentially hazardous – errors. It is crucial for uncertainty estimation to be integrated into every stage of the modeling process of an AI system. This includes providing developers with feedback on the model, identifying potential biases in the data, and enabling control over how the model handles out-of-distribution data. For end-users, having access to a solution for uncertainty estimation and explainability is essential for building trust and increasing acceptance of AI algorithms. A completely error-free algorithm is not required, as long as the machine learning model can provide a reliable estimation of its own uncertainty, indicating its limits and why it returns a certain decision.

Nonetheless, uncertainty estimation alone will not lead to Trustworthy AI. None of the methods available today are perfect. They introduce additional complexity and may lead to new and potentially more subtle errors that need to be understood and addressed as well. Furthermore, developers of AI systems need to guarantee or at least reason about certain properties – such as robustness, fairness, privacy – depending on the application domain and specific context. AI is most successfully applied where the nature of the task is too complex to develop a standard algorithm, making this a very challenging endeavor. Unfortunately, established methods from software testing cannot be directly transferred to AI systems either - for a number of reasons: 1) Specification by examples instead of a formal description leads to specification uncertainty. 2) High-dimensional input spaces lead to a state-space explosion. 3) The unpredictable behavior of (large) neural networks given small input changes makes it impossible to perform representative tests of the entire input space with a limited number of test cases. Therefore, any argumentation around trustworthiness and safety of an AI system will require qualitative and quantitative reasoning as well as an iterative process incorporating an improved understanding of the mechanisms at play over time [7].

Future work should address the shortcomings of existing uncertainty quantification methods as well as complementary contributions towards trustworthy AI. While the accuracy of the uncertainty estimation itself could obviously be improved, our understanding of its limits appears to be the more important challenge, i.e., when can we trust those estimates, what are underlying assumptions and how to test them. As an example, deep ensembles are a well established method to estimate uncertainty but they rely on the assumption that there is *enough diversity* among the ensemble members. However, it is still an open research question which kind of diversity is required for optimal performance and robustness [29]. Nevertheless, ensuring trustworthiness for AI systems will require a holistic approach combining methods from different fields such as explainable AI, formal verification methods, out-of-distribution detection and classical system monitoring. However, the individual contribution of each building block towards trustworthy and safe AI has to be well understood. Looking at the current research landscape in machine learning, there is an abundance of proposed methods and metrics with their number increasing by the day. Unfortunately, most of them are evaluated on artificial benchmarks and with simplistic evaluation metrics leaving a large gap towards real world applications. Furthermore, it is important to acknowledge that most of those methods introduce additional sources of uncertainty that have to be accounted for as well.

In addition to an exhaustive set of trustworthy AI tools and methods, the corresponding development and testing processes have to reflect this requirement as well. More suitable testing and evaluation strategies have to be developed. They should focus on task-specific failure modes and their importance instead of aggregate metrics that heavily rely on a representative test data set. While some high-level proposals already exist [28], there is a strong need for a continuous development and deployment cycle for AI systems similar to the well known *DevOps* in the IT world. The main challenge here is that deep neural networks can change in unpredictable ways with each additional training step rendering every test and analysis done on a previous version meaningless. Ideas borrowed from incremental certification [80] may pave the way for building trustworthy AI systems step-by-step leveraging the work already done while enabling flexibility to further improve performance and functionality based on new data and insights.

References

1. A. Adadi and M. Berrada. Peeking inside the black-box: A survey on explainable artificial intelligence (XAI). *IEEE Access*, 6:52138–52160, 2018.
2. A. Ashukha, A. Lyzhov, D. Molchanov, and D. Vetrov. Pitfalls of in-domain uncertainty estimation and ensembling in deep learning. *arXiv preprint arXiv:2002.06470*, 2020.
3. J. Athavale, A. Baldovin, R. Graefe, M. Paulitsch, and R. Rosales. Ai and reliability trends in safety-critical autonomous systems on ground and air. *2020 50th Annual IEEE/IFIP International Conference on Dependable Systems and Networks Workshops (DSN-W)*, pages 74–77, 2020.
4. C. Beckham, S. Honari, V. Verma, A. M. Lamb, F. Ghadiri, R. D. Hjelm, Y. Bengio, and C. Pal. On adversarial mixup resynthesis. In H. Wallach, H. Larochelle, A. Beygelzimer, F. d'Alché-Buc, E. Fox, and R. Garnett, editors, *Advances in Neural Information Processing Systems*, volume 32. Curran Associates, Inc., 2019.
5. C. Blundell, J. Cornebise, K. Kavukcuoglu, and D. Wierstra. Weight uncertainty in neural network. In *International conference on machine learning*, pages 1613–1622. PMLR, 2015.
6. L. Breiman. Bagging predictors. *Machine Learning*, 24:123–140, 2004.
7. S. Burton. A causal model of safety assurance for machine learning. *arXiv:2201.05451 [cs]*, Jan. 2022.
8. S. Burton and B. Herd. Addressing uncertainty in the safety assurance of machine-learning. *Frontiers in Computer Science*, 5, 2023.
9. S. Burton and B. Herd. Addressing uncertainty in the safety assurance of machine-learning. *Frontiers in Computer Science*, 5, 2023.
10. K. Bykov, M. M. C. Höhne, K.-R. Müller, S. Nakajima, and M. Kloft. How much can i trust you? – quantifying uncertainties in explaining neural networks, 2020.

11. E. Commission. Proposal for a regulation of the european parliament and of the council laying down harmonised rules on artificial intelligence (artificial intelligence act) and amending certain union legislative acts. COM/2021/20106(COD), 2021.

12. R. Dabhi. casting product image data for quality inspection. https://www.kaggle.com/ravirajsinh45/real-life-industrial-dataset-of-casting-product, 2020. Accessed: 2021-09-07, (CC BY-NC-ND 4.0).

13. E. Daxberger, A. Kristiadi, A. Immer, R. Eschenhagen, M. Bauer, and P. Hennig. Laplace redux-effortless bayesian deep learning. *Advances in Neural Information Processing Systems*, 34:20089–20103, 2021.

14. J. Deuschel, A. Foltyn, L. A. Adams, J. M. Vieregge, and U. Schmid. Benchmarking robustness to natural distribution shifts for facial analysis. In *NeurIPS 2021 Workshop on Distribution Shifts: Connecting Methods and Applications*.

15. A. Foltyn and J. Deuschel. Towards Reliable Multimodal Stress Detection under Distribution Shift. In *Companion Publication of the 2021 International Conference on Multimodal Interaction*, pages 329–333, Montreal QC Canada, Oct. 2021. ACM.

16. S. Fort, H. Hu, and B. Lakshminarayanan. Deep ensembles: A loss landscape perspective. *arXiv preprint arXiv:1912.02757*, 2019.

17. J. H. Friedman. Greedy function approximation: A gradient boosting machine. *Annals of Statistics*, 29:1189–1232, 2001.

18. Y. Gal and Z. Ghahramani. Dropout as a bayesian approximation: Representing model uncertainty in deep learning. In *international conference on machine learning*, pages 1050–1059. PMLR, 2016.

19. J. Gawlikowski, C. R. N. Tassi, M. Ali, J. Lee, M. Humt, J. Feng, A. Kruspe, R. Triebel, P. Jung, R. Roscher, et al. A survey of uncertainty in deep neural networks. *arXiv preprint arXiv:2107.03342*, 2021.

20. Gianluca Quaglio, Scientific Foresight Unit, European Parliament. Artificial intelligence in healthcare: Applications, risks, and ethical and societal impacts, 2022.

21. I. J. Goodfellow, Y. Bengio, and A. Courville. *Deep Learning*. MIT Press, Cambridge, MA, USA, 2016. http://www.deeplearningbook.org.

22. A. Graves. Practical variational inference for neural networks. *Advances in neural information processing systems*, 24, 2011.

23. C. Guo, G. Pleiss, Y. Sun, and K. Q. Weinberger. On calibration of modern neural networks. In *Proceedings of the 34th International Conference on Machine Learning - Volume 70*, ICML'17, page 1321–1330. JMLR.org, 2017.

24. Z. Guo, Z. Wan, Q. Zhang, X. Zhao, F. Chen, J.-H. Cho, Q. Zhang, L. M. Kaplan, D. H. Jeong, and A. Jøsang. A survey on uncertainty reasoning and quantification for decision making: Belief theory meets deep learning. *arXiv preprint arXiv:2206.05675*, 2022.

25. B. Gyevnar and N. Ferguson. Aligning explainable ai and the law: The european perspective, 2023.

26. M. Habibpour, H. Gharoun, M. Mehdipour, A. Tajally, H. Asgharnezhad, A. Shamsi, A. Khosravi, and S. Nahavandi. Uncertainty-aware credit card fraud detection using deep learning. *Engineering Applications of Artificial Intelligence*, 123:106248, 2023.

27. M. Havasi, R. Jenatton, S. Fort, J. Z. Liu, J. Snoek, B. Lakshminarayanan, A. M. Dai, and D. Tran. Training independent subnetworks for robust prediction. In *International Conference on Learning Representations*, 2021.

28. R. Hawkins, C. Paterson, C. Picardi, Y. Jia, R. Calinescu, and I. Habli. Guidance on the Assurance of Machine Learning in Autonomous Systems (AMLAS), Jan. 2021.

29. L. Heidemann, A. Schwaiger, and K. Roscher. Measuring ensemble diversity and its effects on model robustness. In H. Espinoza et. al., editor, *Proceedings of the Workshop on Artificial Intelligence Safety 2021 Co-Located with the Thirtieth International Joint Conference on Artificial Intelligence (IJCAI 2021), Virtual, August, 2021*, volume 2916 of *CEUR Workshop Proceedings*. CEUR-WS.org, 2021.

30. M. Hein, M. Andriushchenko, and J. Bitterwolf. Why relu networks yield high-confidence predictions far away from the training data and how to mitigate the problem. In *Proceedings of the IEEE/CVF Conference on Computer Vision and Pattern Recognition*, pages 41–50, 2019.

31. D. Hendrycks, A. Zou, M. Mazeika, L. Tang, B. Li, D. Song, and J. Steinhardt. Pixmix: Dream-like pictures comprehensively improve safety measures. In *Proceedings of the IEEE/CVF Conference on Computer Vision and Pattern Recognition*, pages 16783–16792, 2022.
32. M. Henne, A. Schwaiger, K. Roscher, and G. Weiss. Benchmarking Uncertainty Estimation Methods for Deep Learning With Safety-Related Metrics. In H. Espinoza et. al., editor, *Proc. SafeAI@AAAI 2020*, volume 2560 of *CEUR Workshop Proceedings*, pages 83–90. CEUR-WS.org, 2020.
33. G. E. Hinton and D. van Camp. Keeping the neural networks simple by minimizing the description length of the weights. In *Proceedings of the Sixth Annual Conference on Computational Learning Theory*, COLT '93, page 5–13, New York, NY, USA, 1993. Association for Computing Machinery.
34. T. W. House. Blueprint for an ai bill of rights. https://www.whitehouse.gov/ostp/ai-bill-of-rights/, 2022.
35. E. Hüllermeier and W. Waegeman. Aleatoric and epistemic uncertainty in machine learning: An introduction to concepts and methods. *Machine Learning*, 110(3):457–506, Mar. 2021.
36. I. Hupont, M. Micheli, B. Delipetrev, E. Gómez, and J. S. Garrido. Documenting high-risk ai: A european regulatory perspective. *Computer*, 56(5):18–27, 2023.
37. L. V. Jospin, H. Laga, F. Boussaid, W. Buntine, and M. Bennamoun. Hands-On Bayesian Neural Networks—A Tutorial for Deep Learning Users. *IEEE Computational Intelligence Magazine*, 17(2):29–48, May 2022.
38. T. Joy, F. Pinto, S.-N. Lim, P. H. Torr, and P. K. Dokania. Sample-dependent adaptive temperature scaling for improved calibration. *arXiv preprint arXiv:2207.06211*, 2022.
39. J. M. Jumper, R. Evans, A. Pritzel, T. Green, M. Figurnov, O. Ronneberger, K. Tunyasuvunakool, R. Bates, A. Žídek, A. Potapenko, A. Bridgland, C. Meyer, S. A. A. Kohl, A. Ballard, A. Cowie, B. Romera-Paredes, S. Nikolov, R. Jain, J. Adler, T. Back, S. Petersen, D. A. Reiman, E. Clancy, M. Zielinski, M. Steinegger, M. Pacholska, T. Berghammer, S. Bodenstein, D. Silver, O. Vinyals, A. W. Senior, K. Kavukcuoglu, P. Kohli, and D. Hassabis. Highly accurate protein structure prediction with alphafold. *Nature*, 596:583 – 589, 2021.
40. D. Kaur, S. Uslu, K. J. Rittichier, and A. Durresi. Trustworthy artificial intelligence: A review. *ACM Comput. Surv.*, 55(2), jan 2022.
41. M. Khan, D. Nielsen, V. Tangkaratt, W. Lin, Y. Gal, and A. Srivastava. Fast and scalable bayesian deep learning by weight-perturbation in adam. In *International conference on machine learning*, pages 2611–2620. PMLR, 2018.
42. J.-H. Kim, W. Choo, and H. O. Song. Puzzle mix: Exploiting saliency and local statistics for optimal mixup. In *Proceedings of the 37th International Conference on Machine Learning*, ICML'20. JMLR.org, 2020.
43. V. Kuleshov and S. Deshpande. Calibrated and sharp uncertainties in deep learning via density estimation. In *International Conference on Machine Learning*, pages 11683–11693. PMLR, 2022.
44. B. Lakshminarayanan, A. Pritzel, and C. Blundell. Simple and Scalable Predictive Uncertainty Estimation using Deep Ensembles. In I. Guyon, U. V. Luxburg, S. Bengio, H. Wallach, R. Fergus, S. Vishwanathan, and R. Garnett, editors, *Advances in Neural Information Processing Systems 30*, pages 6402–6413. Curran Associates, Inc., 2017.
45. Y. LeCun, Y. Bengio, and G. Hinton. Deep learning. *nature*, 521(7553):436, 2015.
46. P. Linardatos, V. Papastefanopoulos, and S. Kotsiantis. Explainable ai: A review of machine learning interpretability methods. *Entropy*, 23(1), 2021.
47. H. Liu, Y. Wang, W. Fan, X. Liu, Y. Li, S. Jain, Y. Liu, A. Jain, and J. Tang. Trustworthy ai: A computational perspective. *ACM Trans. Intell. Syst. Technol.*, 14(1), nov 2022.
48. J. Liu, Z. Lin, S. Padhy, D. Tran, T. Bedrax Weiss, and B. Lakshminarayanan. Simple and principled uncertainty estimation with deterministic deep learning via distance awareness. *Advances in Neural Information Processing Systems*, 33:7498–7512, 2020.
49. D. Macêdo, T. I. Ren, C. Zanchettin, A. L. Oliveira, and T. Ludermir. Entropic out-of-distribution detection. In *2021 International Joint Conference on Neural Networks (IJCNN)*, pages 1–8. IEEE, 2021.

50. W. J. Maddox, P. Izmailov, T. Garipov, D. P. Vetrov, and A. G. Wilson. A simple baseline for bayesian uncertainty in deep learning. *Advances in neural information processing systems*, 32, 2019.

51. C. Molnar. *Interpretable Machine Learning*. 2 edition, 2022.

52. M. S. A. Nadeem, J.-D. Zucker, and B. Hanczar. Accuracy-rejection curves (arcs) for comparing classification methods with a reject option. In *Machine Learning in Systems Biology*, pages 65–81. PMLR, 2009.

53. Z. Nado, N. Band, M. Collier, J. Djolonga, M. W. Dusenberry, S. Farquhar, Q. Feng, A. Filos, M. Havasi, R. Jenatton, et al. Uncertainty baselines: Benchmarks for uncertainty & robustness in deep learning. *arXiv preprint arXiv:2106.04015*, 2021.

54. M. P. Naeini, G. Cooper, and M. Hauskrecht. Obtaining well calibrated probabilities using bayesian binning. In *Proceedings of the AAAI conference on artificial intelligence*, volume 29, 2015.

55. R. M. Neal. *Bayesian Learning for Neural Networks*, volume 118 of *Lecture Notes in Statistics*. Springer New York, New York, NY, 1996.

56. J. Nixon, M. W. Dusenberry, L. Zhang, G. Jerfel, and D. Tran. Measuring calibration in deep learning. In *CVPR workshops*, volume 2, 2019.

57. OpenAI. Gpt-4 technical report, 2023.

58. M. P. Oppelt, A. Foltyn, J. Deuschel, N. R. Lang, N. Holzer, B. M. Eskofier, and S. H. Yang. ADABase: A Multimodal Dataset for Cognitive Load Estimation. *Sensors*, 23(1):340, Dec. 2022.

59. K. Osawa, S. Swaroop, M. E. E. Khan, A. Jain, R. Eschenhagen, R. E. Turner, and R. Yokota. Practical deep learning with bayesian principles. *Advances in neural information processing systems*, 32, 2019.

60. Y. Ovadia, E. Fertig, J. Ren, Z. Nado, D. Sculley, S. Nowozin, J. Dillon, B. Lakshminarayanan, and J. Snoek. Can you trust your model's uncertainty? evaluating predictive uncertainty under dataset shift. *Advances in neural information processing systems*, 32, 2019.

61. J. Platt et al. Probabilistic outputs for support vector machines and comparisons to regularized likelihood methods. *Advances in large margin classifiers*, 10(3):61–74, 1999.

62. J. Quiñonero-Candela, M. Sugiyama, A. Schwaighofer, and N. D. Lawrence. *Dataset Shift in Machine Learning*. Neural Information Processing Series. MIT Press, Cambridge, Mass, 2009.

63. A. Rame and M. Cord. DICE: Diversity in Deep Ensembles via Conditional Redundancy Adversarial Estimation. *arXiv:2101.05544 [cs, math]*, Jan. 2021.

64. H. Ritter, A. Botev, and D. Barber. A scalable laplace approximation for neural networks. In *6th International Conference on Learning Representations, ICLR 2018-Conference Track Proceedings*, volume 6. International Conference on Representation Learning, 2018.

65. R. Rombach, A. Blattmann, D. Lorenz, P. Esser, and B. Ommer. High-resolution image synthesis with latent diffusion models, 2021.

66. G. Scalia, C. A. Grambow, B. Pernici, Y.-P. Li, and W. H. Green. Evaluating Scalable Uncertainty Estimation Methods for Deep Learning-Based Molecular Property Prediction. *Journal of Chemical Information and Modeling*, 60(6):2697–2717, June 2020.

67. A. Schwaiger, P. Sinhamahapatra, J. Gansloser, and K. Roscher. Is Uncertainty Quantification in Deep Learning Sufficient for Out-of-Distribution Detection? In *Proc. AISafety@ IJCAI2020*, volume 2640 of *CEUR Workshop Proceedings*, page 8, 2020.

68. I. Secretary of State for Science and Technology. A pro-innovation approach to ai regulation. https://www.gov.uk/government/publications/ai-regulation-a-pro-innovation-approach, 2023.

69. M. Sensoy, L. Kaplan, and M. Kandemir. Evidential deep learning to quantify classification uncertainty. *Advances in neural information processing systems*, 31, 2018.

70. TensorFlow. Uncertainty-aware deep learning with sngp. https://www.tensorflow.org/tutorials/understanding/sngp#deep_ensemble, 2023. [Online; accessed 04-May-2023].

71. S. Thiebes, S. Lins, and A. Sunyaev. Trustworthy artificial intelligence. *Electronic Markets*, 31, 10 2020.

72. S. Thulasidasan, G. Chennupati, J. A. Bilmes, T. Bhattacharya, and S. Michalak. On mixup training: Improved calibration and predictive uncertainty for deep neural networks. *Advances in Neural Information Processing Systems*, 32, 2019.
73. E. Toreini, M. Aitken, K. Coopamootoo, K. Elliott, C. G. Zelaya, and A. van Moorsel. The relationship between trust in ai and trustworthy machine learning technologies. In *Proceedings of the 2020 Conference on Fairness, Accountability, and Transparency*, FAT* '20, page 272–283, New York, NY, USA, 2020. Association for Computing Machinery.
74. J. van Amersfoort, L. Smith, A. Jesson, O. Key, and Y. Gal. On feature collapse and deep kernel learning for single forward pass uncertainty. *arXiv preprint arXiv:2102.11409*, 2021.
75. J. Van Amersfoort, L. Smith, Y. W. Teh, and Y. Gal. Uncertainty estimation using a single deep deterministic neural network. In *International conference on machine learning*, pages 9690–9700. PMLR, 2020.
76. V. Verma, A. Lamb, C. Beckham, A. Najafi, I. Mitliagkas, D. Lopez-Paz, and Y. Bengio. Manifold mixup: Better representations by interpolating hidden states. In K. Chaudhuri and R. Salakhutdinov, editors, *Proceedings of the 36th International Conference on Machine Learning*, volume 97 of *Proceedings of Machine Learning Research*, pages 6438–6447. PMLR, 09–15 Jun 2019.
77. Y. Wang and S. H. Chung. Artificial intelligence in safety-critical systems: a systematic review. *Industrial Management & Data Systems*, 122(2):442–470, 2022.
78. Y. Wen, D. Tran, and J. Ba. Batchensemble: an alternative approach to efficient ensemble and lifelong learning. In *International Conference on Learning Representations*.
79. F. Wenzel, J. Snoek, D. Tran, and R. Jenatton. Hyperparameter ensembles for robustness and uncertainty quantification. *Advances in Neural Information Processing Systems*, 33:6514–6527, 2020.
80. A. Wilson and T. Preyssler. Incremental certification and integrated modular avionics. In *2008 IEEE/AIAA 27th Digital Avionics Systems Conference*, pages 1.E.3–1–1.E.3–8, 2008.
81. A. G. Wilson and P. Izmailov. Bayesian deep learning and a probabilistic perspective of generalization. *Advances in neural information processing systems*, 33:4697–4708, 2020.
82. S. Yun, D. Han, S. J. Oh, S. Chun, J. Choe, and Y. Yoo. Cutmix: Regularization strategy to train strong classifiers with localizable features. In *Proceedings of the IEEE/CVF international conference on computer vision*, pages 6023–6032, 2019.
83. S. Zaidi, A. Zela, T. Elsken, C. C. Holmes, F. Hutter, and Y. Teh. Neural ensemble search for uncertainty estimation and dataset shift. *Advances in Neural Information Processing Systems*, 34:7898–7911, 2021.
84. H. Zhang, M. Cisse, Y. N. Dauphin, and D. Lopez-Paz. mixup: Beyond empirical risk minimization. In *International Conference on Learning Representations*, 2018.
85. L. Zhang, Z. Deng, K. Kawaguchi, A. Ghorbani, and J. Zou. How does mixup help with robustness and generalization? In *International Conference on Learning Representations*, 2021.
86. Z. Zhong, J. Cui, S. Liu, and J. Jia. Improving Calibration for Long-Tailed Recognition. In *2021 IEEE/CVF Conference on Computer Vision and Pattern Recognition (CVPR)*, pages 16484–16493, Nashville, TN, USA, June 2021. IEEE.
87. T. Zhou, L. Zhang, T. Han, E. L. Droguett, A. Mosleh, and F. T. Chan. An uncertainty-informed framework for trustworthy fault diagnosis in safety-critical applications. *Reliability Engineering & System Safety*, 229:108865, 2023.

Chapter 6
Process-aware Learning

Christian M.M. Frey[1], Simon Rauch[12], Oliver Stritzel[1], Moike Buck[1]

Abstract Processes in companies are diverse and complex. The production of different products, inter- or intra-company logistic processes, or other serial event sequences within companies have one thing in common: they can be traced on the basis of a documentation of the individual process steps. Usually, companies have domain experts for each department's processes who use their experience and knowledge to plan and control these steps. However, with increasing complexity and diversity of processes, efficient planning and control is becoming more difficult or even impossible for human decision makers. In the the focus of *Process-aware Learning*, information documented on the data side, which is contained in the flow and execution of any process, should be integrated into AI-enhanced models. This should be accomplished in a way that is useful and as interpretable as possible for non-expert users. These models are used to identify important factors influencing the process, various process key figures, or anomalies in the process, and, based on these insights, to make forecasts or recommendations for action tailored to the process flows.

Key words: Process-aware Learning, Machine Learning, Bayesian statistics, Predictive Analysis, Process Mining, Process Analytics, Industry 4.0, Process AI

6.1 Introduction

The integration of process information into explainable Machine Learning models is generally associated with a high conceptual effort. Therefore, additional efforts of

---

[1] Fraunhofer Institute for Integrated Circuits IIS, Fraunhofer IIS, Nuremberg, Germany
[2] Ludwig-Maximilians-Universität München, Oettingenstr. 67, 80538 Munich, Germany

Corresponding author: Christian M.M. Frey
e-mail: christian.maximilian.michael.frey@iis.fraunhofer.de

© The Author(s) 2024   117
C. Mutschler et al. (eds.), *Unlocking Artificial Intelligence*,
https://doi.org/10.1007/978-3-031-64832-8_6

the *Process-aware Learning* deal with the realization of an automated algorithmic generation of interpretable models for process prognosis and the thereby possible predictive support of process planning and control. The objective is to create approaches that enable the automatic acquisition of causal network structures from process data. Various disciplines from the field of Process Mining [34, 36, 37], in particular the data-driven creation of process models (*'Process Discovery'*), are generally analyzed for their ability to extract causal relationships between process steps and other process parameters from the input data. The added value of such a procedure is an enormous reduction in the manual effort required to convert process data into models while preserving the comprehensibility and interpretability of the forecasts and model-generated suggestions for process optimization.

An important requirement towards the application of Process Mining and Machine Learning is a qualitatively and quantitatively sufficient basis of training data which is even more important for leveraging Deep Learning models. In the area of processing time series, cross-sectional data or text and image processing, as well as in the field of sequential data on processes, it is of utter utility to explore AI-enhanced methods for highly granular, diverse and, above all, incorporating the sequential characteristics of process data sets. This chapter is primarily dedicated to this research on processes. Over the recent years, a trend in the research community emerged that explores novel models facing the challenges of real world business processes. From a practical point of view, the digitization and digitalization of business concepts underpin the digital transformation, where the analysis and optimization of an institution's operations, strategic decisions and value propositions are decisive factors.

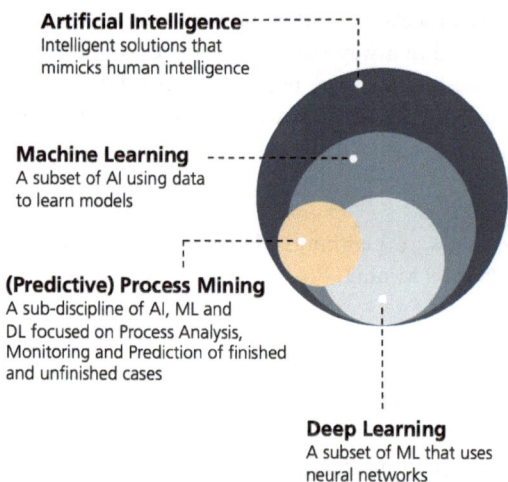

Fig. 6.1: Positioning of (Predictive) Process Mining

Process Mining is a technique designed to analyze and improve business processes as they actually are, not as you think they are. It applies data science to discover, validate, and improve workflows by combining data mining and process analytics.

The process data is extracted from readily available information sources, such as log files, databases or data warehouses, and then analyzed to identify process inefficiencies, bottlenecks, and potential improvements. Using Process Mining, organizations can gain insights into the as-is processes, monitor them in real-time, and optimize them to improve efficiency and quality while reducing costs and resources.

As information systems continue to capture an increasing amount of data about business processes, event logs have become a primary source for Process Mining. Despite the prevalence of event data, organizations often lack a comprehensive understanding of their actual processes. Rather than relying on thorough analysis of event data, management decisions are often influenced by intuitive analysis and judgement of domain experts. While data mining has greatly improved the ability to extract insights from large datasets and support complex decision-making, classical data mining techniques such as classification, clustering, regression, association rule learning, and sequence/episode mining origin from other areas than handling sequential process data (cf. Figure 6.1). Process data is specific to the data generated within a particular process or system, while classical datasets refer to general collections of data used for analysis. As a result, data mining techniques used out of the box might not be effective in uncovering new information within event logs.

Process Mining aims to compare event data, which is actually the observed behavior, with process models that can be created manually or generated automatically. This technology has become available in recent years, and it can be applied to any type of operational process, spanning across various organizations or enterprise-wide systems. It can be utilized for analyzing processes, such as a hospital's treatment processes or material flows in the automotive industry, enhancing the customer service processes of multinational companies, gauging the browsing behavior of individuals while making reservations, diagnosing failures of airline baggage handling systems, or improving the user experience of medical devices, like X-ray machines. These applications usually require dynamic behavior to be linked to process models.

This chapter will provide a starting point and a general overview of Process Mining in Section 6.2. A more thorough discussion on the advances and insights of *Process Intelligence* is given in Section 6.3. A summary and outlook on the challenges of ongoing research conclude this chapter in Section 6.4.

6.2 Overview of Process Mining

This chapter provides an entry point to the main aspects and challenges in the field of *Process Mining*. In Section 6.2.1, we first provide the basic concepts of Process Mining. In the scientific literature, the research works are traditionally divided into three main categories (Process Discovery, Conformance Checking, Model Enhancement) which are presented in more detail in Section 6.2.2. To get a better insight into the procedure of data-centered processes, the characterization of event logs are given in Section 6.2.3 which are mainly used in the field of Process

Mining. In order to evaluate process models, we can define four criteria which are presented in Section 6.2.4 and give examples of processes in Section 6.2.5 to further discuss the challenges.

6.2.1 Process Mining - Basic Concept

Process Mining, which uses techniques to extract knowledge from event logs found in current information systems, offers innovative ways to explore, oversee, and enhance processes across various application fields. Increasing interest in Process Mining stems from two major factors:

1. More and more events are being recorded, thus, providing detailed information about the history of processes;
2. There is a need to improve and support business processes in competitive and rapidly changing environments.

Process Mining serves as a significant connector between Business Intelligence, Business Process Management, Data Mining, and Workflow. It encompasses automated process discovery, conformance checking, organizational/social network mining, construction of simulation models, model extension, model repair, case prediction, and history-based recommendations by extracting process models from an event log as well as monitoring deviations by comparing models and logs [34, 37].

6.2.2 Process Mining - Types

In general, there are three main types of Process Mining [34] as illustrated in Figure 6.2:

1. **Process Discovery.** Here, the task is to output a model by applying discovery techniques on the input event log without any a priori information. A well-known heuristic is provided by the Alpha algorithm [35] which takes an event log as input and produces a process model reflecting the behavior that is stored in the log.
2. **Conformance Checking.** An existing process model is checked if it validates an input event log of the assumed process. Therefore, conformance checking is used to evaluate if the reality, i.e. records in the log, are depicted in the model and vice versa.
3. **Model Enhancement.** Generally, the task is to extend or to further improve an existing process model using the information provided by some event log. In contrast to conformance checking which measures the alignment of the model and the input data, model enhancement focuses on changing or extending the a priori model, e.g., an extension of the model to disclose potential bottlenecks.

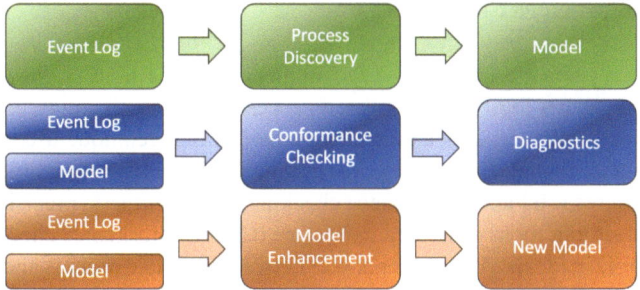

Fig. 6.2: Three types of Process Mining

In the scientific literature, there are also other types of process mining being orthogonal to the ones mentioned above, e.g.:

- The *control-flow perspective*, which focuses on the ordering of activities to find a good characterization of all possible paths.
- The *organizational perspective*, which rather focuses on the resources which are potentially provided within the event log. Here, the goal is to either structure the organization by classifying people in terms of roles and organizational units or to examine the social network underneath.
- The *case perspective*, which focuses on properties of cases.
- The *time perspective*, which focuses on the timing and frequency of events. A sophisticated throughput time analysis provides insight into potential bottlenecks, measures service levels, or can be used to monitor the utilization of resources and predict the remaining processing time of running cases.

6.2.2.1 Process Discovery

Process discovery is one of the three main categories of Process Mining techniques, which use event data to show what people, machines, and organizations are really doing in operational processes. Process discovery involves using event log data to create a process model [35, 43, 42, 12, 15, 38] that describes the actual behavior of the system as recorded in the logs. The goal of process discovery is to find a good process model that best fits the event log data, which can then be used to analyze and improve the operational process. This can provide novel insights that can be used to identify the executional path taken by operational processes and address their performance and compliance problems. Process discovery is essential for any organization that wants to gain a better understanding of their operational processes and find opportunities to optimize them.

6.2.2.2 Conformance Checking

Conformance checking [36, 3] is a technique within the field of Process Mining that compares event logs of a discovered process with an existing reference model to check for compliance. It is a crucial Process Mining technique used to optimize business processes and ensures they are running as efficiently as possible. There are several measures used for conformance checking, such as fitness, precision, generalization, and simplicity as described in Section 6.2.4. Conformance checking plays a significant role in identifying discrepancies between an actual process and a reference model, allowing organizations to make necessary adjustments to improve efficiency and compliance. The technique has been widely used in the field of data science and process management to support the analysis of operational processes based on event logs. For conformance checking, various techniques have been proposed in the literature: rule-based [23, 41], token replay-based [23], and alignment-based techniques [3, 30, 31, 39]. Overall, the aim of conformance checking is to ensure that operational processes are running smoothly and in tune with reference models.

6.2.2.3 Model Enhancement

Model enhancement [36, 4] is a category of Process Mining techniques that compares discovered process models to standard models to identify any errors or deviations. The goal is improve the accuracy and effectiveness of the process models. The heuristics applied for Model Enhancement operate on event data, which is used to discover and analyze operational processes. The discovered process models are then checked against standard models to identify any discrepancies. Based on the result of the checking, the process models can be enhanced with the identified improvements. Generally, model enhancement approaches define a continuous improvement endeavor that strives to enhance the efficiency and effectiveness of operational processes.

6.2.3 Event Log

The basis of the research field of process analysis is data in a special form, so-called event logs. These differ from conventional data structures such as cross-sectional or time series data in that there are usually irregularly distributed data points in the form of executed activities. This makes it difficult to apply classic analysis and forecasting algorithms, however, it offers the opportunity to use methods such as Process Mining to extract process knowledge available on the data side.

Event Logs are a collection of cases, where each element refers to a case, an activity and a point in time (timestamps). Event data sources can be found in various places, such as database systems, transaction logs, ERP systems (*Enterprise Resource*

Planning like SAP or Oracle), message logs, APIs that provide data from websites or social media.

When extracting event logs, a model potentially has to handle one or more of the challenges:

Correlation. Events in an event log are grouped per case. This simple requirement can be quite challenging as it requires event correlation, i.e., events need to be related to each other.

Chronological Ordering. Events belonging to a case need to be chronologically ordered. Typical problems that arise are inaccurate records (only dates), different time zones or delayed logging.

Snapshots. Cases may have a lifetime extending beyond the recorded period, e.g., a case was started before the beginning of the event log.

Scoping. With the execution of processes distributed over different departments, areas or people in an enterprise we need to specify the range of the process analysis and decide which areas and what information, i.e., tables of data, we want to incorporate.

Granularity. The events in the event log are at a different level of granularity than the activities relevant for end users.

Noise. Event logs that have not undergone pre-processing often contain infrequent and rare behavior that is not representative of the typical behavior of the process captured in the event log.

Incompleteness. Logs may contain too few events to discover some of the underlying (control-flow) structures or information in the latent space. Data cleansing methods such as filtering and data mining techniques can be used to extract useful data and alleviate these issues.

Every event log contains certain fields, without which traditional techniques from the field of *Process Mining* will be impossible. These are:

- **Case ID** - instances (objects) being arranged as a sequence of events
- **Activity name** - actions performed within the event log
- **Timestamp** - date and time of recorded events

Optionally, a log file contains additional information which further specifies events:

- **Resource** - additional information about an event's resource
- **Attributes** - additional attributes being recorded for events

Process Mining techniques rely on the assumption that events are recorded sequentially, with each event being associated with a specific activity, or a defined step in a given process that relates to a particular case or process instance. Event logs may also contain supplementary information, such as the resource executing or initiating the activity, or any relevant data elements associated with the event, such as the size of an order, which worker or machine processed a given order or what the outcome of a specific activity was. Such additional information is commonly leveraged by Process Mining techniques whenever possible.

#	case_id	activity	timestamp	costs	resource
0	3	register request	2010-12-30 14:32:00	50	Pete
1	3	examine casually	2010-12-30 15:06:00	400	Mike
2	3	check ticket	2010-12-30 16:34:00	100	Ellen
3	3	decide	2011-01-06 09:18:00	200	Sara
4	3	reinitiate request	2011-01-06 12:18:00	200	Sara
5	3	examine thoroughly	2011-01-06 13:06:00	400	Sean
6	3	check ticket	2011-01-08 11:43:00	100	Pete
7	3	decide	2011-01-09 09:55:00	200	Sara
8	3	pay compensation	2011-01-15 10:45:00	200	Ellen
9	2	register request	2010-12-30 11:32:00	50	Mike
...	... ...		...	...	...

Table 6.1: Running example of an event log

Example. Table 6.1 shows an example of an event log, where the columns *[case_id, activity, timestamp, resource]* refer to the various fields described above. For the case_id 3 there are in total 8 unique activities. Suppose that these activities also hold for all other case_ids in the event log. The column *resource* yields additional information, in this example, the person who accomplished a certain activity at which time. The task of process discovery as described in Section 6.2.2 is to construct a valid process model w.r.t. certain quality criteria as will be discussed in Section 6.2.4. Figure 6.3 shows the process model in the Business Process Modeling Notation (BPMN) [34] of the event log where activities are shown as rectangles whereas specific gateway operations used in BPMN are illustrated as diamonds (here: '+' denote a parallel gateway; '×' denotes an exclusive gateway). The green (orange) markers are used to denote the start (end) of the process.

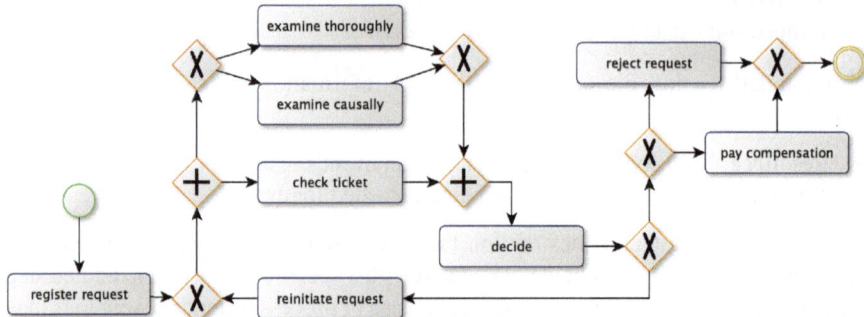

Fig. 6.3: Visualization of the running example shown Table 6.1 in Business Process Modeling Notation (BPMN)

6.2.4 Four Quality Criteria

Completeness and noise refer to qualities of the event log and do not say much about the quality of the discovered model. In fact, there are four competing quality dimensions [36]:

- **Fitness.** The model discovered should be capable of reproducing the behavior observed in the event log. A model with high fitness is able to accommodate most of the behavior observed in the log. A model achieves perfect fitness if it can replay all traces in the log from start to finish.
- **Precision.** The model derived from the event logs must only allow for behavior that is consistent with what was observed in the logs, and should avoid any behavior that is not related to the logs (i.e., underfitting). Underfitting can occur if the model is too general and allows for behavior that is significantly different from what was observed in the logs.
- **Generalization.** How can the discovered model generalize the example behavior seen in the event logs while avoiding overfitting? Overfitting occurs when the model is too specific and only explains the particular sample log given, rather than accounting for the potential variability that may exist in different runs of the same process. This means that the model may not be useful for predicting the behavior of future event logs.
- **Simplicity.** The discovered model should be as simple as possible.

It can be difficult to strike a balance between fitness, simplicity, precision, and generalization when discovering processes. This is why many advanced process discovery techniques offer multiple parameters. To achieve optimal results, it is important to develop enhanced algorithms that can effectively balance these four dimensions of quality. Additionally, any parameters utilized must be user-friendly and easily understood by end-users.

6.2.5 Types of Processes

This chapter concludes with high-level characterization of two different types of processes [37]: *Lasagna Processes* (Section 6.2.5.1) and *Spaghetti Processes* (Section 6.2.5.2).

The spectrum of process types ranges from *structured, semi-structured* to *unstructured*. The characteristic of a structured process (i.e., *Lasagna Process*) encompass work flows where all activities are repeatable and have a well-defined input and output. In semi-structured processes, the requirements of a process are known, and it is possible to sketch the general procedure. However, depending on the specific cases being handled, the work flows might deviate from each other to be more appropriate for the individual case characteristics. In unstructured processes (i.e., *Spaghetti Processes*) the major challenge is to identify proper pre- and post-conditions for the activities being stored in an event log. In practice, the latter is often driven purely by experience, intuition, or trial-and-error routines from domain experts.

6.2.5.1 Lasagna Processes

A process can be classified as a *Lasagna Process* if it is possible to create an established process model with minimal effort, achieving a fitness level of 0.8 or higher, which means that more than 80% of the events occur as intended, and the stakeholders validate the model's correctness. Hence, *Lasagna Processes* are easy to discover, but suffer a loss of information to show the real process as it rather mimics the expectation. Noteworthy, a wide range of heuristics and techniques from the research field of Process Mining can be applied on *Lasagna Processes* where insightful information is predominantly discovered in more sophisticated models aligning the event log and model.

6.2.5.2 Spaghetti Processes

Spaghetti Processes are less structured than *Lasagna Processes*, i.e., only some Process Mining techniques can be applied. For example, techniques examining the operational support activities yield less insights whenever the underlying process inherits too much variability. There are different approaches to get valuable analysis from such kind of processes. As opposed to the imperative paradigm that describes the concrete process flow, we gain more insights into less structured processes by applying tools following the declarative paradigm that describe processes by a set of rules. Another approach would be to leverage *divide-and-conquer* approaches by clustering cases or showing only the most frequent paths and activities.

6.3 Process-Awareness from Theory to Practice

In the following, we present process-aware machine learning (ML) techniques for predictive analysis. For that, Section 6.3.1 provides an entry point into predictive analysis in the scope of Process Mining. The chapter continues with Bayesian Modeling in Section 6.3.2 for solving tasks like activity prediction. In Section 6.3.3, we coin the term of *Process AI* encompassing next to traditional (ML) techniques also state-of-the-art deep learning architectures for solving various downstream tasks.

6.3.1 Predictive Analysis in Process Mining

In practice, information systems provide a more thorough insight into the real work-flow of a business process execution. A key challenge is the proper analysis of the performance-monitoring, which uncovers potential execution patterns. Hence, we can ask questions like *'What activity is the likeliest to follow given a specific historical pattern?'*. From a theoretical point of view, we can model process pattern uncertainty by employing models like *Bayesian Networks* to determine which patterns cause a particular kind of process flow.

Generally, these questions arise in the field of *Predictive Analytics*, i.e., the practice of extracting information from existing data sets in order to determine patterns and predict future outcomes and trends [16, 18, 25, 32, 40]. To generate predictions, existing mining and learning approaches are applied in a business context. Furthermore, predictive analytics is related to business analytics and business intelligence.

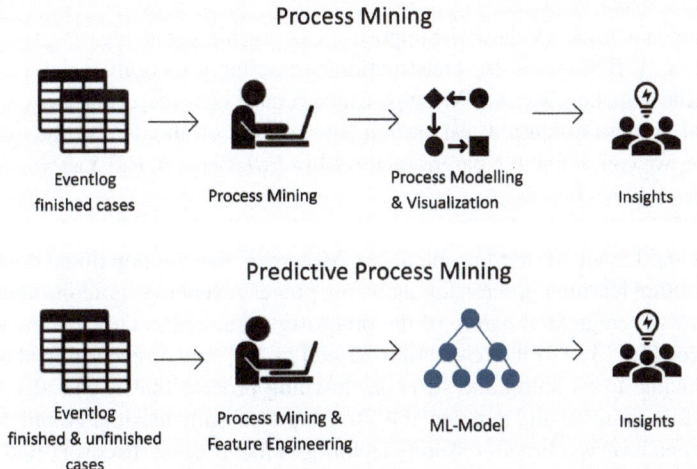

Fig. 6.4: Predictive Process Mining compared to traditional Process Mining

The incorporation of ML-enhanced models used in the scope of *Predictive Process Mining* is illustrated in Figure 6.4. Potentially generated insights from the application of Predictive Process Mining are the prediction of future activities, remaining execution times or other key performance indicators of business processes. The predictive models can aid practitioners in better understanding their processes via the incorporation of Process Mining techniques into the prediction process. Moreover, the assessment of potential uncertainties in processes is aided by (possibly probabilistic) predictions and can be augmented by various recommendations of actions that could nudge running processes towards a more efficient, faster or qualitatively more satisfying result.

6.3.2 Predictive Process Mining with Bayesian Statistics

6.3.2.1 Preliminaries for Bayesian Modeling

First, this section recaps on the general definition of a Bayesian Network (BN). Formally, a Bayesian Network is defined as a directed acyclic graph (DAG) G of tuple (V, A, P), with $V = \{v_1, \ldots, v_n\}$ being a set of vertices of the graph, and $A \subseteq \{a_{ij} | v_i, v_j \in V\}$ representing the interactions among various entities within the graph. Let $X = \{X_1, \ldots, X_n\}$ be a set of random variables such that X_i is a random variable of vertex v_i in the graph. Due to the unidirectional connections of a DAG, we define $Pa(X_i) = \{X_{i1}, \ldots, X_{i,s_i}\}$ to be the parent set, i.e., the predecessors, of the random variable X_i. By applying the chain rule, the joint probability density is defined as

$$P(X_1, \ldots, X_n) = \prod_{i=1}^{n} P(X_i | Pa(X_i)), \tag{6.1}$$

completing our tuple (V, A, P). Suppose a case in the event logs is given as the sequence $< A, B, C, D >$. By construction, we define a random variable for each event in the log, i.e., X_A, X_B, X_C, X_D. Since a case c in the event log contains a sequential process execution, we assume an ordering on the data in the event log. Therefore, we can define the parent relationship $Pa(X_A) = \emptyset, Pa(X_B) = \{X_A\}$, and so on.

It is well-known that learning the graphical structure of a Bayesian Network from a given dataset is an NP-hard problem [6]. Moreover, the computational complexity of the structure learning process for analyzing process event logs is additionally complicated by the sequential nature of the processes themselves. This renders existing score-based [10, 7] as well as constraint-based [24, 26] structure learning algorithms less applicable to an automatic structure learning process that is causally feasible and therefore intuitive and interpretable. To ensure causally feasible graphical structures, we can leverage Process Mining techniques like process discovery and use the extracted activity orderings as input for the structure learning process, as discussed in the following.

6.3.2.2 Quality Criteria for Bayesian Modeling

Having identified or modeled the Bayesian Network according to the given input data, there are further analysis scenarios coming up:

- **Causal Inference.** This type of inference can be used to perform root cause analysis, e.g., investigate the causes of a working delay. Due to the sequential behavior of processes, a single delay in an event can cause a higher throughput time of the overall process (lateness inference).
- **Backward Inference.** With backward inference we can find the probability of the cause's random variable where the affected random variable is known. It is the opposite of causal inference.

- **Explaining-away Inference.** The task of the explaining-away inference is to infer any possible patterns influenced by each state(s) of random variable(s).

6.3.2.3 Context-Aware Structure Learning for Probabilistic Process Prediction

Processes in various domains like manufacturing, customer care, healthcare, or business management processes in general tend to show uncertainties regarding the apparent control-flow [17, 22]. Specific activities might only be occurring in some specific cases or process categories and are absent for the remaining cases. For further analyses and predictions of cases, e.g. the remaining time of a process or the quality of a produced good, it is therefore required to know which process steps have already occurred and which are most likely to be executed in the future. By that, the growing research field of Next Activity Prediction [5, 11, 33, 28] is a crucial step towards feasibly predicting process KPIs for processes with high amounts of uncertainty w.r.t. control-flow. Our current work makes use of Bayesian Networks in combination with process discovery for generating probabilistic graphical process models.

The generation of the Bayesian Network structure does not rely on the existing structure learning algorithms but is achieved by extracting the order of activities directly from the event log. The (directly) following relations between activities – those come into play for different process discovery algorithms like the Alpha Miner [35] or Heuristics Miner [42] – are translated into a DAG that represents a graphical representation of the control-flow similar to a Petri-Net. However, the Bayesian structure needs to infer an ordering that is free of cycles. This is crucial especially for process event log data as cycles can often occur in the recorded activity patterns. The final structure of the Bayesian Network consists of the nodes represented by the activities executed along the process. The connections and transitions between the activities are represented by the edges between the nodes inside the structure.

After the generation of the underlying structural model, the parameters of the model are represented by conditional transition probabilities between the connected activities. The Bayesian model is able to predict the occurrence of all potential activities, the order of the occurring activities (see Figure 6.5) as well as their execution time (see Figure 6.6) in a probabilistic manner.

Possible questions that can be answered by such a context-aware prediction model given the process data at hand are:

...which are the most probable remaining activities of a case given a set of already executed activities?

...what is the remaining execution time of a case given a set of already executed activities?

...which activities can or should be executed to keep the remaining execution time of a case in a certain range with a certain probability?

Making use of backward inference – also known as *What-If* analyses – as formulated in the latter question is one of the significant benefits of Bayesian methods. The Bayesian inference mechanism can be used to generate various recommendations for practitioners that want to control the performance, i.e., execution time of their running processes.

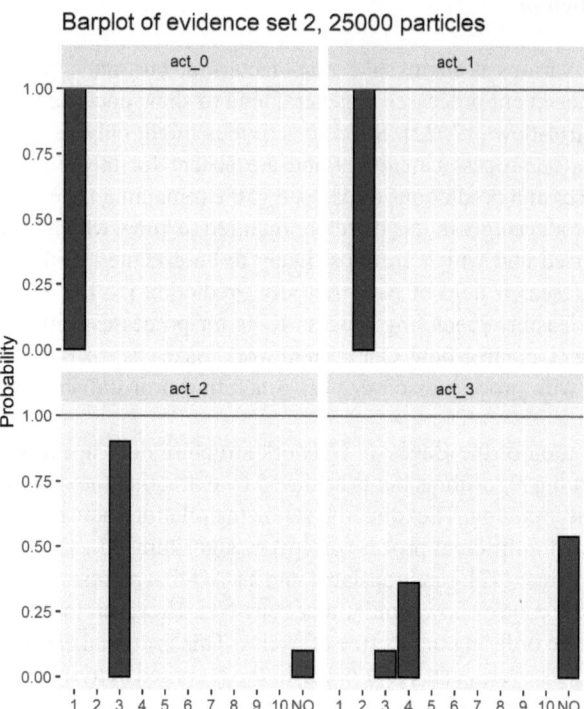

Fig. 6.5: Probabilistic forecast for process activities in a running process for a given trace where the first two activities and their occurrence are known (*act_0, act_1*). The most probable ordering (*Place*) of the remaining activities is inferred in form of occurrence probabilities at the respective places inside the remaining process trace, where *NO* stands for *no occurrence*.

Challenges that have to be overcome in the modeling of a feasible structure with Bayesian Networks are the occurrence of activity cycles in process traces, e.g., partial traces like < ..., $A, B, C, A, ...$ >, and highly complex logs with long traces and many intertwined activities. As Bayesian Network structures need to be free of cycles, multiple occurrences of individual activities would need to be accounted for. There are approaches that are concerned with loop-elimination in Bayesian Networks that could be applied [27, 20]. Also, more complex $1 : N$ relationships of activities drastically increase the overall complexity of the conditional probability tables that need to be calculated for Bayesian inference and prediction for that activity. This

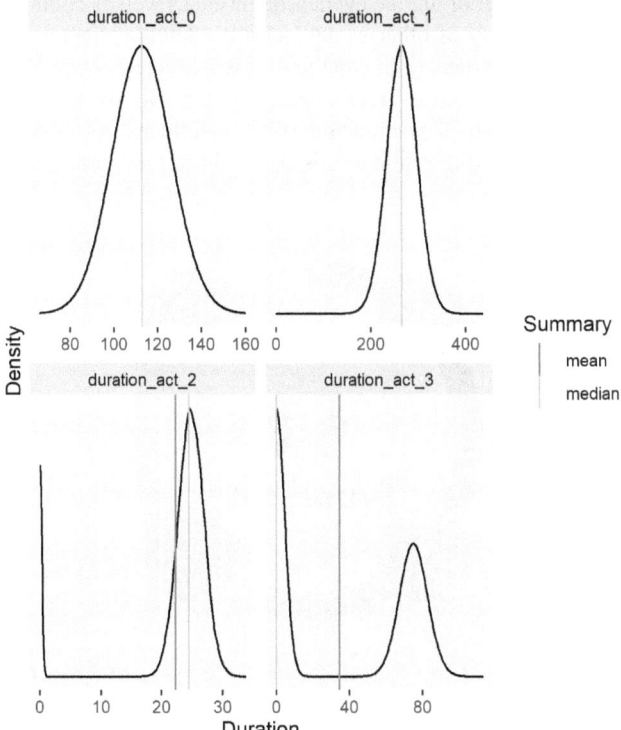

Fig. 6.6: Probabilistic duration prediction based on predicted activities corresponding to the predicted occurrences from Fig. 6.5. The probabilistic occurrence forecast allows for sampling distributions of possible activity durations where either no duration is recorded (for *NO*, i.e. non-occurrence) or a non-zero duration is sampled from the respective activity duration distribution.

displays a potential threat to scalability of the approach which has to be tackled for the applicability of larger and more complex event logs.

6.3.3 Process AI

Process Mining aims at using techniques that focus on both data-driven and process-centric approaches. As described in Section 6.2.2, the techniques are typically grouped into categories such as process discovery, conformance checking, and enhancement. However, one of the major challenges in Process Mining is working with an ever-increasing amount of available data (*Big Data*) accompanied by a high number of dimensions. Real-life event logs may present large amounts of cases, which can represent a diverse set of unique event sequences, as well as contain information on resources and a variety of other event- or case-related attributes. This makes it difficult to featurize event data, or in other words, to extract structured instances from raw event logs, for tasks such as trace clustering or predictive process monitoring being handled with traditional Machine Learning techniques. As a result, such learning tasks can suffer from dimensionality issues and often rely on an ad-hoc method for defining input features.

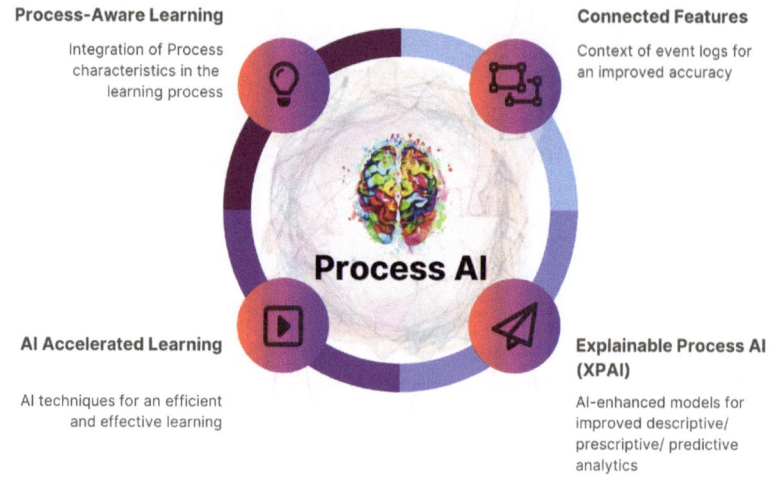

Process-Aware Learning

Integration of Process characteristics in the learning process

Connected Features

Context of event logs for an improved accuracy

Process AI

AI Accelerated Learning

AI techniques for an efficient and effective learning

Explainable Process AI (XPAI)

AI-enhanced models for improved descriptive/ prescriptive/ predictive analytics

Fig. 6.7: Research sphere of Process AI.

In parallel to other disciplines, like natural language processing (NLP), graph analysis, image recognition, tools and techniques from the scope of representation learning have gained a lot of attention outside the NLP community. The essence of representational learning is leveraging the power of neural network models to automatically learn vector representations containing the (contextual) information of the input data in an embedding space. Due to their success in areas such as NLP, image recognition, and graph AI, recent pioneering works employ deep learning architectures also in the scope of business process management [8].

In predictive analytics, neural network-enhanced models have shown superior performance compared to traditional monitoring techniques. The authors of [9] used a recurrent neural network (RNN) to capture the sequential patterns in an event log. In [29], the authors employed a model using Long Short-Term Memory (LSTM) capturing the inherent sequential characteristics of an activity sequence to predict the next event. An extension has been proposed to also predict its timestamp, enabling to predict the remaining time of a partially observed trace. A similar model setup has been employed in [19] for time series classification in the scope of process planning. In [21], the authors propose BINet, a recurrent neural network architecture for real-time multivariate anomaly detection in business process event logs. The model is trained to predict the next event and its attributes.

We coin the terminology of **Process AI** to encompass various AI techniques for solving a diverse set of downstream tasks. As shown in Figure 6.7, Process AI pays attention to *Process-aware learning* heuristics. The tools being leveraged for an *AI-accelerated learning* from complex and high dimensional input data stem from traditional machine learning techniques and range to state-of-the-art deep learning techniques, providing meaningful insights into the behavioral characteristics of processes. Naturally, event logs provide a contextual view of activity sequences and, therefore, provide *connected features* that are used as input for a deep learning model. *Explainable Process AI* (XPAI) plays a crucial role in the practical application and success of Process AI by examining the inference steps of a model in light of the underlying process characteristics. The components focusing on the explainability of models can be further fanned out into *descriptive, prescriptive* [14, 2, 1], and *predictive* [9, 29, 19, 21] (cf. Figure 6.4) factors.

In an ever-increasing connected world with highly complex processes, Process AI has great potential for gaining a more in-depth understanding of processes whilst providing the necessary information for enhanced decision-making.

6.4 Conclusion and Outlook

In modern organizations that manage complex operational processes, the discipline of *Process Mining* (PM) has emerged as an area providing valuable tools for enhanced analyses of processes. Techniques from the scope of *Data Mining* are used to analyze historical data to gain insights into processes, but they tend to be limited in their level of analysis and lack a process-centric approach. PM, on the other hand, uses factual event data rather than relying solely on models. This generates a bridge between *Business Process Management* (BPM) and *Data Mining*. Moreover, PM goes beyond *Process Discovery* and allows for the connection between event logs and process models, which leads to new ways of analysis. This can result in the extension of an existing process model by incorporating insights from various perspectives, or to check whether an existing event log is conforming with a process model. In the era of rising AI tools, one of our main focuses is on Process AI incorporating AI-enhanced modeling tools ranging from traditional machine learning heuristics to modern deep learning heuristics in order to solve a whole range of diverse downstream tasks like predictive analysis. For the latter, we provide an overview into predictive analysis by leveraging the power of Bayesian modeling.

Naturally, log event data provide sequential characteristics where a more in-depth analysis of additional attributive information can improve the accuracy of a downstream task, e.g., the next activity in an incomplete trace. Whereas log events only provide positive samples, i.e., samples that have been observed by the system, there is great potential to look further into negative samples to improve the robustness of process models. A major characteristic of event logs is their temporal annotations.

Therefore, identifying concept drifts in various dimensions (temporal, attributive) in order to optimize business management is essential from a practical point of view in order to adapt to new situations as efficiently as possible. To solve predictive process analysis, Bayesian modeling is of great importance for both, the research community and practitioners. It opens the world for an in-depth analysis of distributional behavior within a process. As the explainable component is of great interest for the practical usage, we focus on XPAI (EXplainable Process AI) in our modeling in order to improve the usability and explainability for non-experts. As practical systems vary tremendously in their scale, another research interest lies on Small Sample Learning (SSL) [13] (a.k.a. Few-shot learning (FSL), or Low-shot learning (LSL)) to potentially adapt pre-learned features most efficiently in a new environment. This raises questions of monotonicity properties, i.e., of the relation between the size of a model and its behavior.

References

1. Z. D. Bozorgi, M. Dumas, M. L. Rosa, A. Polyvyanyy, M. Shoush, and I. Teinemaa. Learning when to treat business processes: Prescriptive process monitoring with causal inference and reinforcement learning, 2023.
2. Z. D. Bozorgi, I. Teinemaa, M. Dumas, M. L. Rosa, and A. Polyvyanyy. Prescriptive process monitoring based on causal effect estimation. *Information Systems*, page 102198, 2023.
3. A. Burattin and J. Carmona. A framework for online conformance checking. In *Business Process Management Workshops*, pages 165–177. Springer International Publishing, 2018.
4. A. Burattin, A. Sperduti, and M. Veluscek. Business models enhancement through discovery of roles. In *2013 IEEE Symposium on Computational Intelligence and Data Mining (CIDM)*. IEEE, Apr. 2013.
5. M. Ceci, P. F. Lanotte, F. Fumarola, D. P. Cavallo, and D. Malerba. Completion time and next activity prediction of processes using sequential pattern mining. In *Discovery Science*, pages 49–61. Springer International Publishing, 2014.
6. D. Chickering, C. Meek, and D. Heckerman. Large-Sample Learning of Bayesian Networks is NP-Hard. *Journal of Machine Learning Research*, 5:1287–1330, 2012.
7. D. M. Chickering. Optimal structure identification with greedy search. *Journal of machine learning research*, 3(Nov):507–554, 2002.
8. P. De Koninck, S. vanden Broucke, and J. De Weerdt. act2vec, trace2vec, log2vec, and model2vec: Representation learning for business processes. In M. Weske, M. Montali, I. Weber, and J. vom Brocke, editors, *Business Process Management*, pages 305–321, Cham, 2018. Springer International Publishing.
9. J. Evermann, J.-R. Rehse, and P. Fettke. Predicting process behaviour using deep learning. *Decision Support Systems*, 100:129–140, 2017. Smart Business Process Management.
10. J. A. Gámez, J. L. Mateo, and J. M. Puerta. Learning bayesian networks by hill climbing: efficient methods based on progressive restriction of the neighborhood. *Data Mining and Knowledge Discovery*, 22(1-2):106–148, May 2010.
11. B. R. Gunnarsson, S. vanden Broucke, and J. D. Weerdt. A direct data aware LSTM neural network architecture for complete remaining trace and runtime prediction. *IEEE Transactions on Services Computing*, pages 1–13, 2023.
12. C. W. Günther and W. M. P. van der Aalst. Fuzzy mining – adaptive process simplification based on multi-perspective metrics. In G. Alonso, P. Dadam, and M. Rosemann, editors, *Business Process Management*, pages 328–343, Berlin, Heidelberg, 2007. Springer Berlin Heidelberg.

13. M. Käppel, S. Schönig, and S. Jablonski. Leveraging small sample learning for business process management. *Information and Software Technology*, 132:106472, Apr. 2021.

14. K. Kubrak, F. Milani, A. Nolte, and M. Dumas. Prescriptive process monitoring: Quo vadis? *PeerJ Computer Science*, 8:e1097, Sept. 2022.

15. S. J. J. Leemans, D. Fahland, and W. M. P. van der Aalst. Discovering block-structured process models from event logs - a constructive approach. In J.-M. Colom and J. Desel, editors, *Application and Theory of Petri Nets and Concurrency*, pages 311–329, Berlin, Heidelberg, 2013. Springer Berlin Heidelberg.

16. F. M. Maggi, C. Di Francescomarino, M. Dumas, and C. Ghidini. Predictive monitoring of business processes. In M. Jarke, J. Mylopoulos, C. Quix, C. Rolland, Y. Manolopoulos, H. Mouratidis, and J. Horkoff, editors, *Advanced Information Systems Engineering*, pages 457–472, Cham, 2014. Springer International Publishing.

17. F. M. Maggi, M. Montali, and R. Peñaloza. Temporal logics over finite traces with uncertainty. *Proceedings of the AAAI Conference on Artificial Intelligence*, 34(06):10218–10225, Apr. 2020.

18. A. E. Marquez-Chamorro, M. Resinas, and A. Ruiz-Cortes. Predictive monitoring of business processes: A survey. *IEEE Transactions on Services Computing*, 11(6):962–977, Nov. 2018.

19. N. Mehdiyev, J. Lahann, A. Emrich, D. Enke, P. Fettke, and P. Loos. Time series classification using deep learning for process planning: A case from the process industry. *Procedia Computer Science*, 114:242–249, 2017. Complex Adaptive Systems Conference with Theme: Engineering Cyber Physical Systems, CAS October 30 – November 1, 2017, Chicago, Illinois, USA.

20. C. Moreira, E. Haven, S. Sozzo, and A. Wichert. Process mining with real world financial loan applications: Improving inference on incomplete event logs. *PLOS ONE*, 13(12), Dec. 2018.

21. T. Nolle, A. Seeliger, and M. Mühlhäuser. Binet: Multivariate business process anomaly detection using deep learning. In M. Weske, M. Montali, I. Weber, and J. vom Brocke, editors, *Business Process Management*, pages 271–287, Cham, 2018. Springer International Publishing.

22. M. Pegoraro and W. M. van der Aalst. Mining uncertain event data in process mining. In *2019 International Conference on Process Mining (ICPM)*. IEEE, June 2019.

23. A. Rozinat and W. Aalst, van der. Conformance checking of processes based on monitoring real behavior. *Information Systems*, 33(1):64–95, 2008.

24. R. Scheines, P. Spirtes, C. Glymour, C. Meek, and T. Richardson. The TETRAD project: Constraint based aids to causal model specification. *Multivariate Behavioral Research*, 33(1):65–117, Jan. 1998.

25. A. Senderovich, C. D. Francescomarino, C. Ghidini, K. Jorbina, and F. M. Maggi. Intra and inter-case features in predictive process monitoring: A tale of two dimensions. In *Lecture Notes in Computer Science*, pages 306–323. Springer International Publishing, 2017.

26. B. Sun, Y. Zhou, J. Wang, and W. Zhang. A new PC-PSO algorithm for bayesian network structure learning with structure priors. *Expert Systems with Applications*, 184:115237, Dec. 2021.

27. R. A. Sutrisnowati, H. Bae, and M. Song. Bayesian network construction from event log for lateness analysis in port logistics. *Computers & Industrial Engineering*, 89:53–66, Nov. 2015.

28. B. A. Tama and M. Comuzzi. An empirical comparison of classification techniques for next event prediction using business process event logs. *Expert Systems with Applications*, 129:233–245, Sept. 2019.

29. N. Tax, I. Verenich, M. La Rosa, and M. Dumas. Predictive business process monitoring with lstm neural networks. In K. Pohl and E. Dubois, editors, *Advanced Information Systems Engineering : 29th International Conference, CAiSE 2017, Essen Germany, June 12-16, 2017. Proceedings*, LNCS, pages 477–492, Germany, 2017. Springer. 29th International Conference on Advanced Information Systems Engineering, CAiSE 2017, CAiSE 2017 ; Conference Date: 12-06-2017 Through 16-06-2017.

30. F. Taymouri and J. Carmona. A recursive paradigm for aligning observed behavior of large structured process models. In *International Conference on Business Process Management*, 2016.

31. F. Taymouri and J. Carmona. Model and event log reductions to boost the computation of alignments. In *Lecture Notes in Business Information Processing*, pages 1–21. Springer International Publishing, 2018.
32. I. Teinemaa, M. Dumas, F. M. Maggi, and C. D. Francescomarino. Predictive business process monitoring with structured and unstructured data. In *Lecture Notes in Computer Science*, pages 401–417. Springer International Publishing, 2016.
33. E. Tello-Leal, J. Roa, M. Rubiolo, and U. M. Ramirez-Alcocer. Predicting activities in business processes with LSTM recurrent neural networks. In *2018 ITU Kaleidoscope: Machine Learning for a 5G Future (ITU K)*. IEEE, Nov. 2018.
34. W. van der Aalst. *Process Mining*. Springer Berlin Heidelberg, 2016.
35. W. van der Aalst, T. Weijters, and L. Maruster. Workflow mining: discovering process models from event logs. *IEEE Transactions on Knowledge and Data Engineering*, 16(9):1128–1142, Sept. 2004.
36. W. M. P. van der Aalst. *Process Mining - Discovery, Conformance and Enhancement of Business Processes*. Springer Berlin Heidelberg, 2011.
37. W. M. P. van der Aalst. *Process Mining: Data Science in Action*. Springer, Heidelberg, 2 edition, 2016.
38. W. M. P. van der Aalst, A. K. A. de Medeiros, and A. J. M. M. Weijters. Genetic process mining. In G. Ciardo and P. Darondeau, editors, *Applications and Theory of Petri Nets 2005*, pages 48–69, Berlin, Heidelberg, 2005. Springer Berlin Heidelberg.
39. S. van Zelst, A. Bolt Iriondo, and B. van Dongen. Tuning alignment computation : an experimental evaluation. In W. van der Aalst, R. Bergenthum, and J. Carmona, editors, *Algorithms and Theories for the Analysis of Event Data 2017. Proceedings of the International Workshop on Algorithms & Theories for the Analysis of Event Data (ATAED 2017),Zaragoza, Spain, June 26–27, 2017*, CEUR Workshop Proceedings, pages 6–20, 2017. Workshop on Algorithms and theories for the analysis of event data (ATAED2017), 26-27 June 2017, Zaragoza, Spain, ATAED2017 ; Conference date: 26-06-2017 Through 27-06-2017.
40. I. Verenich, M. Dumas, M. L. Rosa, F. M. Maggi, and C. D. Francescomarino. Complex symbolic sequence clustering and multiple classifiers for predictive process monitoring. In *Business Process Management Workshops*, pages 218–229. Springer International Publishing, 2016.
41. M. Weidlich, A. Polyvyanyy, N. Desai, J. Mendling, and M. Weske. Process compliance analysis based on behavioural profiles. *Information Systems*, 36(7):1009–1025, Nov. 2011.
42. A. Weijters, W. Aalst, and A. Medeiros. Process Mining with the Heuristics Miner-algorithm. BETA Working Paper Series, WP 166, Eindhoven University of Technology, Eindhoven, 2006.
43. A. Weijters and W. van der Aalst. Rediscovering workflow models from event-based data using little thumb. *Integrated Computer-Aided Engineering*, 10(2):151–162, May 2003.

Chapter 7
Combinatorial Optimization

Jan Krause[1], Tobias Kuen[2], Christopher Scholl[2]

Abstract Optimization is a fundamental topic in mathematics that deals with finding the best solution to a problem from a set of possible solutions. This chapter provides an overview of mathematical optimization, its main objectives, and the methods used to solve optimization problems. It also introduces basic problems such as the modeling of binary decision trees, the pooling problem, the clique problem, and flow models. It concludes with an outlook on online optimization and learning optimization methods, which represent promising areas of research in the field. Overall, this chapter serves as a useful introduction to mathematical optimization, its basic problems, and their practical applications. The chapter is based on a white paper from Bärmann et al. [2].

Key words: optimization, mixed-integer programming, branch-and-bound, tree classifiers, pooling problem, graph theory, cliques, maximum flow problem

7.1 Introduction

While the colloquial use of the word "optimization" is often used for "doing something better" or, in early phases of problem understanding, simply for "doing something differently", mathematical optimization means the targeted search for an optimal solution to a well-defined problem. For this purpose, the planning rules and objectives valid in the company are translated into a mathematical model that takes into account all conceivable solutions for the task at hand. The solution of this model then requires, on the one hand, the availability of data that is considered as a basis

---

[1]University of Technology, UTN, Nuremberg, Germany
[2] Fraunhofer Institute for Integrated Circuits IIS, Fraunhofer IIS, Nuremberg, Germany

Corresponding author: Tobias Kuen
e-mail: tobias.kuen@iis.fraunhofer.de

© The Author(s) 2024
C. Mutschler et al. (eds.), *Unlocking Artificial Intelligence*,
https://doi.org/10.1007/978-3-031-64832-8_7

for decision-making (e.g., data on raw material availability or expected customer demand). On the other hand, powerful algorithms are needed that can solve the optimization problem within the required response time.

A mathematical optimization model as described above consists of one or more objective criteria (so-called objective functions), decision variables and constraints, and is usually based on a concrete planning decision that has to be made in the company. The decision variables describe the adjusting screws of the planning. They often represent small, individual decisions that, when put together, result in the whole plan for the issue. The constraints describe operational, legal or other rules that must apply to a valid overall plan. In particular, they exclude undesirable solutions and effects and describe the interaction and interdependence of the different atomic decisions. Finally, the objective function evaluates a solution according to specified criteria and thus makes different solutions comparable: Solution A is better, worse or equally good as solution B. Based on such a model, a mathematical optimization procedure is able to search for the best possible solution to the posed problem.

While the basic direction of the optimization problem is based on the planning problem, the available data often have a decisive influence on the detailed design of the concrete modeling. On the one hand, the availability, quality and completeness of data influence the speed with which a software solution can be developed; on the other hand, these three criteria significantly determine how trustworthy the solution to the optimization problem is. In practice, the lack of decision-relevant and well-prepared data is often one of the biggest obstacles. Therefore, clean data management is significantly responsible for the success of an optimization project.

If data is available in sufficient quantity and quality for the optimization problem, an optimization algorithm is needed to solve the model. Not all optimization models can be solved in practice on the first try. Therefore, it is often useful to involve experts in mathematical optimization who can advise on the modeling as well as on the selection or, if necessary, the tailored design of solution procedures.

The quality and efficiency of the chosen solution procedure are important aspects. The field of mathematical optimization has developed rapidly over the last 100 years and year by year the methods become progressively powerful. More and more complicated models can be solved, which were considered unsolvable years ago.

In the following, various solution algorithms are presented and their advantages and disadvantages are discussed on the basis of the well-known Traveling Salesman Problem (TSP). The problem arises from the typical problem of planning a tour: It is about planning a round trip to different cities so that the traveling salesman or the truck arrives back at the starting point (the home town or the depot) at the end and has covered a minimum distance. While in 1954 only one instance with 49 cities could be solved optimally [6], in 2006 an instance with 1.9 million places in the world was provably solved almost optimally (to within 0.01%).

7.2 Solving Methods

7.2.1 Heuristics

The simplest method to solve the TSP is the brute force method. Here, the entirety of all possible tours is enumerated, the length is calculated for each individual tour and the best one is selected at the end. This method already reaches its limits for relatively small-sized problems. The reason for this is what in mathematics is called the combinatorial explosion. As the number of cities to be visited increases, the number of tours grows exponentially. While with 5 cities still a manageable number of 12 tours are possible (equivalent solutions combined), with 30 cities there are already more tours than there are stars in the universe. Even if one could calculate 1 million tours per second, one would not get an optimal solution in one's lifetime. Even a conceivably faster computer would not bring any significant progress here. So in practice, after a fixed time, the brute force approach would have to be aborted, without any statement about the quality of the obtained solution compared to the optimal solution.

A second class of methods are so-called greedy heuristics. They rank the decisions and then always choose the local best decision. These local decisions are often motivated by the objective function or "best practice" rules. One of the well-known representatives of this class is the nearest neighbor heuristic. In it, starting from the starting location, one constructs the solution stepwise by always visiting the closest city to the current location next and returning to the starting location at the end. While a solution is obtained quickly by these methods, in practice the solutions are often clearly suboptimal. The reasons for this are, on the one hand, the algorithm's too local view of the decisions made and, on the other hand, the lack of an overall picture for the interactions between these decisions.

A third class of methods are local improvement heuristics (LIH). In contrast to the greedy heuristic, which builds an admissible solution to the optimization problem, LIH already starts with an admissible solution and then tries to improve it by making local changes. A well-known representative for the TSP is the 2-opt algorithm. This iterates over each pair of route segments between two cities and checks whether the 4 considered cities can be connected shorter and, if this is the case, exchanges the routes. This method is also often combined with a greedy heuristic to improve the resulting initial solution. Other well-known methods of this class are Simulated Annealing [9], Tabu Search, Genetic Algorithms [7] and Ant Colony Optimization [8].

An overview of well-known heuristics can be found in [13]. The presented heuristic methods have the advantage that they are easy to implement and use. They are also often already implemented in program libraries. Many generic heuristics have names from nature, as they are inspired by natural phenomena, such as the ant algorithm or genetic algorithms. Their operation is typically easy for management to understand in their descriptiveness, and their use is therefore easy to justify. However, the performance of these generic algorithms often lags behind that of problem-specific algorithms. Another disadvantage is that they provide no way to detect whether an optimization problem is infeasible, making it difficult to maintain and verify the correctness of the optimization model. They also do not provide a way to estimate the quality of the solution found compared to the best possible solution. This gap, i.e., the wasted potential, can be significant even on small instances.

7.2.2 Exact Methods

While a large number of different algorithms have been developed in the field of heuristics, the branch-and-bound method [10] has been the most popular among the exact methods. At its core, the method, like the brute force method, is based on enumerating many decisions. However, the amount of solutions to be tested is significantly reduced by the clever exploitation of so-called primal and dual methods, so that the required runtimes are achieved in more and more applications.

The procedure starts with solving a simplification of the original problem. In doing so, certain constraints are softened to make it a linear optimization problem. For this class, there are exact optimization methods, like the simplex method, which usually provides an optimal solution for the simplified problem in a short time. This solution is usually not admissible for the original problem, but it provides an estimate for the quality of the optimal solution to the original problem. Now, if an admissible solution has been generated using any primal heuristic, the value of this solution together with the value of the reduced problem can be used to provide an estimate of how good the solution found is compared to the best possible solution. With the help of this estimation, it can be decided whether it is worthwhile to search further for better solutions and estimations. If the estimate corresponds to the value of the solution, a mathematically provable global optimal solution for the optimization problem has been found. If the global optimal solution has not yet been found, the procedure starts with the branching step. Here, for example, a binary decision is selected in the problem and it is decomposed into two subproblems. In each of the two subproblems, a different outcome of the decision was selected, resulting in two smaller optimization problems. Now, in turn, new improved estimates can be computed in both subproblems and heuristics can be used to search for new solutions. This procedure can now be continued successively. Each branching step creates two new subproblems and thus a search tree, the so-called branch-and-bound tree, which grows exponentially. The bounding step tries to prevent exactly this and sub-branches of the tree are truncated to reduce the search space. If the estimation of a subproblem is worse than the best solution found, this part of the tree can be removed, because no better solution can be found there even in further branching steps.

The branch-and-bound tree is shown schematically in Figure 7.1. The nodes of the tree are successively worked through and improved barriers and solutions are searched for there. Figure 7.2 shows the progress of the method over time. At each point in time, the difference between the best barrier and solution shows the remaining optimization potential. As the computation time progresses, this difference decreases and the method finally converges to an optimal solution.

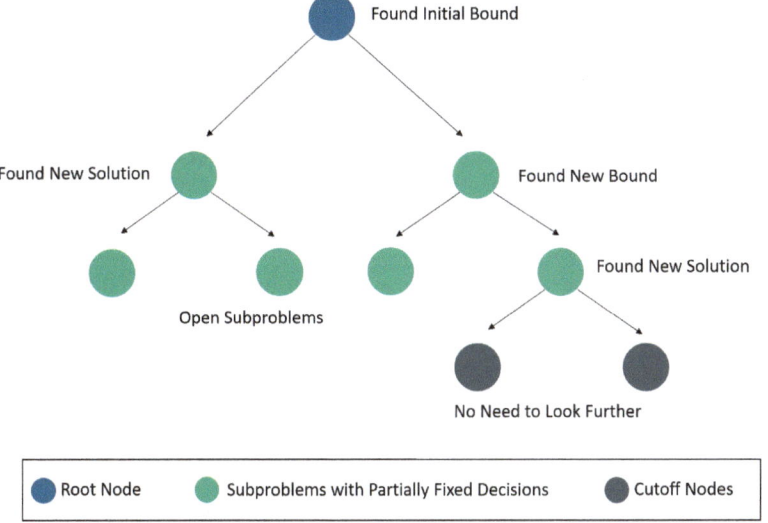

Fig. 7.1: Graphical representation of a branch-and-bound tree method.

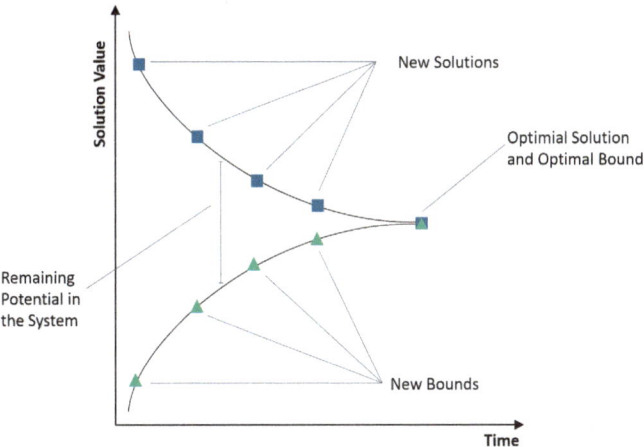

Fig. 7.2: Graphical representation of a convergence optimization method.

There are many different improvements to the basic procedure described. For example, in practice, primal heuristics are used at the nodes that are specifically designed for branch-and-bound and to search for better admissible solutions. Furthermore, there is an extension to the branch-and-cut procedure used in practice, in which additional dual heuristics are used that attempt to provide better estimates to reduce the search space even faster.

The exact methods (especially branch-and-bound type methods) offer the advantage of providing an estimate of the quality of the solution found compared to the optimal solution at any time. Thus, an informed decision can be made at any time whether further computational and time resources should be invested in finding a better solution. Over the runtime, both the estimates and the solutions found are improved, so that it can always be decided whether further computational and time capacity is worthwhile. As a result, simple instances in particular can be solved much faster because there is a clear termination criterion. The procedures enable a significantly improved analysis capability of the problem, since the estimation can be used to evaluate the potential in the system for further developments or different scenarios. The procedures also offer the possibility to identify infeasibilities in the system with pinpoint accuracy.

While heuristic methods are already frequently used due to their lower entry barriers and free availability, exact methods are still less common. However, due to the enormous further development of mathematical methods in the last decades, exact methods are opening up more and more fields of application. Especially for applications with a high economic potential, an evaluation and, if necessary, a switch to exact methods is worthwhile due to their more versatile applicability and more powerful analysis capabilities. Practical problems often lead to complicated optimization problems for which the choice of the optimization approach is crucial. Here, the development of problem-specific algorithms can make the difference between satisfactorily solving optimization problems in practice and maximizing the utility of the developed software tools. In summary, optimization enables companies to make simpler, better and faster decisions. These lead to cost savings and reduce economic risk.

7.3 Modeling Techniques

In this section, we give an overview of basic techniques in mathematical optimization that were used for the different applications which will be described later in this book (14.3.4, 15.6, 16.1, 16.2).

7.3.1 Graph Theory

In mathematical optimization, graphs are used to model the relationship between variables and constraints in an optimization problem. A graph consists of nodes and edges, where nodes represent variables and edges represent the relationships between variables. There are two main types of graphs used in optimization: directed and undirected graphs. Directed graphs have edges that are directed from one node to another, indicating a specific relationship between the two nodes. Undirected graphs have edges that do not have a specific direction, indicating a more general relationship between the nodes. A subset of nodes that are all pairwise connected by an edge is called a clique in graph theory.

7.3.1.1 Clique Problems

Clique problems have applications in a variety of fields, including computer science, operations research, and social network analysis. One common clique problem is the maximum clique problem, which involves finding the largest possible cliques in a given graph. An example of a maximum clique is given in Figure 7.3. The decision version of this problem is to determine whether a given graph contains a clique of a certain size. The maximum clique problem is known to be NP-complete, meaning that no efficient algorithm is known for solving it in general. There are several

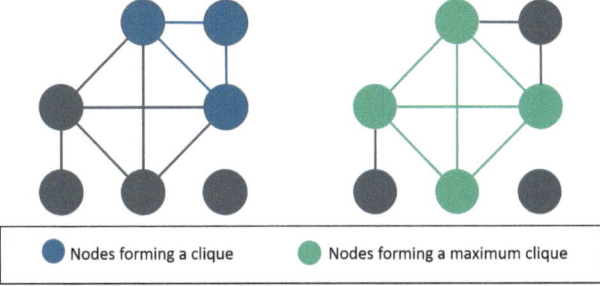

Fig. 7.3: Examples of cliques inside a graph.

strategies to solve clique problems in graph theory. Exact algorithms are designed to solve the clique problem optimally by finding the largest clique in a graph. Examples of exact algorithms include the Bron-Kerbosch algorithm [3], and the brute-force algorithm. These algorithms guarantee an optimal solution but can be computationally expensive for large graphs. Heuristics are approximate algorithms that provide a suboptimal solution to the clique problem. Examples of heuristics include the Tabu search algorithm, the genetic algorithm [7], and the greedy algorithm [5]. These algorithms are generally faster than exact algorithms and can be useful for finding good solutions to large graphs. One can also model the clique problem as a mixed-integer optimization problem and solve it with the branch-and-bound method. Assume we are given an undirected graph $G = (V, E)$ where V is the set of nodes and E the set of edges. We introduce one binary variable x_v for each node. The variable is set to one, if it is included in the clique. The constraints that ensure that the nodes in the clique are pairwise connected are

$$x_u + x_v \leq 1 \quad \forall (uv) \notin E. \tag{7.1}$$

Approximation algorithms are also a popular strategy for solving clique problems. These algorithms are designed to find a solution to the maximum clique problem that is guaranteed to be within a certain factor of the optimal solution. The goal is to find a good balance between accuracy and computational efficiency.

7.3.1.2 Flow Models

Flow models are mathematical models that represent the flow of goods, information, or resources through a network. These models are widely used in operations research, transportation planning, and logistics, among other fields. Flow models typically involve two types of nodes: source nodes and sink nodes. The source nodes are where the flow originates, while the sink nodes are where the flow is consumed. The edges in the graph represent the routes or paths through which the flow can travel. There are several types of flow models in optimization. Maximum flow models aim to determine the maximum amount of flow that can be sent from the source node to the sink node subject to certain constraints. The constraints may include limitations on the capacities of the edges or nodes in the network. Assume for example we want to transport as much goods as possible via train from Station A to Station B. Each edge in the directed graph G represents a freight train between two stations. The set of stations and routes is illustrated in Figure 7.4. The problem can be represented by

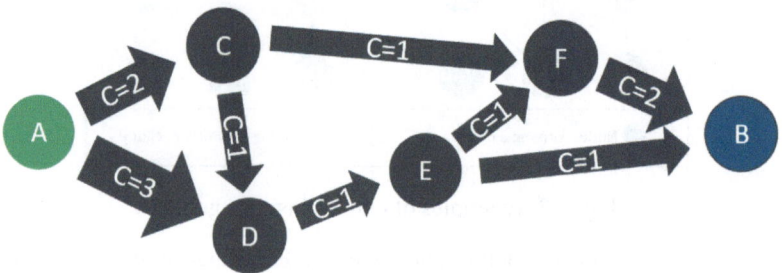

Fig. 7.4: Example flow graph.

the following linear program.

$$\max f_{AC} + f_{AD} \tag{7.2a}$$

$$\text{subject to:} \quad f_{AC} \leq 2,$$
$$f_{AD} \leq 3,$$
$$f_{CD} \leq 1,$$
$$f_{CF} \leq 1, \tag{7.2b}$$
$$f_{DE} \leq 1,$$
$$f_{EF} \leq 1,$$
$$f_{EB} \leq 1,$$
$$f_{FB} \leq 2,$$

$$f_{AC} = f_{CF} + f_{CD},$$
$$f_{AD} + f_{CD} = f_{DE}, \tag{7.2c}$$
$$f_{DE} = f_{EF} + f_{EB},$$
$$f_{CF} + f_{EF} = f_{FB},$$

$$f \in \mathbb{R}_+^8,$$

where f_{UV} is the amount of goods we send via the train from station U to station V. There are two types of constraints. Each freight train has a certain capacity which cannot be exceeded (7.2b). Additionally, we have to ensure flow conservation (7.2c). At each station the amount of goods that arrives must equal the amount that departs. Maximized is the flow we send from station A (7.2a). The optimal value for this example is 2. In scheduling applications one often uses multi-commodity flow models. These determine the optimal flow of multiple commodities through a network. The commodities may have different origins, destinations, and demands, and the network may have limitations on the capacities of the edges or nodes. There exist several algorithms to solve flow problems in optimization. These algorithms vary in terms of their computational complexity, accuracy, and applicability to different types of flow problems. Some of the most common algorithms for solving flow problems include the Ford-Fulkerson algorithm, Edmonds-Karp algorithm and the push-relabel algorithm [5]. These algorithms have been extensively studied and optimized over the years, and many practical implementations are available in optimization software packages.

7.3.2 Mixed Integer Programs and Connections to Machine Learning

Decision trees are a type of machine learning model that uses a tree-like structure to make predictions based on a set of features. Each node in the tree represents a test or decision on one or more features, and each branch represents an outcome of the test.

The leaf nodes represent the final prediction or class label for the input data. Decision trees can be modeled as mixed integer programs (MIP) A mixed integer program consists of a description of the set of feasible points and an objective function. It can be stated as

$$min\{c^T x \mid Ax \leq b, \ x \in \mathbb{Z}^{(p-q)} \times \mathbb{R}^q\}$$

with vectors c and b and a matrix A of appropriate dimensions. In some cases, some constraints on x are algorithmically intractable. In order to decide whether a given solution is fulfilling these constraints, tree classifiers are used assuming that the chosen trained classifier is accurate enough to trust its output. The tree classifier is modeled with linear constraints which are then added to the description of the feasibility set.

7.3.2.1 Modeling Logic

Let $x_1, x_2 \in 0, 1$ be two binary decision variables. Several dependencies between the two variables can be modeled by linear constraints:

- Conflict between $x_1 = 1$ and $x_2 = 1$:

$$x_1 + x_2 \leq 1$$

- $x_1 = 1$ implies $x_2 = 1$:

$$x_1 \leq x_2$$

- $x_1 = 1$ if and only if $x_2 = 1$:

$$x_1 = x_2$$

- $x_1 = 1$ implies $x_2 = 0$ or vice versa:

$$x_1 + x_2 = 1$$

A collection of linear constraints describe the set of feasible points in a mixed-integer program.

7.3.2.2 Binary Decision Trees

We are given some input data as a vector of binary variables $s \in \{0, 1\}^n$. The goal is to decide for each configuration of s, whether this point is feasible for our optimization problem. To label this point either feasible or infeasible, we use a binary decision tree. Decision trees consist of decision nodes $v \in V$ and label nodes $v \in U$. In each node, binary tests are applied on s for example *or* ($\vee$), *not* ($\neg$) or *and* ($\wedge$). Starting at the root node of the tree, the result of the test in one node implies the edge one has to follow until a leaf node (label node) is reached. A small example is given in Figure 7.5. To determine the path that our input data is following, we introduce

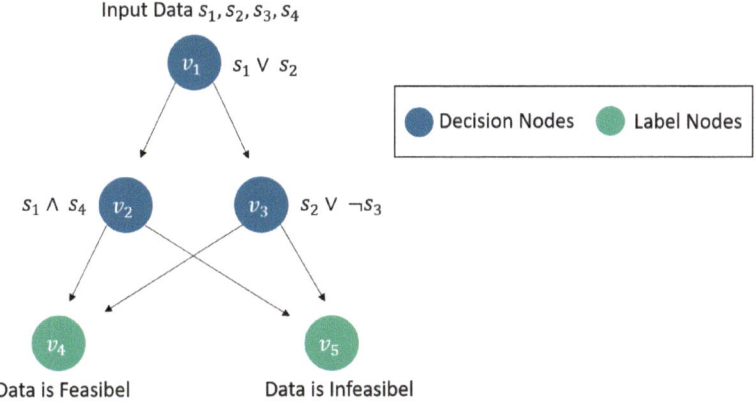

Fig. 7.5: Examples of a binary decision tree.

binary variables x_v for each node $v \in V \cup U$ and binary variables $y_{v_i v_j}$ for each edge between these nodes (E), indicating if the path crosses these elements. To translate this to constraints in the MIP, we use the methods to model logic described above. An edge $v_i v_j$ is active if and only if both its adjacent nodes v_i and v_j are active which can be represented by the inequalities

$$x_{v_i} + x_{v_j} \leq y_{v_i v_j} + 1, \tag{7.3a}$$

$$y_{v_i v_j} \leq x_{v_i}, \tag{7.3b}$$

$$y_{v_i v_j} \leq x_{v_j}, \tag{7.3c}$$

and the binary test at the node v_i implies the activation of the edge. An edge $v_i v_j$ is either implied by a TRUE or a FALSE result for the binary test at v_i. In the example in Figure 7.5, the edge $v_1 v_2$ can only be active, if $s_1 \vee s_2$ is TRUE. This is modeled by the inequality

$$s_1 + s_2 \geq y_{v_1 v_2}. \tag{7.3d}$$

The other binary test inequalities for the example are given by

$$0.5(1 - s_1) + 0.5(1 - s_2) \geq y_{v_1 v_3}, \tag{7.3e}$$

$$0.5 s_1 + 0.5 s_4 \geq y_{v_2 u_1}, \tag{7.3f}$$

$$(1 - s_1) + (1 - s_4) \geq y_{v_2 u_2}, \tag{7.3g}$$

$$s_2 + (1 - s_3) \geq y_{v_3 u_1}, \tag{7.3h}$$

$$0.5(1 - s_2) + 0.5 s_3 \geq y_{v_3 u_2}. \tag{7.3i}$$

Lastly, an active node implies the activation of exactly one of its child nodes.

$$x_{v_i} \geq \sum_{v_j | v_i v_j \in E} x_{v_j}. \tag{7.3j}$$

Now, to decide whether s is feasible for our optimization problem we just have to check if $x_{v_4} = 1$.

7.3.3 Pooling

In the following, we will introduce the pooling problem as it is stated in [1]. Although it originally comes from the petrochemical industry, Chapter 15.6 shows how it could be applied in the food-industrial context. Let $G = (N, A)$ be a simple acyclic-directed graph whose node set is partitioned as $N = I \cup L \cup J$. Here, I denotes the set of inputs, L the set of pools, and J the set of outputs. We assume that $A \subseteq (N \setminus J) \times (N \setminus I)$, i.e., every directed arc originates at a non-output node and terminates at a non-input node. Note that we explicitly allowed the presence of arcs between pools. Traditionally, instances with $A \cup (L \times L) = \emptyset$ are referred to as *standard pooling problems*; otherwise, they are referred to as *generalized pooling problems*. For every $l \in L$, I_l denotes the subset of inputs that have a directed path to l in G. An example for G is illustrated in Figure 7.3.3.

For every $i \in I$, let u_i be the total available supply for this input. Let u_l resp. u_j denote the flow capacities for each pool $l \in L$ resp. each output $j \in J$. The upper bound on arc flows is denoted by u_{ij} for $(i, j) \in A$. Typically, we have $u_{ij} = \min \{u_i, u_j\}$. Further, let K denote the set of quality specifications that are tracked across the problem. For $i \in I$ and $k \in K$, λ_{ik} denotes the level of specification k in raw material at input i. Likewise, $\mu_{jk}^{\min}$ and $\mu_{jk}^{\max}$ are the lower and upper bound requirements on level of specification k and output j.

Let y_a be the flow variable for arc $a \in A$. Non-negative flows originate at inputs and the assumed structure of A implies that each pool receives flows from inputs or other pools and each output receives flows from inputs or pools. For notational simplicity, we will always write equations using the flow variables y_{ij} with the understanding that y_{ij} is defined only for $(i, j) \in A$. Besides the flow variables, we have proportion variables q_{il} for $l \in L$ and $i \in I_l$ that describe the fraction of incoming flow to pool l that originated from some input i. We do not distinguish between flows that started at i and reached l along different paths. With this notation we can now state the constraints describing the pooling problem. Here, we refer to the well-known *q-formulation*:

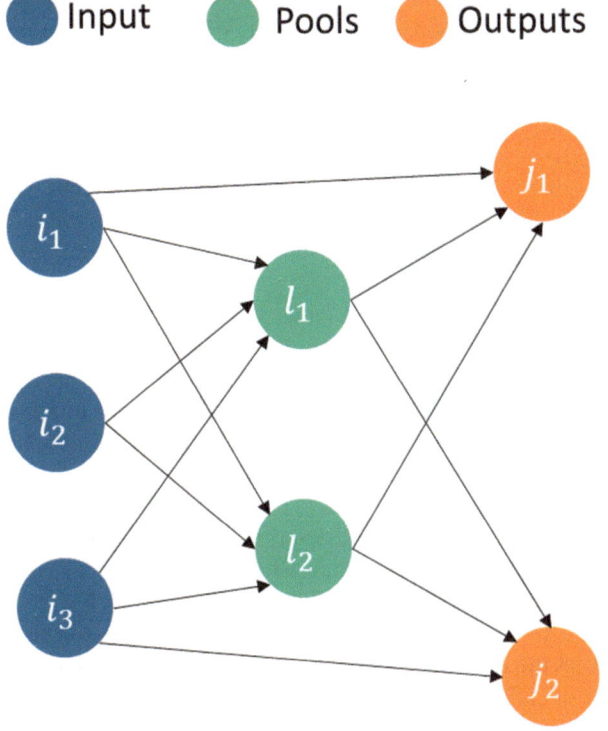

Fig. 7.6: Examples of a pooling graph G with inputs $I = \{i_1, i_2, i_3\}$, one pool $L = \{l_1, l_2\}$ and two outputs $J = \{j_1, j_2\}$.

$$\sum_{i \in I \cup L} y_{il} = \sum_{j \in L \cup J} y_{lj} \quad \forall l \in L \tag{7.4}$$

$$\sum_{j \in L \cup J} y_{ij} \le u_i \ \forall i \in I, \quad \sum_{j \in L \cup J} y_{lj} \le u_l \ \forall l \in L, \quad \sum_{i \in I \cup L} y_{ij} \le u_j \ \forall j \in J \tag{7.5}$$

$$0 \le y_a \le u_a \quad \forall a \in A \tag{7.6}$$

$$q_{il} \ge 0 \ \forall l \in L, i \in I_l, \qquad \sum_{i \in I_l} q_{il} = 1 \ \forall l \in L \tag{7.7}$$

$$y_{il} + \sum_{l' \in L : i \in I_{l'}} q_{il'} y_{l'l} = q_{il} \sum_{j \in L \cup J} y_{lj} \quad \forall l \in L, i \in I_l \tag{7.8}$$

$$\sum_{i \in I} \lambda_{ik} y_{ij} + \sum_{l \in L, i \in I_l} \lambda_{ik} q_{il} y_{lj} \ge \mu_{jk}^{\min} \sum_{i \in I \cup L} y_{ij}, \quad \forall j \in J, k \in K \tag{7.9a}$$

$$\sum_{i \in I} \lambda_{ik} y_{ij} + \sum_{l \in L, i \in I_l} \lambda_{ik} q_{il} y_{lj} \le \mu_{jk}^{\max} \sum_{i \in I \cup L} y_{ij}, \quad \forall j \in J, k \in K \tag{7.9b}$$

Equation (7.4) guarantees flow conservation at each pool. Inequalities (7.5) and (7.6) describe the flow capacities for each node and arc in the network and the non-negativity of flows. Constraint (7.7) is the simplex constraint for the proportion variables, which means that they are non-negative and sum up to 1 at each pool. (7.8) links the flow and proportion variables and describes the amount of flow in pool l that originated from input i. Finally, (7.9a) and (7.9b) accomplish the lower and upper bounds for the quality requirements. With constraints (7.8), (7.9a), and (7.9b) bilinear terms come into play such that the pooling problem belongs to the class of non-convex bilinear optimization problems, whose optimization is in general NP-hard.

Despite the inherent complexity of the pooling problem, the strong effort in improving the algorithms since the introduction in 1978 has made this problem computationally tractable. To that end, a variety of solution methods have been developed where spatial Branch & Cut algorithms stand out as especially promising as seen in [1] and [12]. In particular, the research in in this line follows two aims: To create primal methods for quickly finding good feasible solutions [14] and to find dual methods for obtaining tight bounds [11].

An extension of this basic pooling description is then applicable to multi-level mixture processes in the production chain of the food industry. When it comes to a single-level mixture process such that the inputs are directly blended to obtain the final products, we mathematically consider the simplified blending model, which can be obtained from the model above by setting $L = \emptyset$. In particular, in this case only simplifications of (7.5), (7.6), (7.9a), and (7.9b) are needed. The resulting blending model is a linear program and can thus be solved in practice in a reasonable amount of time by state-of-the-art solver software.

7.4 Conclusion and Outlook

A major current research task is the mutual integration of AI learning methods and mathematical optimization methods. On the one hand, learning methods are usually based on optimization problems: one searches for the best balancing function, the best classifier, or – more generally – the best functional relationship between input and labels to explain the data. Conversely, decision support has long taken place as a purely sequential process between statistics/AI and optimization. For example, to estimate the parameters in an optimization problem, regression techniques were used to predict future trends. After estimating these parameters, optimization took place based on them. In the future, it will be crucial to develop algorithms for optimization problems based on dynamically changing data, such as those found in real-time optimization. Here, it is no longer sufficient to keep the runtime of the algorithms used small (which is a challenge in itself). Now, quality guarantees for the computed solutions under time-varying data must also be maintained. For the often highly dynamic boundary conditions of logistics chains, the availability of such methods will be of particular importance.

Latest projects show that it is possible to derive priorities and cost functions in planning problems from observed past decisions [4]. This methodological approach to logistic problems is particularly relevant because many planning rules are implicit, in the sense of not being formalized and difficult to automate. The ability to derive explicit planning rules from past decisions, thus, ensures a higher degree of automation of the logistics chain and the objectivity of logistical planning decisions. The further development of this approach in the context of learning optimization methods will be of particular importance in future research. The practical application of these methods in an industrial context will constitute a decisive competitive advantage for the user.

References

1. S. Akshay Gupte. Relaxations and discretizations for the pooling problem. *Journal of Global Optimization*, 68(3):631–669, 2017.
2. A. Bärmann, J. Mehringer, C. Menden, U. Neumann, J. Schemm, O. Schneider, B. Sonnleitner, and M. Weissenbäck. Data analytics in der supply chain, 2021.
3. C. Bron and J. Kerbosch. Algorithm 457: finding all cliques of an undirected graph. *Communications of the ACM*, 16(9):575–577, 1973.
4. A. Bärmann, A. Martin, A. Müller, and D. Weninger. A column generation approach for the lexicographic optimization of intra-hospital transports. *OR Spectrum*, Volume 46:607–631, 2024.
5. T. H. Cormen, C. E. Leiserson, R. L. Rivest, and C. Stein. *Introduction to Algorithms*. MIT press, 3rd edition, 2009.
6. G. B. Dantzig, D. R. Fulkerson, and S. M. Johnson. Solution of a large-scale traveling-salesman problem. *Operations Research*, 2(4):393–410, 1954.
7. L. Davis. *Handbook of Genetic Algorithms*. Van Nostrand Reinhold, 1991.
8. M. Dorigo, G. Di Caro, and L. M. Gambardella. Ant algorithms for discrete optimization. *Artificial life*, 5(2):137–172, 1999.
9. P. J. M. Laarhoven and E. H. L. Aarts. *Simulated Annealing: Theory and Applications*, volume 37 of *Mathematics and Its Applications*. Springer Dordrecht, 1987.
10. E. L. Lawler and D. E. Wood. Branch-and-bound methods: A survey. *Operations Research*, 14(4):699–719, 1966.
11. J. Luedtke, C. d'Ambrosio, J. Linderoth, and J. Schweiger. Strong convex nonlinear relaxations of the pooling problem. *SIAM Journal on Optimization*, 30(2):1582–1609, 2020.
12. R. Misener and C. A. Floudas. Advances for the pooling problem: Modeling, global optimization, and computational studies. *Applied and Computational Mathematics*, 8(1):3–22, 2009.
13. C. Rego, D. Gamboa, F. Glover, and C. Osterman. Traveling salesman problem heuristics: Leading methods, implementations and latest advances. *European Journal of Operational Research*, 211(3):427–441, 2011.
14. C. Schulz and B. Uçar, editors. *An Adaptive Refinement Algorithm for Discretizations of Nonconvex QCQP*, volume 233 of *Leibniz International Proceedings in Informatics (LIPIcs)*, Dagstuhl, Germany, 2022. Schloss Dagstuhl – Leibniz-Zentrum für Informatik.

Chapter 8
Acquisition of Semantics for Machine-Learning and Deep-Learning based Applications

Thomas Wittenberg[1,2], Thomas Lang[2,4], Thomas Eixelberger[1,2], Roland Gruber[2,3]

Abstract For the development, training, and validation of machine learning (ML) and deep learning (DL) based methods, such as, e.g., image analysis, prediction of critical events, extraction or reconstruction of information from disrupted data streams, searching similarities in data collections, or planning of procedures, a lot of data is needed. Additionally to this data (images, bio-signals, vital-signs, text records, machine states, trajectories, antenna data, ...) adequate supplementary information about the meaning encoded in the data is required. Only with this additional information – the meaning or knowledge – a tight relation between the raw data and the human-understandable concepts – the *semantics* – from the real world can be established. Nevertheless, as the amount of data needed to develop robust ML or DL methods is strongly increasing, the assessment and acquisition of the related knowledge becomes more and more challenging. Within this chapter, an overview of concepts of knowledge acquisition applied to the different examples of applications is described and evaluated. Six main groups of knowledge acquisition related to AI-based technologies have been identified, namely (1) manual annotation methods, (2) data augmentation, (3) generative networks or simulation techniques, (4) synchronized sensors, (5) Active Learning approaches, and (6) explicit knowledge modeling using semantic networks.

Key words: Semantics, Knowledge, Labeling, Simulation, Augmentation, Active Learning, Reference Systems, Semantic Networks

---

[1]Fraunhofer Institute for Integrated Circuits IIS, Smart Sensing and Electronics, Erlangen, Germany
[2]Fraunhofer Institute for Integrated Circuits IIS, Development Center X-Ray Technology, Fürth, Germany
[3]Friedrich-Alexander-Universität Erlangen-Nürnberg, Erlangen, Germany
[4]FORWISS, Universität Passau, Passau, Germany

Corresponding author: Thomas Wittenberg
e-mail: thomas.wittenberg@iis.fraunhofer.de

© The Author(s) 2024
C. Mutschler et al. (eds.), *Unlocking Artificial Intelligence*,
https://doi.org/10.1007/978-3-031-64832-8_8

8.1 Introduction

For the development, training, and validation of machine-learning (ML) and deep-learning (DL) based procedures, such as, e.g., the automatic segmentation and analysis of 2D or 3D images or videos, the analysis of multi-modal temporal data streams, the prediction of (critical) events (e.g., for autonomous driving or driver surveillance), searching of similarities in huge data collections, or the planning of procedures (as, e.g., gaming tactics and reactive strategies), a plethora of data is needed [62]. In addition to this data $D \in \mathcal{D}$ of any origin adequate additional information about the meaning M of the data is essential. This additional information about the meaning is usually encoded in some types of labels M_i, describing the content of the data with respect to a certain task. Nevertheless, for some data there can always exist a set of different "meanings" $M_1, ..., M_n \in \mathcal{M}$, each one related to one specific task at hand. Thus, if some data is used for a different task, probably a new set of labels with a complete different meaning M_j is required. Together, the data D and its meanings M_i yield the semantics $S = (D, M_i)$, a research field originally based in linguistics.

In 1983 Niemann [49] has postulated this connection of 'data' and 'meaning' as: "In order to collect information about a problem domain Ω, a representative collection of samples $\omega = \{(\mathbf{f}_1(\mathbf{x}), \mathbf{y}_1), ..., (\mathbf{f}_i(\mathbf{x}), \mathbf{y}_i), ..., (\mathbf{f}_N(\mathbf{x}), \mathbf{y}_N)\} \subset \Omega$ is available. In this context $\mathbf{f}_i(\mathbf{x})$ denotes the $i - th$ 'pattern' (the 'data' D) from the problem domain Ω and $\mathbf{y}_i$ some additional information (the 'meaning' M) about the data.[1]

This idea about 'data' and 'meaning' (resp. 'knowledge') is briefly illustrated in Figure 8.1. On the left side a 24×24 patch of numbers (the raw data $D(x, y)$) can be seen, which correspond to the top 24×24 pixels of the image on the right side. The right side depicts an MRI slice of the author's (TW) left wrist, where the radix (1) and the ulna (2) bones have been delineated manually. These delineations describe the locations, geometry, extensions, textures, labels and relation of these two bones to each other within the image contents and hence its 'meaning' M with respect to delineating these bones with ML or DL approaches. Without the image annotation M, the data $D(x, y)$ is more or less meaningless and could not be used to train or parameterize any common supervised machine learning or deep learning method.

Within the addressed application domain of machine learning, deep learning or artificial intelligence, the broad goal of any such system is to use meaningfully-labeled reference data (D, M) for the definition, training, validation and (hyper-) parameter optimization of the developed applications A. This developed application system A can then be applied to and evaluated on yet unknown and unseen data $\tilde{D}$ in order to extract the embedded meaning $\tilde{M}$ from it, thus $A(\tilde{D}) \rightarrow \tilde{M}$, while simultaneously minimizing some loss function $L(M, M')$ between the obtained meaning M' and the intended meaning M [63].

---

[1] Even though, Niemann [49] defined the additional information $y_i \in \{1, ..., k\}$ in a more restricted sense as a set of distinct class labels, the fundamental concept of 'meaning' remains the same.

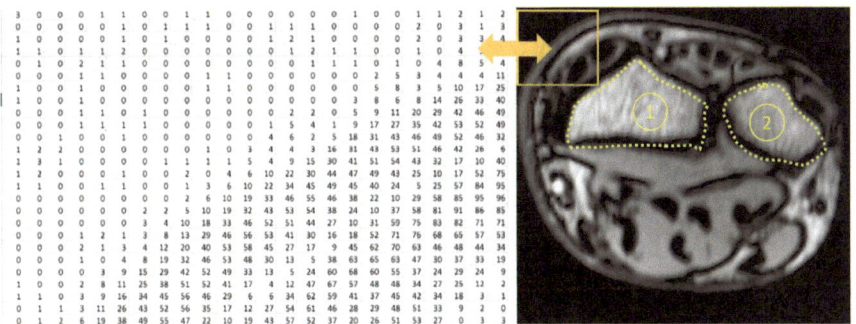

Fig. 8.1: Left side: A 24×24 patch of numbers (the data $D(x, y)$) corresponding to the top left 24×24 pixels of the image on the right side. Right side: An MRI slice of the author's (TW) left wrist, where the radix (1) and the ulna (2) bones have manually been delineated, hence providing information M about the bones position, geometry, extension, texture and labels. Additional information about the skin, muscles, nerves, arteries, veins could be be added. Together, data D (left side) and the annotations M (right side) yield the semantics S of this image.

However, as the amount of data needed to develop adequate and robust AI-based methods is strongly increasing, the simultaneous assessment and acquisition of the related semantics becomes more and more challenging [12, 63, 64]. Specifically, *Nikolenko, 2021* [52] describes this situation as *"Still, any machine learning practitioner will tell you that it is exactly the 'Data' and for some problems especially [the] 'annotation' phases that take upwards of 80% of any real data science project where standard open datasets are not enough. "*

Thus, in this contribution different approaches of knowledge acquisition related to machine and deep learning approaches and applications are identified, described, and evaluated. Furthermore, hybrid approaches are deducted, supporting a shift from human-in-the-loop approaches to the machine-in-the-loop methods for the acquisition of semantics.

8.2 Approaches to Acquire Semantics

In order to obtain an overview of currently used knowledge acquisition methods for AI-based technologies, interviews with members of various research groups of the ADA-Lovelace-Center at the Fraunhofer IIS and also contributors to this book have been conducted and supplemented with adequate literature from the field. Among these applications such as AI for digital pathology (Chapter 12), biosignal analysis and affective sensing (Chapter 11), as well as XXL-CT dataset segmentation (Chapter 18) will serve as examples.

The identified approaches of knowledge acquisition have then been clustered. As result five main groups of knowledge acquisition approaches related to Machine Learning and AI-based technologies have been identified, namely the:

- manual annotation and labeling of data (Section 8.2.1),
- data augmentation techniques (Section 8.2.2),
- simulation of data (Section 8.2.3),
- use of synchronized (expensive) sensors to obtain reference data, (Section 8.2.4),
- active learning (Section 8.2.5), and
- knowledge modeling using semantic networks (Section 8.2.6).

In the following sections these approaches shall be exemplified and illustrated based on application examples described in the Application Part of this book as well as some additional research by the authors.

8.2.1 Manual Annotation and Labeling

For image data (as, e.g., 2D, 3D, 2D+t, 3D+t, or point cloud data) mainly manual annotation methods are used, where experienced users delineate or draw depicted entities in the image data and label them using a predefined set of classifications. For example, the labels for the original ImageNet [13] data collection with currently over 14 million indexed images are based on the WordNet-Ontology [19]. In Figure 8.2a, the complete volumetric data of the fuselage of an historic airplane [23] with a spatial extension of approximately $6,100 \times 15,000 \times 5,200$ voxels and in Figure 8.2b a 512^3-voxel sub-volume as 3D image rendering to be annotated can be seen (see also Chapter 18). Figure 8.2c shows the manual annotation and labeling processing using a graphics tablet (having been found to be precise and intuitive [11]), yielding a labeled data set, see Figure 8.2d. Similarly, trajectories of moving objects in video streams, see Figure 8.5, and temporal single or multi-modal bio-signals of all types (electrocardiogram (ECG), electromyography (EMG), electroencephalogram (EEG), SpO2, RR, emotions, . . .) (see also Chapter 11), are usually manually labeled. In Figure 8.3, important events such as the so-called 'R-peaks' (blue boxes) in ECG data (dotted red line) along the timeline are manually marked and related to a predefined class.

Despite the fact that manual data labeling is currently referred to as the 'gold standard' for complex data [2, 65], the required resources are quite high with respect to delineation time and experienced staff. These expenses remain high, even if specialized annotation pipelines and dedicated annotation tools can be applied to this task, allowing data-guided annotation, proofreading of inference results and model refinements. Consequently, to minimize the costs of experts needed for manual or interactive data labeling, so-called 'crowd-sourcing' approaches have been established in the past [42, 10]. These 'crowd-sourcing' approaches are performed by so-called 'click-workers'. This distributed approach with very low costs per segmentation has in the past enabled the possibility to annotate and label challenging data multiple

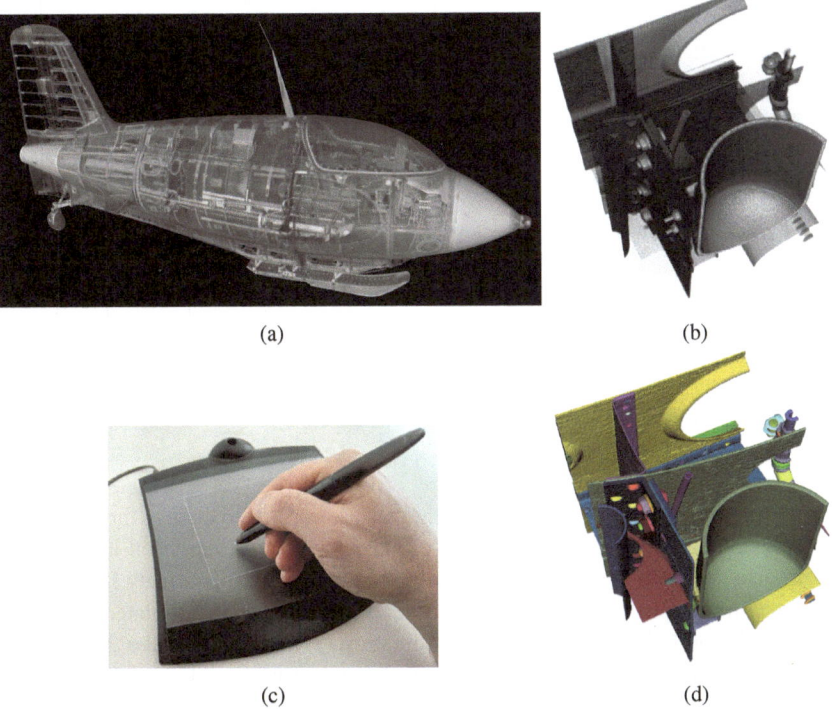

(a)

(b)

(c)

(d)

Fig. 8.2: 8.2a Rendering of the reconstructed fuselage of a completely scanned aircraft (see also Ch. 18, and 8.2b detail located at approximately the midpoint of the fuselage between the nose and the tail of the aircraft, see green box. 8.2c Manual labeling of the detail chunk using a high-resolution graphics tablet and a digital pen, 8.2d resulting a labeled data set. (Images with courtesy from R. Gruber [23]).

times and thus improve the segmentation accuracy by using the 'collective intelligence' of the annotators. Nevertheless, to be efficient, crowd-sourcing also profits strongly from the availability of distributed data via the internet such as Amazon's Mechanical Turk [56], adequate online annotation tools as well as soft skills of the annotators. However, besides the organizational, legal and logistic overhead, one drawback of crowd-sourcing can be the limited understanding of the click-workers about the annotation problem at hand and the complexity of the data depicting the various types of entities.

8.2.2 Data Augmentation Techniques

Even though data augmentation techniques (DA) do not provide explicit (new) knowledge embedded in the data (such as, e.g., manual annotation (Section 8.2.1) or phys-

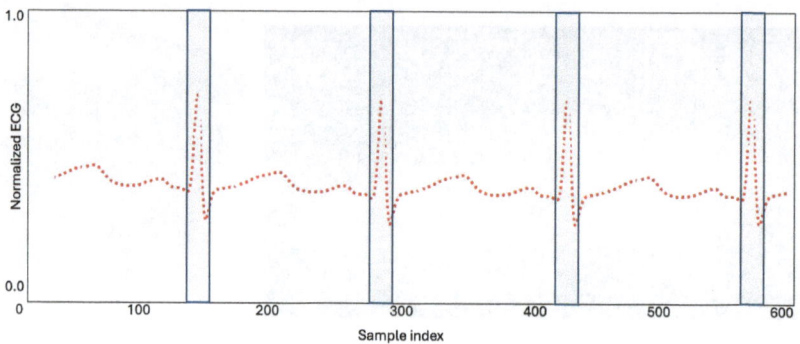

Fig. 8.3: Example of manually labeled R-Peaks (blue boxes) from ECG data (red dotted lines)).

ical simulation (Section 8.2.3), DA techniques are applied to change the appearance of the available data based on some a-priori known information on how the data might also appear within a certain domain, while keeping the already known labels aka 'meaning' constant. Specifically, these changes are mainly related to rotation (R), translation (T), scaling (S) and illumination (I) in order to make the dedicated applications independent to these RTSI-influences, if needed. Even though no new knowledge is gained by DA, the amount of training data with known labels can thus be extended, which can be quite useful for data generalization as, e.g., needed for unsupervised training or auto-encoders. Besides extending the data collections for a certain domain or task, data augmentation is also widely used to achieve a so-called domain generalization, addressing the idea how to take knowledge acquired from an arbitrary number of related domains and apply it to previously unseen domains [47].

For machine learning and deep learning-based image analysis systems using spatial, volumetric or temporal image data, typical augmentation techniques include geometric (rotation, flipping, cropping, shifting, zooming, scaling, shearing) and photometric changes (change of color, brightness, contrast, gamma), the addition of noise (erasing, noise injection) as well as kernel based alterations (using, e.g., linear and non-linear high- and low-pass, or FIR filters) [3]. Specifically an alteration of color can be achieved by principle component analysis (PCA), change of hue, saturation, gray-level equalization, random contrast, auto-contrast, contrast limited adaptive histogram equalization (CLAHE), solarization, or color jitter (where brightness, contrast and saturation of an image are randomly changed). Many of these DA techniques – such as the geometrical, noise and kernel based changes – can also be modified for the use of temporal data, such as, e.g., vital signs (EMG, ECG, EEG, temperature, ...) or state data from machines such as robots or cars [53].

For more specific applications, e.g., as the development of DL-based computer-assisted diagnosis (CAD) system within the field of digital pathology using whole-slide-images (see Chapter 12), more task-specific augmentation techniques are

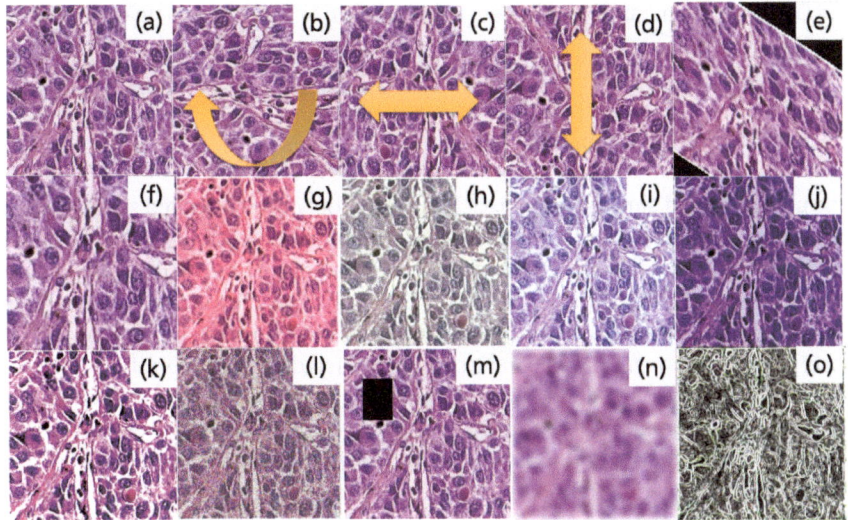

Fig. 8.4: Examples for image augmentation for HE-stained tissue in the field of digital pathology (see Chapter 12): (a) original image patch; (b) rotation, (c+d) flipping, (e) shearing, (f) zooming, (g-j) change of color stain, (k) change of brightness, (l) addition of noise, (m) random erasing, (n) Gaussian blur, and (o) Sobel based filtering. The semantics of the new image patches (b – o) is not changed as they provide the same information, but in a different peculiarity.

needed, which can consider possible variations of the hematoxylin and eosin (HE) staining [33] as well as the influence of different scanning systems [4].

Figure 8.4 gives an example of a small image patch of HE-stained tissue from the field of digital pathology, where the appearance of original image (a) has artificially been changed with respect of rotation and flipping (b,c,d), shearing and zooming (e,f), change of color (g-j), change of brightness (k), the addition of noise (l) and random erasing (m), and the use of a Gaussian and and Sobel kernel filter (n+o). As can be seen, the semantics of all the new image patches (b – o) is not changed as they provide the same information, but in a different peculiarity.

8.2.3 Simulation and Generation

If no sufficient data is available for the development and evaluation of ML and AI-based applications or manual data labeling (see Section 8.2.1) becomes too tedious, complex or expensive, (physical) data simulation as well as (deep learning-based) data generation techniques can be used. These approaches are able to yield adequate data D and its related meaning M at the same time. Two approaches, namely physical modeling (motivated from computer games and using such visualization engines), and DL-driven generation of new data using generative adversarial networks (GANs), are described here.

8.2.3.1 Physical Modeling

The physics-based simulation approaches usually only need a limited set of physical rules to yield realistic data. For instance, for the physical simulation of radiographs (images produced by X-rays, gamma rays, or similar radiation) with different content, Monte-Carlo-Simulations (MCS) can be applied, as they include the treatment of many known interaction mechanisms contributing to image formation, such as absorption, incoherent and coherent scattering including bonding effects and pair formation. However, the description of the object geometries can be challenging. In the biomedical domain Monte-Carlo-Simulated radiographs can for example be used to generate realistic breast tomosynthesis images [5] or chest radiographies [45, 51] from a voxelized geometry model to represent the patient anatomy. In non-medical applications, Monte-Carlo-Simulations can be used used to simulate radiographs of technical objects [28, 44, 59].

Besides the physical modeling of light-matter interactions to yield simulated radiographs, also object-to-object interactions can be modeled in order to simulate the movement of soccer players during a soccer match.

Application Example – Searching Soccer Scenes:

Finding critical or interesting scenes in video recordings of sports matches (such as soccer, basketball, or ice hockey games) and evaluating these scenes is a fundamental step with match- and game-strategy analysis for sports clubs, associations and societies as well as media reporting. Currently, the status quo is the manual annotation of the video providing adequate labels during or after a match or a time-consuming search in video material after a game. The use of these labels is intended to provide added value in terms of more effective training management and talent scouting, e.g., through the automatic identification of weak points or the automatic compilation of well or poorly solved scenes of a player's game.

Hence, to support and provide scene search for soccer games (and other sports matches) an AI-based search engine has been developed [39, 40], which is able to detect similar situations in a databases of past games by means of positional data. Besides finding similar game situations or constellations in the data collection, in a second step, these scenes should be evaluated and analyzed with respect to their contribution to the game and the outcome. Also, it is envisioned that it should be possible to suggest gaming strategies and solutions to optimize a game situation [43].

Nevertheless, as the complete manual labeling of full matches (or only key relevant scenes) to obtain an adequate labeled data collection is quite expensive with respect to time, personnel and other resources (see Section 8.2.1), a soccer game simulator was applied in order to provide a sufficient set of interesting visual scenes (the data

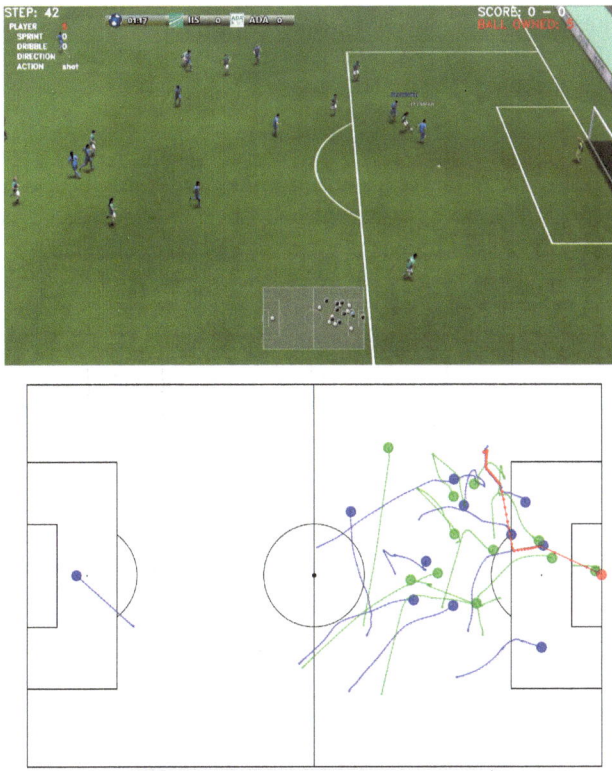

Fig. 8.5: Top: one single frame of a simulated soccer scene; Bottom: the corresponding ground truth (the meaning M, related to) the individual trajectories of the soccer players and the ball during a goal scene.

D) together with the trajectories of the players and the ball. These trajectories of the players and the ball relate to the 'meaning' M embedded within the video data and which are needed for the automated scene analysis and game evaluation [40, 43]. Figure 8.5 depicts on the top a simulated visual soccer scene, while the bottom shows the related simulated ground truth (the meaning M) in form of the individual trajectories of the soccer players and the ball during a goal. More details about the specific use of the thus generated and labeled game trajectories can be found in the works of Loeffler et al. [40] as well as Marzilger at al. [43].

8.2.3.2 Generative Adversarial Networks

Generative adversarial networks (GANs) [21] refer to the class of unsupervised deep learning approaches which are able to automatically extract and characterize the structure and patterns of the incoming data in such a way, that the model can then be used to automatically generate new data in the same domain and similar to the original data.

The basic approach of GANs relates to a pair of deep neural networks, namely the generator G – usually an autoencoder – and a discriminator D. From a vector of latent variables, the generator produces new data, while the task of the discriminator is to decide whether this data is real or fake. Usually the generator receives noise as input and produces new data as output, while the discriminator takes the newly generated data as well as real reference data as input and provides a probability value as output, indicating the chance of the generated data being real or fake. Furthermore, as a positive side effect, the necessary meaning M (as, e.g., labels, locations, contours, shapes, ...) about the content of the generated synthetic data D can simultaneously be provided.

Typical examples for generative adversarial networks being used to provide large scale (labeled) data collections are, e.g., the generation of photographs of human faces [27, 25], synthetic pictures of polyps in colonoscopy images [57], synthetic fundus images [6], image data for crack detection in electromagnetic nondestructive testing, [60], or defect detection in steel blocks from ultrasonic B-scan images [54].

Application Example – Generating Synthetic Data for NDT:

Specifically in the field of non-destructive testing (NDT), the use of GANs has recently received high interest. Especially during the production of devices with task-critical functionalities in the automotive domain (as, e.g., electric circuits, car wheels, tires, pistons ...) non-destructive testing based on various imaging modalities such as X-ray, CT, ultrasound or hyper-spectral imaging during the production has become mandatory. To analyze the thus acquired large scale image data many different machine and deep learning approaches have been proposed and investigated in the past. Nevertheless, as the occurrence of real defects during the production of these devices is in the range of one percent and below, the collection and acquisition of adequate reference data for the development of appropriate ML and DL methods remains challenging. Hence, to this end, GANs have recently been investigated to augment real NDT image data with artificially generated defects. Using artificially generated defects in aluminum cast wheels as an example [8, 48], Figure 8.6 provides a detail (rim and two spokes) of a GAN generated X-ray of an aluminum wheel including an artificial defect (red box).

Nevertheless, these deep learning-based approaches to provide realistic X-ray image data have the disadvantage, that a plethora of adequate training data – including sufficient images with and without defects – is needed. Hence the training of such GANs itself becomes challenging.

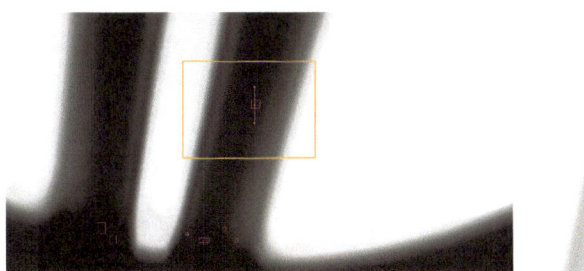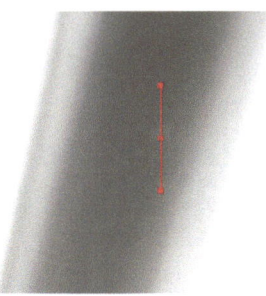

Fig. 8.6: Left side: Detail of an artificially generated X-ray of an aluminum wheel (here the connection between the rim and two spokes) with an artefact (red box); Right side: detail of the artificial generated artifact.

8.2.4 High-End Reference Sensors

In ML and DL applications, where expensive sensors (e.g., high-end cameras or ECG-sensors) are replaced by portable low-cost devices (e.g., smartphones), which then make use of AI-procedures to enhance the captured raw data, the high-grade data from the synchronized (expensive) sensors can directly be used as the needed reference knowledge for the training of the AI-methods [63]. In the field of biomedical engineering examples for this approach are the use of professional high end medical grade sensor devices for the wire-bound acquisition of high-quality physiological data such ECG, EMG, EEG, blood oxygen saturation (SpO2), ...) as reference for contact-less optical heart and respiratory rate assessment on smartphones, or high-end optical tracking or kinematic systems as reference for smartphone-based gait analysis and localization [17, 20, 31].

Application Example – Reference data for indoor localization

Radio-based localization and tracking of objects or people (e.g., by their smartphones) is a key component of many indoor applications and industrial environments. Typically, a set of synchronized antennas receives fixed frequency radio signal oscillations from mobile transmitting units to exploit values such as run-time measurement, time-of-arrival, and difference in time-of-arrival to estimate a position. For the (self) localization process, deep neural networks, and more specifically so-called long short-term memory (LSTM) networks can be applied [16, 18], extracting the exact location from the mentioned time-depending variables. Nevertheless, multi-path propagation of high-frequency signals is one of the most common sources of error that adds dynamic drift to the measurements. On top of that, the objects and people to be located and tracked usually move dynamically and non-linearly in real-world applications (e.g., in sports) and even change their motion behavior both in the short and long term, e.g., they stop abruptly or run in circles for a long

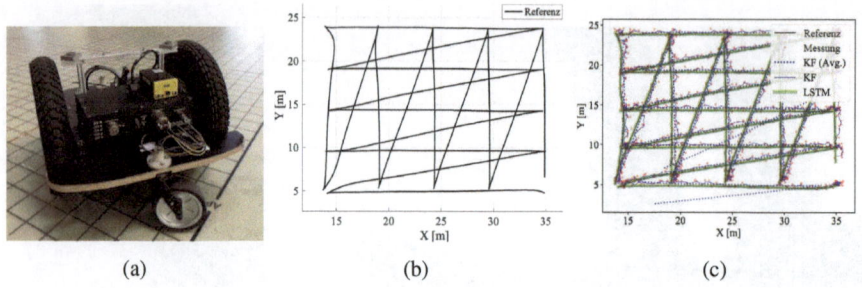

(a)   (b)   (c)

Fig. 8.7: (a) Mobile robot to capture reference data for radio-based location and tracking; (b) reference zig-zag path obtained with the robot; (c) Results of two path estimators using a Kalman filter and an LSTM network architecture to reconstruct the path from radio based data.

time. Therefore, the challenge is twofold: First to compensate for multi-path effects, second to distinguish between the actual motion and the drift of the measurements.

To address both challenges and provide reliable and accurate localization and tracking, reference positions and reference trajectories are employed (being the meaning M). In this case several synchronized high-end reference systems were used to collect this data. On one hand, an accurate indoor optical laser-based positioning system was applied providing positions at 30 Hz with an average (vertical and horizontal) mean error of less than 1 mm. On the other hand, a mobile robot was used to collect highly dynamic motion data with a maximum speed of 30 km/h and an acceleration of 2 m/s^2. These positions were then used as a reference data M for the development and training of various deep neural network approaches. It has to be noted, that the synchronization process between the devices is a challenge by itself, but by this approach smart-phone-based tracking can be enhanced.

Figure 8.7 shows the mobile platform used to collect the reference data (8.7a), the reference zig-zag path obtained from the mobile platform (8.7b), and the direct comparison of the ground-truth (semantics) with two approaches (Kalman filter and an LSTM network architecture) to estimate the paths from radio-based data (8.7c). For more datails, see [15].

8.2.5 Active Learning

With the exception of DA, the described approaches and techniques to acquire and collect (implicit or explicit) knowledge (or meanings M_i) about some data D are all depending on a strong involvement of human experts, either for data labeling (Section 8.2.1), structured knowledge modeling (Section 8.2.6), programming and training, designing and implementing adequate high-end simulations (Section 8.2.3), or applying synchronized high-grade devices (Section 8.2.4) [63].

To reduce these expensive workloads, more intelligent and hybrid approaches are needed, shifting the focus from human-in-the-loop to machine-in-the-loop. Such approaches could for example be reinforcement learning (RL) [46, 38] (see Section 3.2), active learning (AL) [7, 29, 34, 55] (see Section 4.3), or boot-strapping (BS) where already available (machine learning) methods can be used for the augmentation of data with labels (knowledge), and where human experts are only needed for verification or corrections.

All named approaches (RL, AL, BS) share the assumption that there exists a subset of data samples $\tilde{D} \subset D$, which can be used to train or optimize an ML-approach and will yield a better – but not yet optimal – performance than training on the complete dataset. Additionally, for RL and AL it is assumed that there exists a selection method that finds this subset $\tilde{D} \subset D$ faster than random sampling, whereas the methods to select the most adequate training data are mainly based on heuristics and try to employ insights of the application task or the ML-model. Nevertheless, the main motivation to use AL approaches is to reduce the expensive human resources to generate annotations.

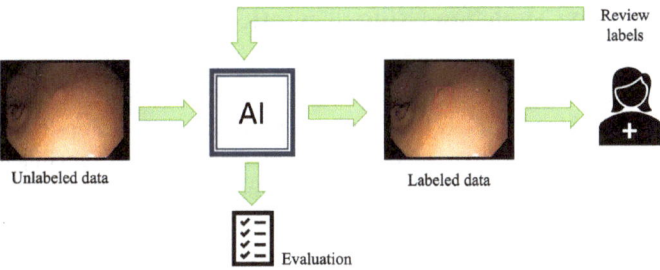

Fig. 8.8: Human-in-the-loop approach to generate training data.

Application Example – Boot-Strapping:

For example, Eixelberger at al.+[14] recently proposed a human-in-the-loop boot-strapping approach, which is used to iteratively increase the labeled training data collection, see Figure 8.8. In this case a deep neural network N_0 was initially pre-trained using labeled data L_0 from a publicly available repository, yielding a network model N_1. This network N_1 was then used to pre-label a yet unlabeled private image data-set U_1, whose labels were then reviewed, checked and partially corrected by a human expert (also known as the "oracle"). By this, a correctly labeled data-set L_1 was created, and added to the learning data. Then the network model was retrained using the additional training data (L_0, L_1), hence yielding a new network N_2. By this iterative approach with increasingly new labeled training data, the outcome of the network was constantly improved, while simultaneously the amount of labeled training data increased.

Furthermore, this machine-in-the-loop approach reduced the time required for labeling new training data. The involved experts only needed to review and correct the suggested labels instead of drawing each bounding box on their own. Based on the work by Su et al. [58] the median annotation time of a click-worker for setting a bounding box is 42 seconds (mean: 88 seconds), yielding a duration of nearly twelve hours for labeling 1,000 images. In this case the implemented network N_i needs approximately 25 ms per frame for the prediction and pre-labeling of the new data, thus, 1,000 yet unseen images can be processed within 250 seconds or less than 5 minutes. A user can review these predictions with one frame per second, thus needing approximately 17 minutes. If 25% of the images need to be corrected, and an experienced user needs 10 seconds to correct the bounding box and set a label, this sums up to 42 minutes for the necessary intervention. Hence, in total the user can check and correct 1,000 images per hour. Compared to the twelve hours, this is a time reduction of over 90% [14].

8.2.6 Knowledge Modeling Using Semantic Networks

In the above-described data-driven resp. bottom-up approaches the acquired data D is linked and combined directly on the pixel-, voxel-, or time-stamp-level with their related meanings M. For images (volumes, videos, point clouds, ...) this means that a set of spatially adjacent pixels (voxels, surface points, ...) is usually combined and labeled, likewise for uni- or multi-modal data a set of synchronized and temporally adjacent and successive set of data points are combined and labeled.

In contrast to theses methods, also classic rule-based methods can be applied. Here the factual and procedural knowledge of human experts about the domain, context and content of the data is translated into machine-understandable information (also known as ontologies), and then implemented in adequate data analysis measures. Examples for such a rule-based description of the content of the acquired and observed data are scene-graphs [9] or semantic networks, as proposed by Niemann et al. [50].

Such a semantic network or scene graph can be defined as a labeled, directed, and acyclic graph $G = (\mathcal{V}, \mathcal{E})$ consisting of the two sets $\mathcal{V}$ and $\mathcal{E}$. The set $\mathcal{V} = \{v_1, \ldots, v_N\}$ consists of vertices v_i (nodes) representing concepts, ideas, physical or conceptual objects, or features of objects. The set $\mathcal{E} = \{e_1, \ldots, e_M\}$ is a subset of $\mathcal{V} \times \mathcal{V}$ and describes edges e_l connecting ordered pairs of vertices (v_i, v_j) and thus relations between two objects [64]. The most important types of relations (edges) used within semantical networks are relations between classes and sub-classes of objects such as 'is a', 'has a' or 'has feature', as well as instance relations between object instances and object classes. For machine learning and deep learning driven approaches, such semantic networks or scene graphs can simultaneously be applied to organize and describe (a) very complex inter-relationships and the inter-dependencies of the involved objects (e.g., such as all the parts and their relationships of a car or the human anatomy) and hence the knowledge or meaning M about the data, and

(b) the needed methods to extract the related information from the data [32, 24]. Nevertheless, in practice not every node v_i needs to relate to a feature or an image processing functionality. In consequence, the system does not further process such a node and its sub components.

As an example for such a semantic network G, Figure 8.9 provides a coarse semantic network (or scene graph) of a car. The main node (on the top) denotes the car itself with its associated label ('car') and a set of important parts (as, e.g., four wheels, a windshield and the motor). On the next level, the object 'wheel' is a combination of a 'rim' and and a 'tire' object, where once again the rim may consist of the hub, some spokes and the horn. All these individual parts or objects and their relationships ('has a' or 'part-of', 'connects-to', ...) are denoted in black. Each object furthermore incorporates some descriptive features (denoted in green), such as, e.g., a unique label ('car', 'wheel', 'rim', 'spoke', ...) and the amount of its occurrences within the next level object ('4 wheels', '8 spokes'). Also attached to an object node can be a reference to some image analysis method (depicted in yellow) and its necessary parameters, which can be applied to detect and segment types of this entity in the related data. For example, 'wheel' objects (rim and tire) of vehicles can be identified in video-data using the generalized Hough transform [22], which is a well understood image analysis module to detect and describe parametric curves such as circular or elliptical objects from edge images. Depending on the structure, size, geometry, extension or complexity of an object at node v_i, besides the generalized Hough transform for circles or ellipses, any other image detection and segmentation module can be used, either being a deep neural network or a method from classical image analysis. Similar approaches have been proposed for the analysis of volumetric image data in medicine as, e.g., the segmentation of chest CT data using an anatomical model as semantic network.

Application Example – Instance Segmentation from Industrial CT:

For the automatic segmentation of various vehicle parts from large scale XXL-CT data Lang et al. [35] recently proposed the use of a semantic network based on the 'standard anatomy' of a car, usually having three to four wheels, a large windshield, a huge solid car body and so forth. Hence, in a first step the main components of a car are modeled as a scene graph, where in a second step the main object nodes are related to image adequate analysis methods. For the detection and hierarchical segmentation of the windshield the XXL-CT data D is combined with the meaning M encoded in the scene graph of the car. Based on the volumetric data (see Figure 8.10a) and the related semantic network (as, e.g., depicted in Figure 8.9), a bounding box of an object $v_{Windshield}$ is detected (see Figure 8.10b). Using this bounding box as a reduced search space, the included windshield components are iteratively analyzed and fused, hence yielding a complete segmentation of the windshield (see Figure 8.10c). Using this hierarchical, top-down, knowledge-driven approach, the scene graph G of the car can iteratively be traversed from node to node, and the related methods applied to the

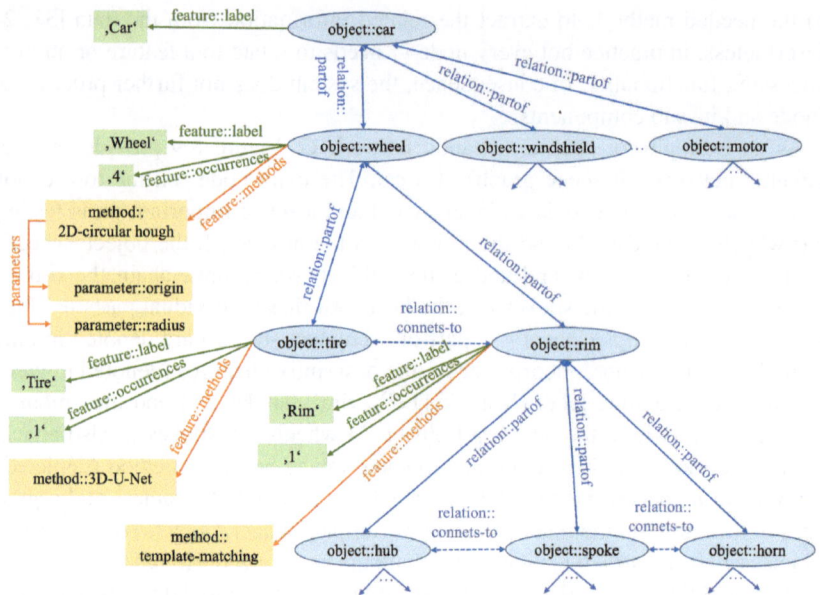

Fig. 8.9: Coarse semantic network describing a car (top node) with some of its parts (four 'wheels', 'windshield', 'motor',...), parts of the wheels ('tire' and 'rim' objects) and parts of the rims ('hub', 'spokes', 'horn', ...). Object nodes and their relationships ('part-of', 'has-a', 'connected-to', ...) are provided in black, feature nodes ('label', 'occurrences', ...) are shown in green, while method nodes and their parameters are displayed in yellow.

volumetric data. Hence, node-by-node the individual parts of the car can piece-wise be detected and segmented.

8.2.7 Discussion

A qualitative comparison and summary of the described methods for the acquisition of 'knowledge' resp. 'meaning' M of some data D (together referred to as the semantics S) for machine and deep-learning driven methods can be found in Table 8.1. In the table the listed methods are subjectively evaluated with respect to

- the amount of involved human interaction to relate the knowledge to the data,
- the programming effort needed to establish the method or handle the data,
- the quality of the yielded knowledge resp. meaning M, and
- if new and additional data D can be generated by this method.

With the exception of data augmentation (Section 8.2.2), all described approaches yield the needed meaning resp. knowledge in the expected quality. For **data aug-**

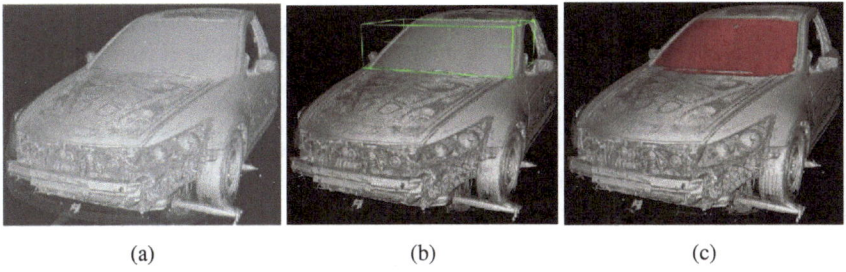

(a)   (b)   (c)

Fig. 8.10: Example for the hierarchical segmentation of a windshield in XXL-CT data using a semantic network. Based on the volumetric data of a car (a) and a related semantic network (as, e.g., depicted in Figure 8.9), the volume inside a bounding box of an object $v_{Windshield}$ is searched for. Using the bounding box, the included components are iteratively analyzed and fused yielding a segmentation of the windshield. (Images ©2023 by IEEE. Reprinted, with permission from T. Lang et al. [35]).

mentation, the 'quality' of the knowledge acquisition is rated as 'medium', as the already available data D is simply modified within a-priori known ranges related to the application.

Both the necessary human interaction (e,.g., for data labeling or system parametrization) and the efforts with respect to some required programming task can be related to costly human resources. Whereas the programming effort to establish, design, implement, host, and maintain an (online) annotation tool for data management and **manual data annotation** (Sections 8.2.1 and 8.2.5) is straightforward and manageable, the (crowd-sourcing) effort for the actual object delineation and labeling can be quite high, and the achieved results depend strongly on the expertise and availability of the labelers and annotators.

In contrast, the **augmentation of data** (Section 8.2.2) is quite cheap, both with respect to human interaction (actually none) and programming efforts. As data augmentation is currently already integrated in many available ML and DL frameworks, the programming part only involves the selection and integration of adequate augmentation strategies for the data involved. It should be noted that DA does not really create new data or knowledge, but extends the available data with new representations.

For the **simulation and generation** of data (Section 8.2.3) the programming effort is quite high, both for defining and establishing an adequate and realistic physical simulation and appearance of the data (such as a soccer game or the cell spread simulators), or for designing and training a generative adversarial network to generate new data. The human interaction for data simulation and data generation can be considered as quite low and refers most likely to the selection and adjustment of the necessary parameters. In contrast to all other methods, only the (physical) simulation and (deep learning-based) generation approaches actually yield (within some limits) new data together with a meaningful description, where all other methods only support the connection of available data D to the related meaning M. Nevertheless, it should be noted, that due to distribution shifts not all generated resp. simulated data depict the 'real world' and it must be carefully looked to it, which of this data is adequate to be used as training data.

For the use of (expensive) synchronized and **correlated sensors** (Section 8.2.4) the programming as well the human interaction efforts can be considered as medium, as for the synchronization and handling of the various (multi-modal) data sources some engineering, programming as well as human interaction and engagement is needed. If some data D can automatically be collected together with its reference (the meaning M) and some other related data within real life scenarios, such as mobile sensors, robots, or cars, this connection could be considered as semantic data harvesting [36, 37]. Compared to the other named approaches, data harvesting requires no expensive preparation step (e.g., to design and implement a data generator) and no post-processing step (for data annotation or labeling).

Within **active learning** (Section 8.2.5), the expensive human resources are tried to be minimized, shifting the workload from *human-in-the-loop* to *machine-in-the-loop*. Yet, this also implicates, that with a reduced need of the *domain expert* a shift to the *data scientist* occurs, who are now responsible to provide the infrastructure to evaluate and select the most adequate and yet unlabeled data for the domain expert to annotate or correct. Overall, this combination of man and machine is able to yield high quality labeled data.

Finally, defining, implementing and establishing adequate **semantic networks** or **scene graphs** (Section 8.2.6) for a new task or application involves a high load of human interaction. However, the actual design and programming of the inference loop to traverse the scene graph is subjectively considered with a medium workload, under the assumption that many DL and ML libraries for various standard (image) analysis components as part of the graphs nodes are already available and can directly be accessed. Certainly, if an image data analysis methods for a certain node v_i is not available yet, the workload related to the design, implementation and training of this missing method must be considered as 'high'.

8.3 Conclusion and Outlook

Independently of their final application, the design, development, training, and evaluation of new machine or deep learning methods is tightly related to the availability of data D in combination with an adequate and machine-readable description about the content of the data, the meaning M. While in the ages of huge data [26, 61] big data [41], or even massive data [1] the availability and collection of almost any type of data has become easy – also in combination with social networks, where a plethora of different data can be downloaded – the connection of the raw data (images, videos, vital data, texts, messages, machines states) to their meaning remains still challenging. Nevertheless, if such immense annotated data is available and can be used to train domain-independent machine learning or deep learning methods (as,

Method	Human Interaction	Programming Effort	Quality of Knowledge M	Generation of new data D
Manual Labeling	High	Low	High	No
Augmentation	None	Low	Medium	No
Simulation	Low	High	High	Yes
Generation (GANs)	Low	High	High	Yes
Expensive Sensors	Medium	Medium	High	No
Active Learning	Low	Medium	High	No
Semantic networks	High	Medium	High	No

Table 8.1: Qualitative comparison of knowledge acquisition methods (adapted and extended from [63], licensed under CC-BY 4.0) with respect to necessary human interaction, programming effort, quality of the obtained knowledge as well the possibility additionally generate new data.

e.g., for image segmentation or event detection), the development of powerful new software tools and related applications can be expected [30].

Using application examples described later in this book as well as some research by the authors, different approaches to collect and obtain the semantics $S = (D, M)$ for the different tasks and various data types (antenna data, trajectories of soccer players, XXL-CT volumes, whole slide images, multimodal bio-signals, ...) have been investigated and reviewed. Amongst them completely different approaches were identified, including manual data annotation and crowd-sourcing, active learning, data augmentation, data simulation and generation, data harvesting using additional sensors, as well as knowledge modeling using semantic networks or scene graphs. As all these approaches are strongly related to the involvement of human resources for annotation or implementation or other tasks, more intelligent and hybrid approaches have to be considered in the future, shifting the workload from human-in-the-loop to machine-in-the-loop.

Acknowledgments:
We want to thank the following colleagues from the Fraunhofer IIS for supporting this contribution with image material: Tobias Feigl (Section 8.2.4), Bishwajit Mohan Gosswami (Section 8.2.3), Robert Marzilger and Nicolas Witt (Section 8.2.3).

References

1. J. Abello, P. M. Pardalos, and M. G. Resende. *Handbook of massive data sets*, volume 4. Springer, 2013.
2. M. Aljabri, M. AlAmir, M. AlGhamdi, M. Abdel-Mottaleb, and F. Collado-Mesa. Towards a better understanding of annotation tools for medical imaging: a survey. *Multimedia Tools and*

Applications, 81(18):25877–25911, July 2022.

3. K. Alomar, H. I. Aysel, and X. Cai. Data augmentation in classification and segmentation: A survey and new strategies. *Journal of Imaging*, 9(2), 2023.

4. M. Aubreville, N. Stathonikos, C. A. Bertram, R. Klopfleisch, N. ter Hoeve, F. Ciompi, F. Wilm, C. Marzahl, T. A. Donovan, A. Maier, J. Breen, N. Ravikumar, Y. Chung, J. Park, R. Nateghi, F. Pourakpour, R. H. Fick, S. Ben Hadj, M. Jahanifar, A. Shephard, J. Dexl, T. Wittenberg, S. Kondo, M. W. Lafarge, V. H. Koelzer, J. Liang, Y. Wang, X. Long, J. Liu, S. Razavi, A. Khademi, S. Yang, X. Wang, R. Erber, A. Klang, K. Lipnik, P. Bolfa, M. J. Dark, G. Wasinger, M. Veta, and K. Breininger. Mitosis domain generalization in histopathology images — the midog challenge. *Medical Image Analysis*, 84:102699, 2023.

5. A. Badal, D. Sharma, C. G. Graff, R. Zeng, and A. Badano. Mammography and breast tomosynthesis simulator for virtual clinical trials. *Computer Physics Communications*, 261:107779, 2021.

6. V. Bellemo, P. Burlina, L. Yong, T. Wong, and D. Ting. Generative adversarial networks (gans) for retinal fundus image synthesis. In G. Carneiro and S. You, editors, *Computer Vision – ACCV 2018 Workshops - 14th Asian Conf. on Computer Vision*, pages 289–302. Springer, 2019.

7. S. Budd, E. C. Robinson, and B. Kainz. A survey on active learning and human-in-the-loop deep learning for medical image analysis. *Medical Image Analysis*, 71:102062, 2021.

8. R. K. Chanda. Synthetic x-ray image dataset generation for machine learning-based nondestructive testing – transferring methods from medical radiography. Masters thesis, Friedrich-Alexander-University Erlangen-Nürnberg, April 2023.

9. X. Chang, P. Ren, P. Xu, Z. Li, X. Chen, and A. Hauptmann. A comprehensive survey of scene graphs: Generation and application. *IEEE Transactions on Pattern Analysis and Machine Intelligence*, 45(1):1–26, 2023.

10. C. Chen, P. W. Woźniak, A. Romanowski, M. Obaid, T. Jaworski, J. Kucharski, K. Grudzien, S. Zhao, and M. Fjeld. Using Crowdsourcing for Scientific Analysis of Industrial Tomographic Images. *ACM Transactions on Intelligent Systems and Technology (TIST)*, 7:1 – 25, 2016.

11. C. Dach, C. Held, R. Palmisano, T. Wittenberg, and S. Friedl. Evaluation of input modalities for the interactive image segmentation of fluorescent micrographs. In *Biomed Tech*, volume 56, page S1, 2011.

12. S. Dasiopoulou, E. Giannakidou, G. Litos, P. Malasioti, and Y. Kompatsiaris. Survey of Semantic Image and Video Annotation Tools. In *Knowledge-Driven Multimedia Information Extraction and Ontology Evolution: Bridging the Semantic Gap*, pages 196 – 239, 2011.

13. J. Deng, W. Dong, R. Socher, L.-J. Li, K. Li, and L. Fei-Fei. Imagenet: A large-scale hierarchical image database. In *2009 IEEE Conference on Computer Vision and Pattern Recognition*, pages 248–255, 2009.

14. T. Eixelberger, G. Wolkenstein, R. Hackner, V. Bruns, S. Mühldorfer, U. Geissler, S. Belle, and T. Wittenberg. YOLO networks for polyp detection: A human-in-the-loop training approach. *Current Directions in Biomedical Engineering*, 8(2):277–280, 2022.

15. T. Feigl. *Datengetriebene Methoden zur Bestimmung von Position und Orientierung in funk- und trägheitsbasierter Koppelnavigation*. doctoralthesis, Friedrich-Alexander-Universität Erlangen-Nürnberg (FAU), 2021.

16. T. Feigl, S. Kram, P. Woller, R. H. Siddiqui, M. Philippsen, and C. Mutschler. A bidirectional lstm for estimating dynamic human velocities from a single imu. In *Proc's Int. Conf. on Indoor Positioning and Indoor Navigation (IPIN)*, pages 1–8, 2019.

17. T. Feigl, S. Kram, P. Woller, R. H. Siddiqui, M. Philippsen, and C. Mutschler. RNN-aided human velocity estimation from a single IMU. *Sensors*, 20(13):3656, Jun 2020.

18. T. Feigl, T. Nowak, M. Philippsen, T. Edelhäußer, and C. Mutschler. Recurrent neural networks on drifting time-of-flight measurements. In *Proc's Int. Conf. on Indoor Positioning and Indoor Navigation (IPIN)*, pages 206–212, 2018.

19. C. Fellbaum. *WordNet: An Electronic Lexical Database*. Bradford Books, 1998.

20. M. Gjoreski, M. Luštrek, M. Gams, and H. Gjoreski. Monitoring stress with a wrist device using context. *Journal of Biomedical Informatics*, 73:159–170, 2017.

21. I. Goodfellow, J. Pouget-Abadie, M. Mirza, B. Xu, D. Warde-Farley, S. Ozair, A. Courville, and Y. Bengio. Generative adversarial nets. In Z. Ghahramani, M. Welling, C. Cortes, N. Lawrence, and K. Weinberger, editors, *Advances in Neural Information Processing Systems*, volume 27. Curran Associates, Inc., 2014.

22. N. Gothankar, C. Kambhamettu, and P. Moser. Circular hough transform assisted cnn based vehicle axle detection and classification. In *2019 4th Int. Conf. on Intelligent Transportation Engineering (ICITE)*, pages 217–221, 2019.

23. R. Gruber, N. Reims, A. Hempfer, S. Gerth, M. Salamon, and T. Wittenberg. An annotated instance segmentation xxl-ct dataset from a historic airplane. *arXiv*, 2022.

24. G. Heidemann, F. Kummert, H. Ritter, and G. Sagerer. A hybrid object recognition architecture. In C. von der Malsburg, W. von Seelen, J. C. Vorbrüggen, and B. Sendhoff, editors, *Artificial Neural Networks — ICANN 96*, pages 305–310, Berlin, Heidelberg, 1996. Springer Berlin Heidelberg.

25. F.-T. Hong, L. Zhang, L. Shen, and D. Xu. Depth-aware generative adversarial network for talking head video generation. In *Proc's IEEE/CVF Conf. on Computer Vision and Pattern Recognition (CVPR)*, pages 3397–3406, June 2022.

26. P. J. Huber. Huge data sets. In R. Dutter and W. Grossmann, editors, *Compstat*, pages 3–13, Heidelberg, 1994. Physica-Verlag HD.

27. T. Karras, S. Laine, and T. Aila. A style-based generator architecture for generative adversarial networks. In *Proceedings of the IEEE/CVF Conference on Computer Vision and Pattern Recognition (CVPR)*, June 2019.

28. S. Kasperl. *Qualitätsverbesserungen durch referenzfreie Artefaktreduzierung und Oberflächennormierung in der industriellen 3D-Computertomographie*. Phd thesis, Friedrich-Alexander-Universität Erlangen-Nürnberg (FAU), 2005.

29. T. Kim, K. H. Lee, S. Ham, B. Park, S. Lee, D. Hong, G. B. Kim, Y. S. Kyung, C.-S. Kim, and N. Kim. Active learning for accuracy enhancement of semantic segmentation with CNN-corrected label curations: Evaluation on kidney segmentation in abdominal CT. *Scientific Reports*, 10(1):366, Jan. 2020.

30. A. Kirillov, E. Mintun, N. Ravi, H. Mao, C. Rolland, L. Gustafson, T. Xiao, S. Whitehead, A. C. Berg, W.-Y. Lo, P. Dollár, and R. Girshick. Segment anything, 2023.

31. R. Koch, N. Pfeiffer, N. Lang, M. Struck, O. Amft, B. Eskofier, and T. Wittenberg. Evaluation of HRV estimation algorithms from PPG data using neural networks. *Current Directions in Biomedical Engineering*, 6(3):505–509, 2020.

32. S. J. Kopetzky and M. Butz-Ostendorf. From Matrices to Knowledge: Using Semantic Networks to Annotate the Connectome. *Front Neuroanat*, 12:111, 2018.

33. P. Kuritcyn, C. I. Geppert, M. Eckstein, A. Hartmann, T. Wittenberg, J. Dexl, S. Baghdadlian, D. Hartmann, D. Perrin, V. Bruns, and M. Benz. Robust Slide Cartography in Colon Cancer Histology. In C. Palm, T. M. Deserno, H. Handels, A. Maier, K. Maier-Hein, and T. Tolxdorff, editors, *Bildverarbeitung für die Medizin 2021*, pages 229–234, Wiesbaden, 2021. Springer Fachmedien Wiesbaden.

34. T. Lang and T. Sauer. Geometric active learning for segmentation of large 3d volumes. *ArXiv*, abs/2210.06885, 2022.

35. T. Lang, T. Sauer, T. Wittenberg, S. Gerth, and N. Uhlmann. Ontoseg - segmentation of large volumetric datasets using semantic knowledge. In *2023 IEEE 17th International Conference on Semantic Computing (ICSC)*, pages 65–72, 2023.

36. U. Lee, E. Magistretti, M. Gerla, P. Bellavista, P. Lió, and K.-W. Lee. Bio-inspired multi-agent data harvesting in a proactive urban monitoring environment. *Ad Hoc Networks*, 7(4):725–741, 2009. I. Bio-Inspired Computing and Communication in Wireless Ad Hoc and Sensor Networks II. Underwater Networks.

37. U. Lee, E. Magistretti, B. Zhou, M. Gerla, P. Bellavista, and A. Corradi. Efficient data harvesting in mobile sensor platforms. In *Fourth Annual IEEE International Conference on Pervasive Computing and Communications Workshops (PERCOMW'06)*, pages 5 pp.–356, 2006.

38. R.-Z. Liu, H. Guo, X. Ji, Y. Yu, Z.-J. Pang, Z. Xiao, Y. Wu, and T. Lu. Efficient reinforcement learning for starcraft by abstract forward models and transfer learning, 2021.

39. C. Loeffler, W.-C. Lai, B. Eskofier, D. Zanca, L. Schmidt, and C. Mutschler. Don't get me wrong: How to apply deep visual interpretations to time series, 2022.
40. C. Löffler, L. Reeb, D. Dzibela, R. Marzilger, N. Witt, B. M. Eskofier, and C. Mutschler. Deep siamese metric learning: A highly scalable approach to searching unordered sets of trajectories. *ACM Trans. Intell. Syst. Technol.*, 13(1), 2021.
41. S. Lohr. The age of big data. *New York Times*, 11(2012), 2012.
42. L. Maier-Hein, S. Mersmann, D. Kondermann, S. Bodenstedt, A. Sanchez, C. Stock, H. G. Kenngott, M. Eisenmann, and S. Speidel. Can Masses of Non-Experts Train Highly Accurate Image Classifiers? In P. Golland, N. Hata, C. Barillot, J. Hornegger, and R. Howe, editors, *Medical Image Computing and Computer-Assisted Intervention – MICCAI 2014*, pages 438–445, Cham, 2014. Springer International Publishing.
43. R. Marzilger, F. Hirn, R. A. Alvarez, and N. Witt. Sports scene searching, rating and solving using AI. In D. Krumm, S. Schwanitz, and S. Odenwald, editors, *spinfortec2022 : Tagungsband zum 14. Symposium der Sektion Sportinformatik und Sporttechnologie der Deutschen Vereinigung für Sportwissenschaft (dvs)*, 2022.
44. S. Melnik. *Artefaktkorrektur für die 3D-Röntgenbildgebung auf Basis synthetischer Röntgenprojektionen.* PhD thesis, Technische Universität Berlin, 2020.
45. H. R. Mendes, J. C. Silva, M. Marcondes, and A. Tomal. Optimization of image quality and dose in adult and pediatric chest radiography via monte carlo simulation and experimental methods. *Radiation Physics and Chemistry*, 201:110396, 2022.
46. V. Mnih, K. Kavukcuoglu, D. Silver, A. A. Rusu, J. Veness, M. G. Bellemare, A. Graves, M. Riedmiller, A. K. Fidjeland, G. Ostrovski, S. Petersen, C. Beattie, A. Sadik, I. Antonoglou, H. King, D. Kumaran, D. Wierstra, S. Legg, and D. Hassabis. Human-level control through deep reinforcement learning. *Nature*, 518(7540):529–533, Feb. 2015.
47. K. Muandet, D. Balduzzi, and B. Schölkopf. Domain generalization via invariant feature representation. In S. Dasgupta and D. McAllester, editors, *Proceedings of the 30th International Conference on Machine Learning*, volume 28(1) of *Proceedings of Machine Learning Research*, pages 10–18, Atlanta, Georgia, USA, 17–19 Jun 2013. PMLR.
48. D. Neufeld, T. Würfl, R. Gruber, T. Schön, and A. Maier. Realistic image synthesis of imperfect specimens using generative networks. In *Proc's 9th Conf. on Industrial Computed Tomography (iCT), e-Journal of Nondestructive Testing*, volume 24(3), 2019.
49. H. Niemann. *Klassifikation von Mustern*. Springer Verlag, 1983.
50. H. Niemann, G. Sagerer, S. Schröder, and F. Kummert. Ernest: A semantic network system for pattern understanding. *IEEE Trans. Pattern Anal. Mach. Intell.*, 12:883–905, 1990.
51. K. Nikolaidis, S. Kristiansen, V. Goebel, T. Plagemann, K. Liestøl, and M. Kankanhalli. Augmenting physiological time series data: A case study for sleep apnea detection. In U. Brefeld, E. Fromont, A. Hotho, A. Knobbe, M. Maathuis, and C. Robardet, editors, *Machine Learning and Knowledge Discovery in Databases*, pages 376–399, Cham, 2020. Springer International Publishing.
52. S. I. Nikolenko. *Introduction: The Data Problem*, pages 1–17. Springer International Publishing, 2021.
53. N. Nonaka and J. Seita. Data augmentation for electrocardiogram classification with deep neural network, 2020.
54. L. Posilović, D. Medak, M. Subašić, M. Budimir, and S. Lončarić. Generative adversarial network with object detector discriminator for enhanced defect detection on ultrasonic b-scans. *Neurocomputing*, 459:361–369, 2021.
55. H. A. Qadir, J. Solhusvik, J. Bergsland, L. Aabakken, and I. Balasingham. A framework with a fully convolutional neural network for semi-automatic colon polyp annotation. *IEEE Access*, 7:169537–169547, 2019.
56. J. Ross, L. C. Irani, M. S. Silberman, A. Zaldivar, and B. Tomlinson. Who are the crowdworkers?: shifting demographics in mechanical turk. *CHI '10 Extended Abstracts on Human Factors in Computing Systems*, 2010.
57. Y. Shin, H. A. Qadir, and I. Balasingham. Abnormal colon polyp image synthesis using conditional adversarial networks for improved detection performance. *IEEE Access*, 6:56007–56017, 2018.

58. H. Su, J. imagenet, and L. Fei-Fei. Crowdsourcing annotations for visual object detection. In *Human Computation - Papers from the 2012 AAAI Workshop, Technical Report*, AAAI Workshop - Technical Report, pages 40–46, Dec. 2012.
59. F. Sukowski and N. Uhlmann. Monte Carlo simulations in NDT. In S. Mordechai, editor, *Applications of Monte Carlo Method in Science and Engineering*, chapter 1. IntechOpen, Rijeka, 2011.
60. L. Tian, Z. Wang, W. Liu, Y. Cheng, F. E. Alsaadi, and X. Liu. An improved generative adversarial network with modified loss function for crack detection in electromagnetic nondestructive testing. *Complex & Intelligent Systems*, 8(1):467–476, Feb. 2022.
61. E. J. Wegman. Huge data sets and the frontiers of computational feasibility. *Journal of Computational and Graphical Statistics*, 4(4):281–295, 1995.
62. T. Wittenberg. The need of annotation for reference image data sets. *International Congress Series*, 1281:453–458, 2005. CARS 2005: Computer Assisted Radiology and Surgery.
63. T. Wittenberg, M. Benz, A. Foltyn, R. Hackner, J. Hetzel, V. Wiesmann, and T. Eixelberger. Acquisition of Semantics for AI-based Applications in Medical Technologies: An overview. *Current Directions in Biomedical Engineering*, 7(2):515–518, 2021.
64. T. Wittenberg, M. Elter, and R. Schulz-Wendtland. Complete Digital Iconic and Textual Annotation for Mammography. In A. Horsch, T. M. Deserno, H. Handels, H. P. Meinzer, and Tolxdorff, editors, *Proc's Bildverarbeitung für die Medizin*, pages 91–95, 2007.
65. L. Wißler, M. Almashraee, D. Monett, and A. Paschke. The Gold Standard in Corpus Annotation. In *Proc's 5th {IEEE} Germany Student Conference*, Passau, 2014.

Part II
Applications

Chapter 9
Assured Resilience in Autonomous Systems – Machine Learning Methods for Reliable Perception

Gereon Weiss, Jens Gansloser, Adrian Schwaiger, Maximilian Schwaiger

Abstract Machine learning in the form of deep neural networks provides a powerful tool for enhanced perception of autonomous systems. However, the results of such networks are still not reliable enough for safety-critical tasks, like autonomous driving. We provide an overview of common challenges when applying these methods and introduce our approach for making the perception more robust. It includes utilizing uncertainty quantification based on ensemble distribution distillation and an out-of-distribution approach for detecting unknown inputs. We evaluate the approaches for object detection tasks in different autonomous driving scenarios with varying environmental conditions. The results show that the additional methods can support making the perception task of object detection more robust and reliable for future usage in autonomous systems.

Key words: uncertainty estimation, out-of-distribution detection, autonomous driving

9.1 Introduction

Over the past decade, advances in machine learning (ML) and high-performance computing have led to a huge increase in available methods for improved perception of autonomous systems. These new approaches significantly outperform many conventional, previously used methods in many specific tasks such as 2D/3D object detection ([22], [18], [24], [29]), 3D depth estimation, image recognition ([14]) or semantic segmentation ([3]). Deep neural networks (DNNs) turned out to be one of the biggest contributors to this new wave of innovation due to their ability to solve

Fraunhofer Institute for Cognitive Systems IKS, Fraunhofer IKS, Munich, Germany

Corresponding author: Gereon Weiss
e-mail: gereon.weiss@iks.fraunhofer.de

© The Author(s) 2024
C. Mutschler et al. (eds.), *Unlocking Artificial Intelligence*,
https://doi.org/10.1007/978-3-031-64832-8_9

179

highly complex problems with a high degree of accuracy. DNNs are now part of every software stack for autonomous driving vehicles. However, it has been shown that deep learning models often lack in giving robust and calibrated predictions ([8], [6]), making it hard to have confidence in the reliability of their outputs. One important aspect for providing increased robustness is to monitor and to reason about the predictions at test time as well as to check if a given input during operating time differs strongly from the samples in the training distribution. Calibrated confidence estimations combined with out-of-distribution (OOD) detection integrated in safety-critical applications like autonomous driving systems can provide valuable additional information for the correctness of predictions for the situational awareness. More-over, they can reduce the risk of hazards resulting from functional insufficiencies by decreasing the number of unknown unsafe scenarios, which is a critical part for the so-called safety of the intended functionality (SOTIF) [1] of such systems.

In this work, we review, develop, and evaluate new methods for quantifying uncertainty in DNNs and OOD detection related to perception tasks as well as to monitor perception systems' performance. The main focus is on safety-critical applications by example of autonomous driving to which these methods provide crucial knowledge for avoiding high-risk behavior and increasing overall reliability.

9.1.1 The Perception Challenge

An autonomous system must be able to observe and interpret its environment to safely operate in it – the so-called *perception*. Accurate perception of the surroundings is a key part for many autonomous systems where an agent is required to interact with its environment. Some of the systems operate in a controlled setting where the area of operation (operational design domain) can be constrained to fit the needs required for them to work appropriately, e.g., by geo-fencing.

However, even though these application can also benefit from more robust predic-tions through uncertainty quantification, the major target cases for these methods are in open-world environments. For these applications, a huge amount of possible con-figurations of the input data is possible and likely occur if the system is in operation. As the inputs for their perception modules mostly consist of very high dimensional sensor data, like LIDAR point clouds or (stereo) camera images, and the applied deep learning models have a large amount of trainable parameters, a formal verifica-tion which guarantees correct predictions is almost impossible even for constrained sub-problems due to the huge state space for both, inputs and parameters.

9.2 Approaches to reliable perception

There are many aspects which must be tackled in order to make a machine learning based system reliable. We provide an outline and brief summary of the requirements, challenges and approaches to design reliable systems using ML components.

9.2.1 Choice of Dataset

The training and test datasets need to be representative for the chosen area of deployment. To do so, a thorough understanding of the target domain is required. It is important to define under which conditions the ML system should operate, e.g., which kinds of objects are in the environment or which kinds of weather conditions are expected. The main challenge here is how to measure and quantify suitable datasets. Examples of metrics are coverage of the application domain, relevance, equivalence of cases/situations or coverage of positive and negative examples. Following standard practice, it is important to train the model independently of the test dataset, e.g., no hyperparameter tuning should be done on test data. Another difficulty is, how to ensure that the training data covers all relevant information (i.e., semantics) of the data which the model will face during deployment. To mitigate the negative effect of distributional shift, the training dataset can be continuously improved by gathering data after deployment.

A general critical issue in this case is to capture all relevant edge cases required for the expected functionality of the model. A straightforward approach to achieve this is to introduce large amounts of data to increase the generalization capabilities of the trained model. This can be implemented by:

- Augmentation of training data to introduce more variation.
- Extending the training dataset with synthetic data or data from generative models.
- Finding underrepresented classes and mitigating this class imbalance.
- Improving the model iteratively during operation by pseudo-labeling or active learning.

9.2.2 Unexpected Behavior of ML Methods

Currently, it is a hard challenge to verify larger DNNs and to make sure that their behavior is always as expected. In particular, providing proof that the network actually learns the semantics of the problem and has a sufficient understanding of the system it should model is challenging. Additionally, it is important that a model is adversarially robust, i.e., small perturbations in the input should not change the predictions drastically. In addition to that, an ML model should also be robust against noise in the input space. Noisy input data can be faulty sensors or changing weather conditions.

Interpretability is an additional important building block towards a safe ML system. A model should be capable of explaining its predictions in a human understandable way. This increases trust in the model and is a helpful tool for finding problems like wrong or biased predictions. Another important step towards safer systems is that a DNN should know when it cannot make reliable predictions. This is related to OOD detection, where the system can, e.g., ask for human support when an input is encountered that is different to the learned concepts. Often labeled training data is very limited, and training a well performing model with few data is difficult as well. A straightforward approach here is to use semi-supervised learning where a small amount of labeled data and large amounts of unlabeled data are used for training. Another option is active learning, where the model uses pseuo-labeling to improve over time. We refer to [27] for an exhaustive list of safety concerns.

The safety-related topics we tackle in our contribution are as follows. **Uncertainty quantification** learns reliable calibrated uncertainties for each input to determine how confident the model is that its prediction is correct. **Redundancy and ensembles**: Redundancy can be applied at many stages of a ML pipeline, e.g., using redundant sensors or different neural network architectures. A common approach for improving the performance of an ML model is to train multiple models independently with different initializations. **Out-of-distribution detection** is a mechanism for detecting novel inputs that have different semantics than those contained in the training dataset. Detecting these is crucial, since standard neural networks often misclassify OOD samples with high confidence.

9.2.3 Reliable Object Detection for Autonomous Driving

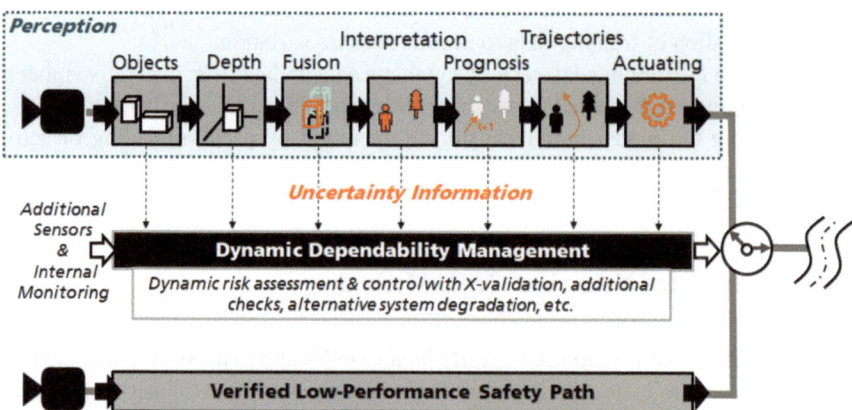

Fig. 9.1: The envisioned perception pipeline, integrating reliability information such as uncertainty estimates which is a metric used for the dynamic dependability management. Additionally, we show a low-performance safety path which is used when the reported uncertainty is too high.

For increasing the trustworthiness of an ML-based perception system for autonomous vehicles, we investigated the use of uncertainty quantification for *object detection*, one of the essential perception tasks of autonomous systems. The additional reliability information gained from the uncertainty quantification can in turn be utilized to increase the overall safety of the system [13]. Figure 9.1 shows the envisioned concept for incorporating the uncertainty information in a complete perception pipeline of an autonomous vehicle. The overall perception system is thereby split into two separate paths, a high-performance path based on ML methods and a fallback path relying on classical, non-data driven approaches. The subsystems in the upper high-performance path are extended with additional reliability information, such as uncertainty estimates. At runtime, a dynamic dependability management system takes this reliability information into account and combines it with additional sensor information and other monitoring systems. In each given situation, it dynamically assesses the reliability of the outputs of the high-performance path and if required switches to the safety path. The safety path is intended to provide basic functionality to bring the autonomous vehicle in a safe state, e.g., performing minimal risk maneuvers which allow coming to a halt on the hard shoulder of a motorway.

Additionally, we extend the object detection with a separate OOD component to detect wrong OOD detections of the object detector. The aim of this component is to provide an additional safeguard that filters out wrong predictions of the object detection pipeline.

Such a reliable object detector could in the future be integrated into an autonomous driving systems. To demonstrate its general suitability, we performed different evaluations by benchmarking the underlying feature extractors of the object detector, comparing it to other state-of-the-art approaches w.r.t. safety metrics, and finally testing it in a simulation environment. For an overview of generally available methods for uncertainty quantification and out-of-distribution detection we refer the interested reader to [11].

9.2.4 Uncertainty Quantification for Image Classification

To find suitable uncertainty estimation approaches that are further considered for robust object detection, we first performed two benchmarking experiments on the underlying task of image classification. In the first experiments [12], we compared four uncertainty quantification methods for DNNs – *Monte Carlo dropout* [6], *deep ensembles* [16], *learned confidence* [4], and *evidential deep learning* [26] – against the baseline of assuming the outputs of the softmax activations used for classification as confidences. We evaluated their performance across three standard image classification datasets and two network architectures. For that, we used evaluation metrics that also consider safety-related aspects: Network calibration, which measures how well confidences are calibrated, and *Remaining Error Rate* vs. *Remaining Accuracy Rate*, a metric we introduced to capture the trade-off between performance and safety when discarding inputs based on a confidence threshold. Our initial findings show that standard softmax values are usually overconfident and deep ensembles consistently showed the best results (see also Chapter 5 for a more theoretical introduction into the topic). However, due to the increased computational cost in training and inference for deep ensembles, learned confidence as sampling-free approach may also be an interesting approach for further studies.

In our second experiment [25], we investigated whether the same uncertainty quantification methods used in the first benchmark are suitable to detect novel concepts in input images that otherwise would lead to false positives. To that end, for each method we trained three different architectures on three different datasets. For the evaluation, we chose an out-of-distribution dataset for each training dataset or a different split. We investigated if discarding predictions based on the uncertainty allows rejecting novel inputs without impacting the overall performance too much. The results show that deep ensembles consistently showed the best results, closely followed by evidential deep learning, a sampling-free approach. However, the data indicated that for a truly reliable novelty detection approach other more specific measures are required, as a significant portion of out-of-distribution inputs could not be discarded by any method without greatly impacting the overall performance.

The benchmarks were performed on the task of image classification and intended to find suitable approaches to transfer to the downstream task of object detection. To this end, we further investigated the application of deep ensembles, how to transfer them to object detection, and how to minimize their computational complexity.

9.2.5 Ensemble Distribution Distillation for 2D Object Detection

For the integration of deep 2D object detection models into a perception pipeline of safety-critical applications, the lack of awareness about uncertainty in the given predictions represents a common problem for evaluating the trustworthiness of the system. Previous work mostly tackled this problem with sampling-based approaches to produce predictive uncertainty estimates. However, this requires multiple forward passes to create a statistical output which poses a big problem for real-time applications. Sampling-free methods on the other hand, only capture limited estimations of uncertainty which is not sufficient in many cases. A DNN that explicitly parameterizes a distribution is also referred to as a *prior network* which was first introduced for OOD detection by [19]. Therefore, we developed an ensemble distribution distillation approach to train a student model to predict an output distribution that is similar to an ensemble of teacher models with just requiring a single forward pass by using prior networks. We used the Yolov3 [23] architecture for our experiments and set the teacher models trained using maximum likelihood estimation as the baseline.

The behavior of an ensemble can be approximated by minimizing the Kullback–Leibler (KL) divergence between the student model and the expected predictive distribution of the teacher models:

$$L(\theta, T_{ens}) = E_{\hat{p}(x)}\Big[KL\big[E_{\hat{p}(\phi|T)}[P(y|x;\phi)]||P(y,x;\theta)\big]\Big], \tag{9.1}$$

where θ represents the parameters of the student model while T_{ens} is the teacher ensemble parameterized by ϕ for each model respectively. However, for the case that the student model directly learns the teacher outputs, this approach essentially only captures the mean predictions of the ensemble while the diversity is lost. In order to preserve information about the predictive distribution of the ensemble, it was proposed by [20] to let the student learn the parameters of an underlying distribution parameterizing the output distribution of the network:

$$P(y|x^*, \theta) = P(y|\delta), \quad \delta \sim p(\delta|x^*, D). \tag{9.2}$$

In Equation 9.2, δ can be any distribution, but is often chosen to be the conjugate prior of y due to traceability.

Formally, the student model is trained to distill the implicit distribution over distributions of M teacher models from a set of samples into an explicit distribution modeled by a single prior network. The set of distributions from M teacher models $\{P(y|x;\phi^{(m)})\}_{m=1}^{M}$ is approximated by a single distribution $p(\delta|x;\hat{\theta})$ which is supposed to match the teacher ensemble as close as possible.

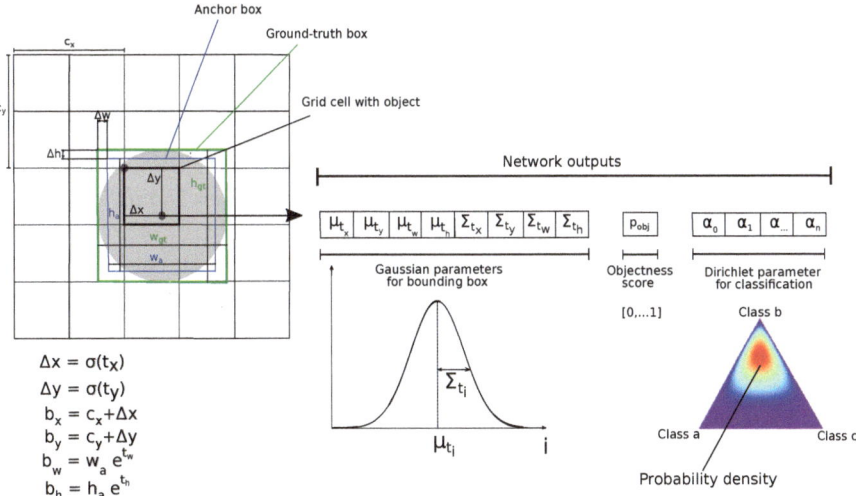

Fig. 9.2: Distilled Yolo prediction head.

For the classification case the conjugate prior to the categorical distribution commonly used to describe the problem would be a Dirchlet distribution resulting in:

$$L_{cls}^{ens} = -\sum_{m=1}^{M}\sum_{j=1}^{W}\sum_{l=1}^{H}\sum_{k=1}^{A} \gamma_{j,l,k} log(Dir(\tau_{j,l,k}^{m}|\hat{\alpha}_{j,l,k})) + \epsilon. \tag{9.3}$$

In this case, we sum over M teacher models with A grid cells containing anchor boxes with W, H widths and heights for each feature map. A small term ϵ is added for numerical stability as the α's are computed by involving an exponential function which can cause issues. The $\gamma_{j,l,k}$ parameter incorporates the objectness and the box scale for the respective ground-truth box $\gamma_{j,l,k} = ((2 - w_{gt}h_{gt})o_{j,l,k})/2$.

Similarly, for the regression of the bounding boxes a Gaussian distribution is assumed:

$$L_{reg}^{ens} = - \sum_{m=1}^{M} \sum_{j=1}^{W} \sum_{l=1}^{H} \sum_{k=1}^{A} \gamma_{j,l,k} log\left(\mathcal{N}\left(t_{i_{j,l,k}}^{m}|\mu_{t_i}, \Sigma_{t_i}\right)\right) + \epsilon, \qquad (9.4)$$

where L_{t_i} denotes the loss for a single bounding box parameter (e.g. t_x). Similar to Equation 9.3, the NLL of the teacher prediction t_i given the predicted parameters μ_{t_i} and Σ_{t_i} is summed over all teacher models and anchor boxes with all dimensions.

To show the effectiveness of the proposed method, the models are compared in terms of the quality of the uncertainty estimates as well as classical performance metrics. For performance on the detection task, *mean average precision (mAP)* and *mean average recall (mAR)* at an intersection over union (IoU) from 0.5-0.95 was used. *Mean average precision* is the most commonly used metric for comparing object detection models. It measures the area under the precision-recall curve for an *average* over different intersection over union thresholds (average precision) defining a true positive as well as building the *mean* over all classes (*mAP*). The *mean average recall (mAR)* measures the recall averaged under all intersection over union thresholds and all classes indicating purely how many relevant objects have been found while *mAP* indicates how many have been found as well as how precise these are with a certain trade-off.

For quantifying the ability of the networks to jointly assign high probability to true positives with respect to categories and bounding boxes, the *probabilistic detection quality (PDQ)* [9] was used.

Four different types of models were evaluated: 1) *Best teacher model* is the best performing model from the ensemble, 2) *Deep Ensemble* is the aggregated result from the ensemble of teachers. The boxes are merged by averaging the confidence scores as well as the bounding box position over all models. The diversity of the models is also considered by calculating the variances for the classification and regression parts for each box, 3) *Distilled student with ground-truth* is a student model trained using soft teacher labels as well as ground-truth label, 4) *Distilled student without ground-truth* is a student model trained with just using the predictions from the ensemble of teachers as targets.

Table 9.1 shows the results of the models on the three different datasets. It can be observed that all approaches outperform the baseline of taking the best performing teacher model from the ensemble for all metrics by a large margin. By taking a closer look between the distilled student models and the teacher ensembles containing 5 different networks one can see comparable for all metrics with the ensemble mostly performing slightly better than the students. However, in some cases the students even perform marginally better on the PDQ metric which is an interesting observation. In addition to that, it is also notable that including the ground-truth labels to the student training do not change the results in a significant way. This might be the case due to

Table 9.1: Evaluation Results

Model	mAP (%)	mAR(%)	PDQ (%)	FPS
KITTI test dataset				
Best teacher model (Baseline)	45.4	50.9	8.4	60
Deep Ensemble (5 models)	**52.9**	**56.5**	41.3	12
Distilled student with gt (ours)	51.7	55.5	40.8	58
Distilled student without gt (ours)	51.9	56.1	**41.5**	58
COCO2017 validation dataset				
Best teacher model (Baseline)	32.5	42.1	2.6	60
Deep Ensemble (5 models)	34.6	**44.1**	**23.3**	12
Distilled student with gt (ours)	**34.6**	43.6	16.5	58
Distilled student without gt (ours)	34.4	43.4	17.2	58
BDD validation dataset				
Best teacher model (Baseline)	23.1	30.2	2.5	60
Deep Ensemble (5 models)	**25.7**	**32.7**	15.5	12
Distilled student with gt (ours)	25.4	32.1	**18.5**	58
Distilled student without gt (ours)	25.1	32.3	17.3	58

the fact that the teachers are already converged to a certain level below the combined ensemble and knowledge discovery mostly takes place by learning the features out of multiple ensemble models as some models tend to find relevant objects which others do not and vice versa due to restrictive feature space boundaries of some local optima.

For the scope of this work, the PDQ metric is particularly interesting as it is the only metric taking class and bounding box uncertainties into account. It is no surprise that the single teacher models alone do not perform well on this metric as the variances are set to zero in this case, which reduces the expected probability inside the box significantly due to a sharp box boundary. The most prominent benefit of the ensemble distribution distillation approach is the huge increase in speed during inference time measured in frames per second (FPS) over the ensemble, as the student models do not require any sampling and thus, compute the output in just one forward pass making it multiple times faster. The speed was tested on a NVIDIA RTX2080 Ti graphic card.

In order to get more insight on the properties of the predicted values they are plotted in Figure 9.3 against the expected precision (top) and the Intersection-over-Union (IoU) (bottom) between predicted and ground-truth box for cases where the IoU is higher than 0.4 with a correct classification.

All results in Figure 9.3 are based on the student model trained on the dataset Common Objects in Context (COCO) from 2017 (COCO2017) from the ensemble only. The first plot shows the relationship between the average confidence and precision. In the context of object detection this can also be referred to as calibration [15]. It can be seen that the model is highly underconfident in its predictions, with even low mean confidence predictions having high precision in many cases. This, how-

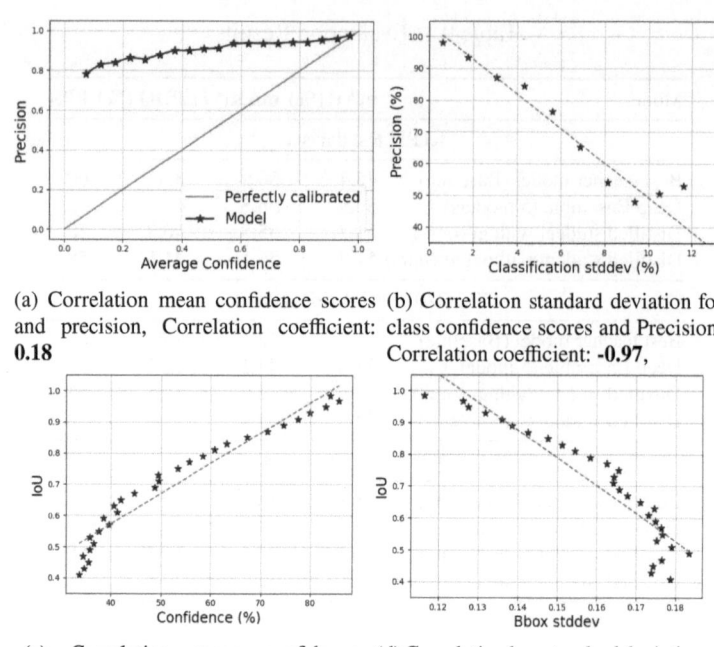

(a) Correlation mean confidence scores and precision, Correlation coefficient: **0.18**

(b) Correlation standard deviation for class confidence scores and Precision, Correlation coefficient: **-0.97**,

(c) Correlation mean confidence scores and IoU, Correlation coefficient: **0.485**

(d) Correlation box standard deviations and IoU, Correlation coefficient: **-0.34**

Fig. 9.3: Correlations for distilled student without ground-truth model for COCO2017 validation dataset.

ever, seems to be a property of the Yolo [23] architecture in general, as the teachers were also observed to show the same behavior which reinforces the need for some more reliable source of information. This can be found in Figure 9.2b) where the standard deviation is plotted on the precision. There is a strong negative correlation between these two values showing that for a more spread Dirichlet distribution over the class label distribution the precision significantly drops. This indicates that the disagreement over the predictive distribution gives a more robust predictor with respect to the likelihood of a prediction to be relevant than just taking the mean confidence. Having a more trustworthy approximation on what examples have a higher chance to be false positives is of high importance in safety-critical applications like in autonomous driving.

The two bottom graphs in Figure 9.3 show the correlation between mean confidence (left) and the predicted bounding box standard deviation (right) and the IoU with the ground-truth. For the confidences we can see a positive and for the standard deviations a negative correlation, which is as expected. However, we expected the correlation from the bounding box deviation to be higher than for the confidences. Nevertheless, these results show that for correct boxes the classification confidence and the bounding box standard deviation can indicate how well a box fits to the actual ground-truth which can be used as a source of information in systems using stereo vision to estimate the exact 3D position of the detected objects.

Overall, we were able to show that our distilled student model significantly outperforms the teacher models by improving mean-average precision (mAP) from 32.5% to 34.6% and PDQ from 2.6% to 16.5% on the COCO2017 evaluation dataset. In addition to that, we showed that the predicted uncertainties of our model correlate well with the quality of the classification and bounding box position predictions.

9.2.6 Robust Object Detection in Simulated Driving Environments

To evaluate the developed robust object detector more in-depth, we chose to create a set of scenarios in the autonomous driving simulator CARLA [5]. This allowed us to systematically investigate the influence of additional parameters, including lighting and weather conditions, occlusion, and object types.

9.2.6.1 Scenarios Setup

Scenario Name	Description
Accident ahead	The ego vehicle drives behind another car that suddenly changes lanes due to an accident site ahead.
Pedestrians crossing	The ego vehicles takes a right turn where a group of three pedestrians is about to cross the street.
Group of runners	The ego vehicle drives along a lane behind another car. On the opposite lane, a group of 10 runners is running behind a safety car.
Occluded pedestrian crossing	The ego vehicle drives along a straight street when suddenly a pedestrian crosses the street between two parked cars.
Roundabout crash	The ego vehicle enters a two lane roundabout when the other car is about to exit it.
Red light violation	The ego vehicle arrives at a crossing and is about to perform a left turn, when a car coming from the left side is violating a red traffic light.
Random items	The ego vehicle slowly drives along a straight street where random items — e.g., a garden gnome, plastic bag or shopping cart — are placed on the side walk.

Table 9.2: A short description of the scenarios generated in the CARLA simulator for the purpose of evaluating the robust object detector developed within this project.

Within CARLA, we created a set of seven scenario types, which are described in Table 9.2. The *ego vehicle* denotes the autonomous driving car incorporating the perception system. Each scenario thereby has 18 variations, allowing for a better investigation of the robustness of the approaches and enabled studying the impact of changes in the environment in a controlled manner. The variations are listed in Table 9.3 and sample images are shown in Figure 9.4. Each scenario in all its variations thereby is executed deterministically using CARLA ScenarioRunner [5].

Time of Day	Rain	Clouds	Fog
{Day, Sunset, Night}	None	None	None
{Day, Sunset, Night}	Light	Light	None
{Day, Sunset, Night}	Heavy	Heavy	None
{Day, Sunset, Night}	Heavy	Heavy	Heavy
{Day, Sunset, Night}	None	Heavy	Light
{Day, Sunset, Night}	None	Heavy	Heavy

Table 9.3: The 18 variations for each scenario given in Table 9.2. It consists of six basic weather conditions, repeated for three different times of day.

The resulting data, comprising the RGB camera output, ground truth for 2D object bounding boxes and additional metadata, are recorded and stored as a dataset to simplify the evaluation pipeline.

Fig. 9.4: Example images taken from the generated scenarios.

9.2.6.2 Methods and Metrics

We evaluated models trained on the datasets Berkely Deep Drive (BDD)) [28], COCO [17] and KITTI [7] with ensemble distribution distillation outlined in Section

9.2.5 and standard softmax training respectively. As a base model architecture we chose a slightly advanced version of Yolov3.

To compare the results between the proposed ensemble distribution distillation method and a standard implementation of a Yolo model in terms of robustness, we compared the results for the different CARLA scenarios on mean average precision (mAP) and mean average recall (mAR), see also explanations in Section 9.2.5.

9.2.6.3 Results

The results for the CARLA scenarios are provided as supplementary material in tabular form [2]. As expected from the results in Section 9.2.5, the distilled model outperforms the standard models in most cases. These differences get more pronounced under difficult conditions like dusk and rain. However, even though the trend clearly shows better performance for the distilled model, in some cases the standard model is better. This is another indication that the quality of a certain object detector is mostly not strictly superior to another model even if the overall results are noticeably better. A further interesting observation is, that there are large differences in performance solely induced by different weather condition for the same street scenario which indicates high bias of models trained on very restricted incomplete datasets. Especially striking is the trend that models tend to struggle with the same scenes and more important, the same weather conditions, even though they are trained independently on different datasets with different methods. This is the case despite the weather conditions vary strongly in the respective training datasets.

Overall, we could achieve a noticeable improvement for object detection over standard models by developing the ensemble distribution distillation method which is capable to give uncertainties for the classes as well as for the bounding box edges. We showed that this method also shows better performance in terms of classical object detection evaluation metrics as mAP and mAR on both, real world and simulated data with vastly different environment conditions. In Section 9.2.5, we also demonstrated the negative correlation between the uncertainties for classes and box edges with the average precision and average intersection over union respectively which strongly aligns with the expectations as more uncertain predictions are indeed more risky to be wrong. Another interesting finding is the simultaneous degradation in performance for all models which where independently trained with different methods and datasets on the same weather condition on the simulated environment, even though some of them are visually of medium difficulty for the human eye.

9.2.7 Out-of-Distribution Detection

The purpose of the OOD detection module is to detect wrong predictions of the object detector, providing an additional layer of robustness. The OOD module should classify bounding boxes of objects learned by the object detector as *in-distribution*

(ID), and bounding boxes of everything else (e.g., background) as OOD as shown in Figure 9.5.

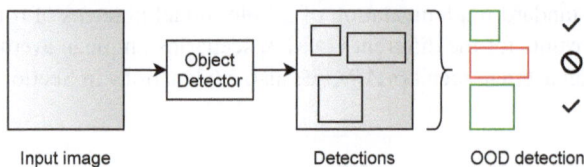

Fig. 9.5: The OOD module classifies the predicted bounding boxes from the object detector as in-distribution (shown in green) or out-of-distribution (shown in red).

The OOD module is trained on the ground truth bounding boxes which are also used to train the object detector. Since in an open-world scenario like autonomous driving input images could contain anything, we only evaluate methods that do not rely on training with OOD data.

We experiment with various simple OOD detection strategies. To evaluate the OOD detection methods, we use all cropped ground truth bounding boxes for cars and pedestrians from the various scenarios described in Section 9.2.6.1 as ID dataset, in the following referred to as carla-id. Note that this dataset also contains noisy images, i.e., images containing fog or rain. As OOD dataset we use the SVHN dataset [21] (referred to as svhn) to represent inputs that strongly differ from the ID data. Additionally, we use random crops from the scenario backgrounds that do not contain known objects which represent wrong predictions of the object detector and are closer to the ID data (referred to as carla-ood). The used datasets are shown in Figure 9.6.

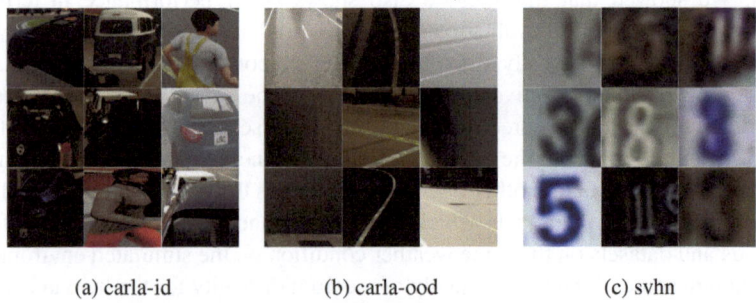

         (a) carla-id   (b) carla-ood   (c) svhn

Fig. 9.6: For the evaluation of the OOD detection methods, carla-id is used as ID dataset. This dataset contains the bounding boxes for cars and pedestrians. The OOD datasets are carla-ood which contains random crops from the scenario backgrounds and svhn which contains images of house numbers.

The described setup can be seen as a simple binary classification problem, which is straightforward to solve. A classifier with a small neural network can easily

Table 9.4: Results in AUROC, AP and FPR at 95% TPR for carla-id as ID dataset and carla-ood and svhn as OOD datasets. The single simple-cnn and resnet models are the best performing models in the ensemble. For the autoencoders, different number of clusters and latent dimensions were tested and the best performing models are reported.

	|AUROC|	AP	|FPR at 95% TPR
	|	carla-ood/svhn	
simple-cnn	0.83/0.94	0.58/0.88	0.65/0.28
resnet	0.75/0.89	0.48/0.69	0.76/0.36
simple-cnn ensemble	0.62/0.67	0.32/0.35	0.73/0.58
resnet ensemble	0.62/0.65	0.36/0.36	0.77/0.67
AE-gmm	0.84/0.96	0.72/0.94	0.54/0.29
AE-kmeans	0.33/0.46	0.20/0.29	0.94/0.90
AE-rec	0.81/0.98	0.73/0.97	0.71/0.10

achieve perfect accuracy. For this reason, more sophisticated OOD methods cannot be applied here. For example, confidence calibration cannot be used in this case since all confidences for ID data are close to 100%. Since the classifier always has a confidence of 100%, it is already perfectly calibrated. Since the ID dataset only contains two classes, simple OOD detection methods work well. We run our experiments with the following OOD detection strategies.

- **Classifier thresholding (simple-cnn, resnet)**: We threshold the maximum class probability of a classifier, all predictions below the threshold are OOD, all above the threshold are ID.
- **Classifier thresholding with ensembles (simple-cnn ensemble, resnet ensemble)**: The same as classifier thresholding but the class probabilities are the averaged predictions of an ensemble.
- **AE-gmm**: We train a Gaussian mixture model (GMM) on the ID latent representations of an autoencoder trained on ID data and threshold the likelihood of samples.
- **AE-kmeans**: We cluster the ID latent representations of an autoencoder trained on ID data and threshold the distance of a sample to the nearest cluster center.
- **AE-rec**: We train an autoencoder on the ID data and threshold the reconstruction error to classify samples as ID or OOD.

Note that the classifier based OOD detection methods require class labels for training. The autoencoders are trained in an unsupervised way. For the classifier based methods, we use a shallow simple convolutional neural network (CNN) architecture as well as the resnet18 architecture [10]. For the autoencoder, we use resnet18 as encoder and a simple convolutional decoder. We evaluate the methods with the following metrics:

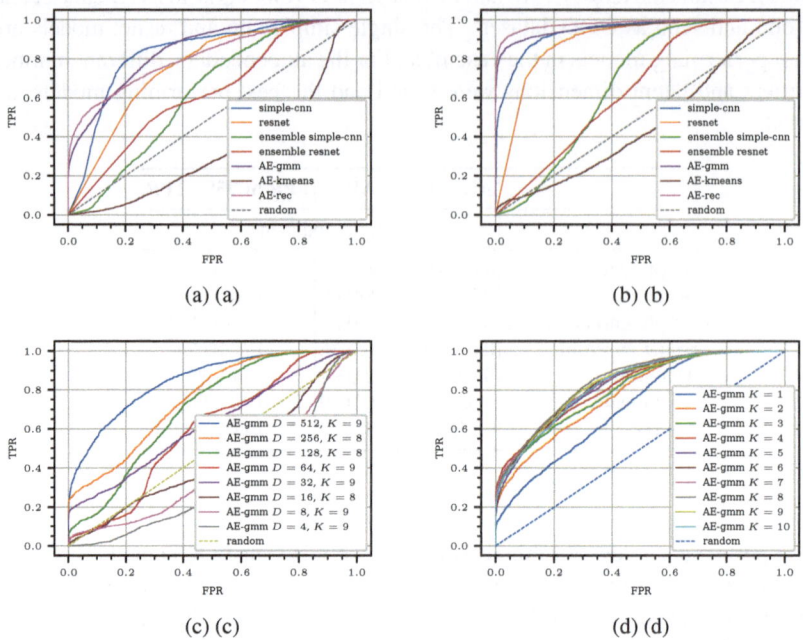

Fig. 9.7: ROC curves for (a) carla-id and carla-ood (b) carla-id and svhn (c) AE-gmm with different latent dimensions D and best number of clusters K on carla-id and carla-ood. (d) AE-gmm with latent dimension 512 and different cluster sizes on carla-id and carla-ood.

-   **The Area Under the Receiver Operating Characteristic Curve (AUROC)** is a threshold independent metric for evaluating classifiers. It computes the area under the receiver operating characteristic (ROC) curve which represents the FPR and TPR evaluated for all possible thresholds. An AUROC of 1 corresponds to an OOD detector that can perfectly distinguish between ID and OOD inputs.
-   **Average Precision (AP)** is also a threshold independent metric similar to AUROC. It computes the area under the precision-recall curve evaluated for all possible thresholds. A high AP means the model has high precision as well as high recall.
-   **FPR at 95% TPR** is the FPR when a threshold is used such that the TPR is 95%. It reflects the performance of the OOD detector for a specific choice of threshold.

We show the experimental results in Table 9.4. Each ensemble contains eight independently trained classifier. For the single classifier results we use the best performing classifier from the ensemble. For the autoencoders, we tried different latent dimensions and used the best performing. Additionally, for AE-gmm and AE-kmeans we tried different numbers of clusters and took the best performing model.

In Figure 9.7a and Figure 9.7b the ROC curves for the respective methods for carla-ood and svhn are displayed. The best performing method for carla-ood is AE-gmm. For svhn as OOD dataset, the best performing method is AE-rec. For both datasets, simple-cnn, AE-gmm and AE-rec are the best performing models. We can see that fitting simple Gaussians to the latent representations does reflect the distribution of ID data in the latent space well. Compared to that, the L2 distance used by K-means does not work at all and shows that the formed clusters have different variance or overlapping convex hulls. We see that for this simple problem the simple-cnn model works always better than the more complex resnet model. Additionally, the best single model works better than the ensemble. Figure 9.7c shows the ROC curve for AE-gmm with different latent dimension sizes evaluated for carla-ood. For each curve, we evaluated different numbers of clusters and report the best performing setup. We see that for small latent dimensions the performance is significantly worse. Figure 9.7d shows AE-gmm with different number of clusters and latent dimension 512, also evaluated for carla-ood. With latent dimension 512 more clusters improves the AUROC score. This shows that the latent space for ID data does not contain only two clusters, each for one class, but is structured based on more complex features.

9.3 Conclusion and Outlook

Machine learning (ML) is a core technology for developing intelligent autonomous systems. It is crucial to reliably integrate ML into safety-critical systems like autonomous cars or collaborative robots. We investigated current state-of-the-art methods and approaches to enhance the reliability and robustness of ML-based perception for autonomous vehicles. Our evaluations of diverse autonomous driving scenarios show the enhancements in ML-based perception using the proposed approaches of uncertainty quantification and out-of-distribution detection. For changing environmental conditions and street scenes, we achieve more reliable detection rates. Nevertheless, from a safety perspective the introduction of ML into safety-critical tasks still requires application-specific solutions. Therefore, further research activities should target specific real-world systems, taking their individual safety requirements on the perception module into account.

References

1. ISO 21448:2022 road vehicles – safety of the intended functionality, 2022.
2. Supplementary material – Carla Scenarios. http://dx.doi.org/10.24406/fordatis/312, 2023.
3. V. Badrinarayanan, A. Kendall, and R. Cipolla. Segnet: A deep convolutional encoder-decoder architecture for image segmentation. *IEEE transactions on pattern analysis and machine intelligence*, 39(12):2481–2495, 2017.
4. T. DeVries and G. W. Taylor. Learning confidence for out-of-distribution detection in neural networks. *CoRR*, abs/1802.04865, 2018.

5. A. Dosovitskiy, G. Ros, F. Codevilla, A. Lopez, and V. Koltun. CARLA: An open urban driving simulator. In *Proceedings of the 1st Annual Conference on Robot Learning*, pages 1–16, 2017.
6. Y. Gal and Z. Ghahramani. Dropout as a bayesian approximation: Representing model uncertainty in deep learning. In *Proc. ICML 2016*, volume 48, pages 1050–1059. PMLR, June 2016.
7. A. Geiger, P. Lenz, C. Stiller, and R. Urtasun. Vision meets robotics: The kitti dataset. *International Journal of Robotics Research (IJRR)*, 2013.
8. C. Guo, G. Pleiss, Y. Sun, and K. Q. Weinberger. On Calibration of Modern Neural Networks. In *Proc. ICML 2017*, pages 1321–1330. JMLR.org, Aug. 2017.
9. D. Hall, F. Dayoub, J. Skinner, H. Zhang, D. Miller, P. Corke, G. Carneiro, A. Angelova, and N. Sünderhauf. Probabilistic object detection: Definition and evaluation. In *The IEEE Winter Conference on Applications of Computer Vision*, pages 1031–1040, 2020.
10. K. He, X. Zhang, S. Ren, and J. Sun. Deep residual learning for image recognition, 2015.
11. M. Henne, J. Gansloser, A. Schwaiger, and G. Weiß. Machine learning methods for enhanced reliable perception of autonomous systems. Technical report, Fraunhofer IKS, 2021.
12. M. Henne, A. Schwaiger, K. Roscher, and G. Weiss. Benchmarking Uncertainty Estimation Methods for Deep Learning With Safety-Related Metrics. In *Proc. SafeAI@AAAI 2020*, volume 2560 of *CEUR Workshop Proceedings*, pages 83–90, 2020.
13. M. Henne, A. Schwaiger, and G. Weiss. Managing uncertainty of ai-based perception for autonomous systems. In *Proceedings of the Workshop on Artificial Intelligence Safety AISafety@IJCAI'19*, volume 2419 of *CEUR Workshop Proceedings*, 2019.
14. A. Krizhevsky, I. Sutskever, and G. E. Hinton. Imagenet classification with deep convolutional neural networks. In *Advances in neural information processing systems*, pages 1097–1105, 2012.
15. F. Kuppers, J. Kronenberger, A. Shantia, and A. Haselhoff. Multivariate Confidence Calibration for Object Detection. In *Proceedings of the IEEE/CVF Conference on Computer Vision and Pattern Recognition Workshops*, pages 326–327, 2020.
16. B. Lakshminarayanan, A. Pritzel, and C. Blundell. Simple and scalable predictive uncertainty estimation using deep ensembles. In *Advances in Neural Information Processing Systems 30*, pages 6402–6413. Curran Associates, Inc., 2017.
17. T.-Y. Lin, M. Maire, S. Belongie, J. Hays, P. Perona, D. Ramanan, P. Dollár, and C. L. Zitnick. Microsoft COCO: Common objects in context. In *European conference on computer vision*, pages 740–755. Springer, 2014.
18. W. Liu, D. Anguelov, D. Erhan, C. Szegedy, S. Reed, C.-Y. Fu, and A. C. Berg. Ssd: Single shot multibox detector. In *European conference on computer vision*, pages 21–37. Springer, 2016.
19. A. Malinin and M. Gales. Predictive uncertainty estimation via prior networks. In *Advances in Neural Information Processing Systems*, pages 7047–7058, 2018.
20. A. Malinin, B. Mlodozeniec, and M. Gales. Ensemble distribution distillation. *arXiv preprint arXiv:1905.00076*, 2019.
21. Y. Netzer, T. Wang, A. Coates, A. Bissacco, B. Wu, and A. Y. Ng. Reading digits in natural images with unsupervised feature learning. In *NIPS Workshop on Deep Learning and Unsupervised Feature Learning 2011*, 2011.
22. J. Redmon, S. Divvala, R. Girshick, and A. Farhadi. You only look once: Unified, real-time object detection. In *Proceedings of the IEEE Conference on Computer Vision and Pattern Recognition (CVPR)*, 2016.
23. J. Redmon and A. Farhadi. Yolov3: An incremental improvement. *arXiv preprint arXiv:1804.02767*, 2018.
24. S. Ren, K. He, R. Girshick, and J. Sun. Faster r-cnn: Towards real-time object detection with region proposal networks. In *Advances in neural information processing systems*, pages 91–99, 2015.
25. A. Schwaiger, P. Sinhamahapatra, J. Gansloser, and K. Roscher. Is Uncertainty Quantification in Deep Learning Sufficient for Out-of-Distribution Detection? In *Proceedings of the Workshop on Artificial Intelligence Safety AISafety@IJCAI'20*, volume 2640 of *CEUR Workshop Proceedings*, page 8, 2020.

26. M. Sensoy, L. Kaplan, and M. Kandemir. Evidential Deep Learning to Quantify Classification Uncertainty. In *Advances in Neural Information Processing Systems 31*, pages 3179–3189. Curran Associates, Inc., 2018.
27. O. Willers, S. Sudholt, S. Raafatnia, and S. Abrecht. Safety concerns and mitigation approaches regarding the use of deep learning in safety-critical perception tasks. In *Computer Safety, Reliability, and Security. SAFECOMP 2020 Workshops*. Springer International Publishing, 2020.
28. F. Yu, H. Chen, X. Wang, W. Xian, Y. Chen, F. Liu, V. Madhavan, and T. Darrell. Bdd100k: A diverse driving dataset for heterogeneous multitask learning. In *IEEE/CVF Conference on Computer Vision and Pattern Recognition (CVPR)*, June 2020.
29. Y. Zhou and O. Tuzel. Voxelnet: End-to-end learning for point cloud based 3d object detection. In *Proceedings of the IEEE Conference on Computer Vision and Pattern Recognition (CVPR)*, 2018.

Chapter 10
Data-driven Wireless Positioning

Maximilian Stahlke, Tobias Feigl, Sebastian Kram, Jonathan Ott, Jochen Seitz, Christopher Mutschler

Abstract Radio-based indoor localization is a crucial enabler for various tasks like robot navigation or quality assurance in industrial assembly. However, especially industrial environments often present challenging propagation conditions, with reflections, diffraction obstruction, and blockage. While traditional, lateration-based positioning algorithms can provide high accuracies in line-of-sight (LOS) conditions, signal blockages cause non-line-of-sight (NLOS) propagation and degrade the performance dramatically. To overcome this problems, recently artificial intelligence (AI)-driven localization algorithms have shown promising results and can provide robust and high localization performance in such challenging environments. In this chapter, we evaluate the performance of AI-models trained on radio fingerprints in various challenging industrial environments. We evaluate the accuracy and the effect of environmental changes on the localization performance and robustness. Our results show that environmental changes significantly degrade the performance, which leads to a high effort in maintenance, i.e., keeping the models up-to-date. We show how to combine classical lateration-based algorithms with data-driven models employing uncertainty estimation to reduce the the effort for initial deployment and maintenance for a more robust localization solution.

Key words: indoor localization, lateration-based positioning, AI-driven localization, hybrid localization.

Fraunhofer Institute for Integrated Circuits IIS, Fraunhofer IIS, Nuremberg, Germany

Corresponding author: Maximilian Stahlke
e-mail: maximilian.stahlke@iis.fraunhofer.de

© The Author(s) 2024
C. Mutschler et al. (eds.), *Unlocking Artificial Intelligence*,
https://doi.org/10.1007/978-3-031-64832-8_10

199

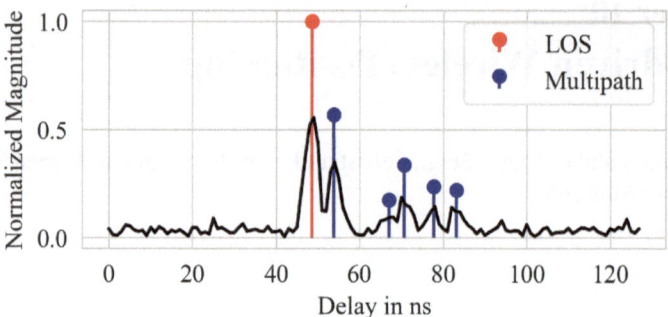

Fig. 10.1: Simulated CIR with 250 MHz bandwidth limitation. The LOS and multipath components are depicted as stems.

10.1 Introduction

Indoor positioning systems achieve reliable localization in industrial environments and enable various tasks such as process monitoring or the navigation of autonomous robots. There are methods based on cameras [13, 12], a combination of inertial sensors and cameras [14], or other modalities like LiDAR [23] and ultrasonic [23], which enable high localization accuracies. However, these methods often lack robustness due to changing light conditions, to dynamics of industrial environments, or interferences in the measured data [22, 15, 3, 7]. Therefore, radio-based localization methods are often employed providing high localization accuracies in such environments. Traditionally, methods such as lateration based on the time-difference-of-arrival (TDOA) are used to estimate positions. While these algorithms achieve localization accuracies in the decimeter range in LOS scenarios, NLOS between the receiver and the transmitter – the access point (AP) – significantly lowers the positioning accuracy. However, in realistic indoor scenarios, often a limited amount of APs is available, which often leads to areas where the minimum number of three APs in LOS cannot be ensured. To enable high localization accuracy in such NLOS dominated areas, instead of classical TDOAs, channel impulse responses (CIRs) can be used as an information source for positioning. These signals are complex time-series that are acquired through correlation with known pseudo-random pulses. They resolve all propagation components and therefore can contain a variety of spatial information.

Fig. 10.1 shows a bandwidth-limited CIR that includes a LOS component (red) as well as five additional signal components (blue). To exploit all the information contained in the high-dimensional CIR, machine learning (ML)-based methods can be used to assist radio localization systems, i.e., NLOS identification [20, 19], time-of-arrival (TOA) error mitigation [5], or multipath-assisted localization [9]. While [5, 9] consider all signal components (red and blue stems in Figure 10.1), the methods proposed by [20, 19] improve the positioning by enhancing LOS information (red stem in Figure 10.1). However, those methods still require few APs in LOS to achieve reliable positioning. To enable localization in NLOS-dominated areas, fingerprint-based localization methods utilizing channel information, such as the bandwidth-limited CIR (black line in Figure 10.1), can be employed [11, 17, 21]. These, so-called direct positioning methods, are independent to the radio propagation conditions and thus do not require a redundant deployment of APs to ensure LOS conditions. However, their main drawback is that they need an expensive life cycle management due environmental changes, including recording and labeling of data, fine-tuning and deployment of positioning models.

In this work, we briefly review AI-assisted localization methods and focus on investigations of the robustness of AI-based direct positioning localization models at different environmental changes, i.e., moving persons or objects or (non-)deterministic LOS blockage and discuss the impact on the life cycle management. Finally, we propose an approach to combine AI models trained on radio fingerprints with traditional localization methods in order to lower the efforts for maintenance and deployment.

10.2 AI-Assisted Localization

AI-assisted localization methods combine classical localization methods, like lateration based methods, with AI models. Their idea is to mitigate errors introduced by the environment like NLOS conditions or exploit additional information, which can not be modeled analytically, like multipath information.

A promising method is to identify NLOS signals and exclude them for localization. In our previous work, we investigated different supervised and unsupervised ML approaches. We used convolutional neural networks (CNNs), which employ the CIR to identify whether a signal has LOS or NLOS [19]. Given CIRs with labels (LOS, NLOS), CNNs can effectively model the temporal correlations in the high dimensional CIR to classify the channel conditions. However, labeling NLOS signals is often challenging in real world environments. While LOS signals are similar to LOS in visible light conditions, radio signals can propagate through various materials, like cardboards or wood and thus beeing in LOS for radio signals, but not for visible light. Thus labeling NLOS signals is often challenging or not possible in certain types of environments. To overcome this problem, we proposed to only use LOS signals for a classification by modeling the distribution of LOS CIRs and identify NLOS signals as out-of-distribution samples [20]. We used a variational autoencoder to model the distribution of in-domain samples, i.e., LOS CIRs, to identify NLOS signals by its lower data likelihood. While we achieve lower accuracies compared to supervised methods, we overcome the problem for labeling NLOS signals.

Another challenge is dense multipath propagation. Due to the limited bandwidth of radio signals, multipath components (MPCs), i.e., reflections from the environment, may not reliable separated in the radio channel. Thus, also under LOS conditions the TOA measurements may be erroneous due to interference of MPCs. Also here AI methods can be used to estimate reliable TOA estimations. We employed a CNN to estimate the TOA in a provided CIR given the true TOA [5]. We have shown that our method works even under dense multipath propagation with restricted bandwidth. Thus we could successfully mitigate the errors of interference and provide reliable TOA measurements for classical lateration based methods. However, the radio channel does not only include the first direct path of arrival (FDPOA), but also all the reflections of the environment. This information can be used by multipath-assisted positioning [6]. Their idea is to learn virtual anchors, represented by reflections of e.g. walls, and use them as additional APs. A crucial component in the learning process is a reliable MPC delay estimation. We employed a U-Net [16], well known for time series segmentation, to estimate the delays of MPC components outperforming classical sub-space methods in accuracy and inference time [9].

10.3 Direct Positioning

AI-assisted localization methods still require APs with LOS conditions to enable localization. For very complex, NLOS dominated, environments, only methods that exploit the radio fingerprint of a radio system estimate very accurate positions. Stahlke et al. [17] have shown that exploiting the channel impulse response (CIR) [2] with the corresponding time-of-arrival (TOA) can achieve high localization accuracy in complex NLOS environments en par with the performance of traditional tracking solutions under ideal LOS conditions. The idea of the data-driven approach is to train a deep learning (DL) model with a labeled dataset of channel impulse responses (CIRs), TDOAs, and the corresponding reference positions. The trained model can then, independent of the propagation conditions of the environment, predict the position utilizing channel information. However, the main disadvantages of the approach are (1) that the training database must cover fine-grained radio fingerprints of the area of interest within the environment to enable accurate localization, and (2) that the radio fingerprint, and thus the database, depend on the specific propagation environment. Hence, changes in the environment, e.g., movements of blocking objects such as cars or shelves, significantly change the fingerprint of the radio device, and consequently, lead to significant localization errors. In this work, we evaluate the performance of data-driven fingerprint-based positioning in various industrial-like environments and the effect of environmental changes on the performance of the models.

10.3.1 Model

We follow the idea of Stahlke et al. [17] to create a DL model, which exploits the raw CIR in the time-domain and the corresponding TDOA, for direct positioning. An example visualization is shown in Figure 10.2. For every burst (set of received

CIRs per timestep), for the calculation of the TDOA, the first (earliest) TOA is used. The CIRs, shown in blue on the left hand side, of all APs are horizontally stacked, while every CIR is padded by the TDOA to get a relative alignment of the signals in the input tensor of the neural network. The idea is to extract and exploit all (spatial) information of a CIR, i.e., diffractions, reflections, and absorptions, to get a unique fingerprint for every position in an environment.

Inspired by the work of Niitsoo et al. [11] we also employ CNNs as they have shown very accurate results in processing image and time-series data. However, Fawaz et al. [4] have shown, that very simple CNN architectures without local pooling can outperform highly complex models with local pooling layers in time-series applications. Hence, in contrast to the CNN of Niitsoo et al. [11], we employ a simple CNN architecture without local pooling to exploit the full information of the time-series data.

10.3.2 Experimental Setup

We used a 5G-compatible software-defined radio system with commercial of the shelf hardware. The system is a downlink TDOA system with 6 APs within an industrial hall. The system has a bandwidth of 100 MHz with a carrier frequency of 3.75 GHz (Lower bandlimit: 3.7 MHz; Upper bandlimit: 3.8 MHz). The TX Power of the APs is 20 dBm. The receiver records bursts (consecutive signals for every AP) at 100 Hz.

The hardware setup of the receiver and the distribution of the APs within the L.I.N.K. application center of Fraunhofer IIS in Nuremberg [1] are shown in Fig-

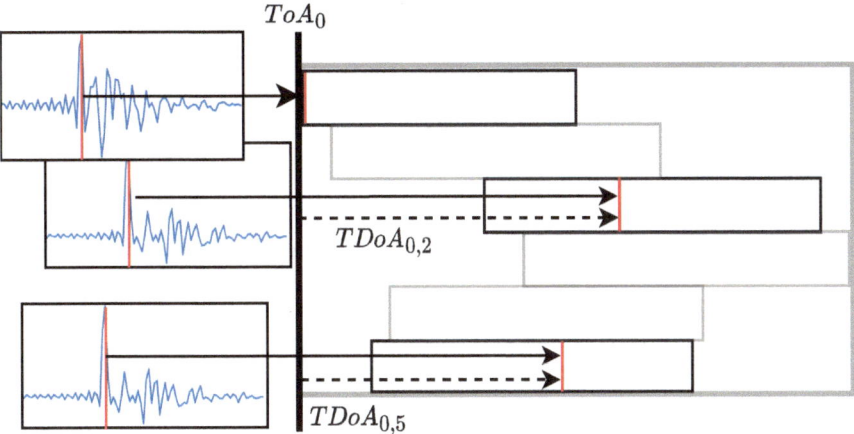

Fig. 10.2: Input embedding for the neural network. The CIRs, shown on the left hand side in blue, are aligned by their TDOA and horizontally stacked.

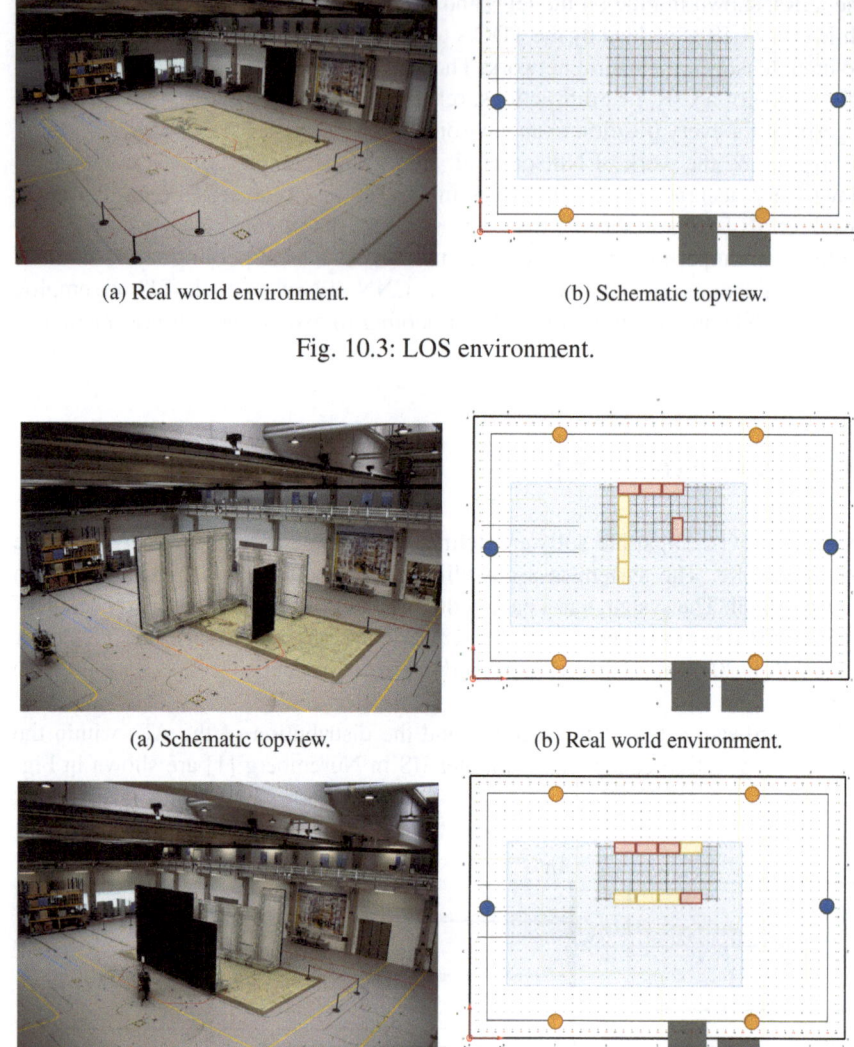

(a) Real world environment.   (b) Schematic topview.

Fig. 10.3: LOS environment.

(a) Schematic topview.   (b) Real world environment.

(c) Schematic topview.   (d) Real world environment.

Fig. 10.4: Deterministic mixed environments with absorber L-shaped (DS1) on the top and absorber corridor (DS2) on the bottom.

ure 10.3. In total 6 APs are available, indicated as dots within the schematic top view on the left-hand side. The blue APs are placed at a height of 6.5 m, while the orange APs are placed at a height of 7.5 m.

 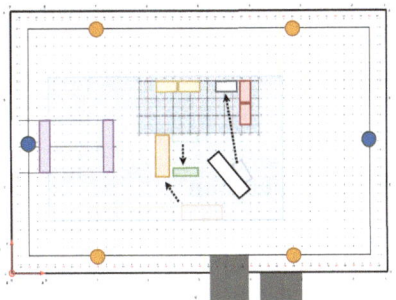

(a) Real world environment.   (b) Schematic topview.

Fig. 10.5: Realistic industrial environment with various industrial objects.

10.3.2.1 Measurement Campaign

In order to demonstrate the accuracy, robustness, and transferability of the proposed data-driven approach an extensive measurement campaign was conducted including three different types of environments. In general, all recordings took part in the L.I.N.K. hall, shown in Figure 10.3 on the left-hand side. For all environments the recording area (28 m × 18 m) remained the same. For each environment, we recorded 1 - 1.5 hours of data covering several people on random trajectories to mimic a realistic data acquisition that still covers the area of interest. With a recording frequency of 100 Hz, we therefore achieve a database size of 2-3 million CIRs per environment with 6 APs. As ground truth we used a Nikon iGPS, an optical reference system with a positioning error of $MAE < 1$ mm.

10.3.2.2 Environments

In total, six different environments were recorded to investigate the effects of various environmental changes. Figure 10.3 shows the LOS environment, while the image on the left-hand side shows the real environment and the figure on the right-hand side shows a schematic top view. It contains no obstacles, which means that all APs have LOS to the receiver rendering it ideally for traditional TDOA-based localization systems. In Figure 10.4, the environments (DS1) and (DS2) are shown. We added walls, which absorb the signals on the outside (black) and reflect them on the inside (metal). In this environments we created severe NLOS conditions blocking the LOS to almost the half of the APs all the time. Due to the different composition of the absorber walls, the fingerprints are different at almost every position rendering a good candidate for a cross environment evaluation.

The last environments are shown in Figure 10.5. We created a typical industrial environment (RS1) with various objects, like a forklift, a van, metal shelves, and a working platform. We introduced realistic changes (RS2), where we moved the van, added an additional work platform, and moved the forklift (3C). Those changes should reflect realistic changes of a dynamic industrial environment. More details about the environments and its changes can be found in [17].

10.3.3 Evaluation

We trained our models in all of the environments and conducted a cross validation, i.e., tested them in the same and in the other environments. The results can be seen in Table 10.1, i.e. mean absolute error (MAE) on the left and the 90th percentile of the cumulative distribution function of the absolute error (CE_{90}) on the right. The models are trained and tested in all environments to see the impact of the environmental change. The diagonal elements show the results, where the model is trained and tested in the same environment, while the off-diagonal elements show the results, where the models are trained and tested in different environments. In general, the performance is high for all environments, with a $CE_{90} < 0.7$ m for all scenarios. The positioning accuracy is the lowest in the LOS environment compared to the other mixed environments. Under LOS conditions the CIR contains less information about the environment, as only few multipath components are present (e.g., floor or ceiling) compared to mixed environments with dense multipath like in the corridor environment (DS2). However, in the other environments the positioning accuracy is very high. The performance degrades for the model in a different composition of the environment, as the radio fingerprint changes due to different multipath propagation and changes of the channel state (i.e., LOS to NLOS and vice versa). If the model is trained in the LOS environment and tested in a mixed scenario, the performance degrades with a $CE_{90} > 2$ m for all environments. This is due to the different propagation conditions for the area. Dense multipath and different channel states (LOS / NLOS) cause changed radio fingerprints, and therefore, lead to a degrading positioning accuracy. If the model is trained in the initial industrial environment (RS1) and evaluated in the changed industrial scenario (RS2), the positioning performance degrades only slightly. This is due to the fact that the environment only changed slightly, which leads to only local anomalies.

Table 10.1: Position accuracy and robustness of our model against various environmental changes ($MAE \mid CE_{90}$ in [m]).

		Evaluated								
		LOS		DS1		DS2		RS1		RS2	
Trained	LOS	0.41	0.68	1.24	2.50	1.71	3.90	1.42	3.00	1.40	2.87
	DS1	1.23	2.22	0.34	0.61	1.54	2.75	2.01	4.00	1.99	3.90
	DS2	1.18	2.16	1.32	2.65	0.26	0.49	1.84	3.59	1.84	3.76
	RS1	0.83	1.59	1.50	2.78	1.21	2.17	0.32	0.58	0.49	0.92
	RS2	1.05	1.89	1.50	2.93	1.20	2.21	0.48	0.92	0.35	0.61

10.3.4 Hybrid Localization

As shown in Section 10.3.3, data-driven models are affected by environmental changes and need a life cycle management to ensure reliable localization. The effort increases with the size of the environment, as the new data has to be recorded and labeled for all changed areas. However, indoor environments are often heterogeneous, which means that there are sections which are dominated by NLOS, while some areas still have enough LOS to enable traditional positioning.

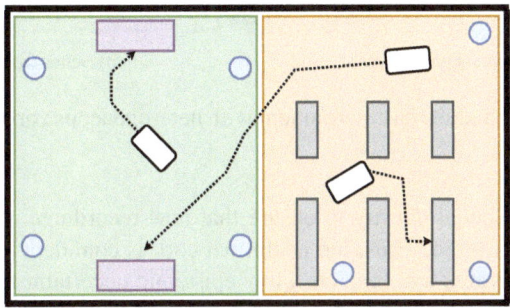

Fig. 10.6: Schematic top view of a typical (dynamic) industrial environment with an ultra-wideband localization system.

Figure 10.6 represents a typical industrial environment, with shelves with goods (grey rectangles) and machines (purple rectangles). An ultra-wideband (UWB) radio system is deployed, shown as blue circles, to localize automated guided vehicles (AGVs), indicated as white rectangles. The area is separated in two sections with different complexities for localization. For the localization in the left area, a machine hall (green), traditional localization approaches can be employed as there is always LOS to at least four transmitters. In contrast, the area on the right-hand side, the LOS is blocked to the majority of the transmitters. To still achieve high localization accuracies, data-driven models have to be deployed, with the overhead for an expensive life cycle management. To enable a cost-efficient, continuous, and robust localization, a combination of AI based and traditional localization methods is needed, which requires the identification of the spatial limitation of the data-driven positioning model.

10.3.5 Zone Identification

To identify the spatial limitation of our data-driven model, we employ the method proposed by Stahlke et al. [18]. Their idea is to model the epistemic uncertainty [8] of the AI model to identify out-of-distribution samples. In fingerprint-based localization, we assume that the received signals are unique for a certain position, as the received signals only depend on the environment. We record the data in the training

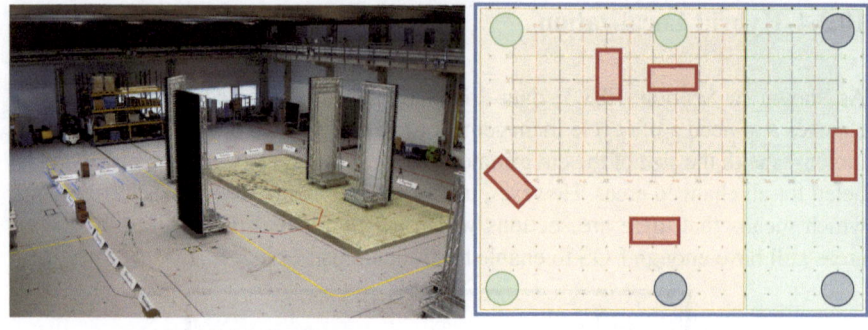

(a) Real world environment.   (b) Schematic topview.

Fig. 10.7: Realistic industrial environment with heterogeneous zones for localization.

area with a high spatial density to ensure that new recordings within the training area are already covered. Thus, the neural network is confident about unseen data within the training area and predicts a low epistemic uncertainty. In contrast, samples recorded outside the training area are different to all recorded samples within the training area. This leads to a high epistemic uncertainty of the neural network, which helps us to identify the spatial limitation of the model. We employed the probabilistic ensemble [10], with a neural network architecture very similar to the one used in Section 10.3.1. More details about the implementation and architecture can be read in [18]. To identify samples, which are outside of the training area, a threshold has to be defined for the uncertainty to detect samples within and outside the training data distribution. Methods like interquartile ranges could be used to detect out-of-distribution samples. However, they might require some fine-tuning. To ensure a reliable threshold, we used a logistic regression classifier, which uses few samples out-of and inside-of the training area.

10.3.6 Experimental Setup

For the evaluation, a UWB radio system with six stationary transceivers is employed with one dynamic transceiver carried by a small robot platform. The system is configured to estimate the round-trip-time (RTT) at the robot platform with a bandwidth of 499.2 MHz at a center frequency of 4 GHz. The recorded data is labeled with an optical ground truth reference system with a recording frequency of 4 Hz.

10.3.7 Environments

Also in this experiment, we created an industrial-like environment including walls, which absorb radio signals on the outside (black) and reflect them on the inner side. The environment is shown in Figure 10.7, where the real-world environment is on the left-hand side and the schematic top view is on the right-hand side. The stationary transceivers are indicated as dots and are placed on the upper and lower part of the recording area. The reflective walls are indicated in red and are placed to block the LOS between the stationary transceivers and the moving robot, which causes ranging errors and thus also to a degradation in the localization performance with traditional positioning approaches. The environment is separated into two areas: The left area (orange), is very cluttered and mostly dominated from NLOS to the transceivers, while the right area (green) is more open and provides always LOS to the transceivers indicated in grey.

10.3.8 Evaluation

The uncertainty, estimated by our probabilistic ensemble, can be seen in Figure 10.8. The dashed black line separates the training area (left) and the unseen area (right). It can clearly be seen that the uncertainty identifies the spatial limitations of the direct

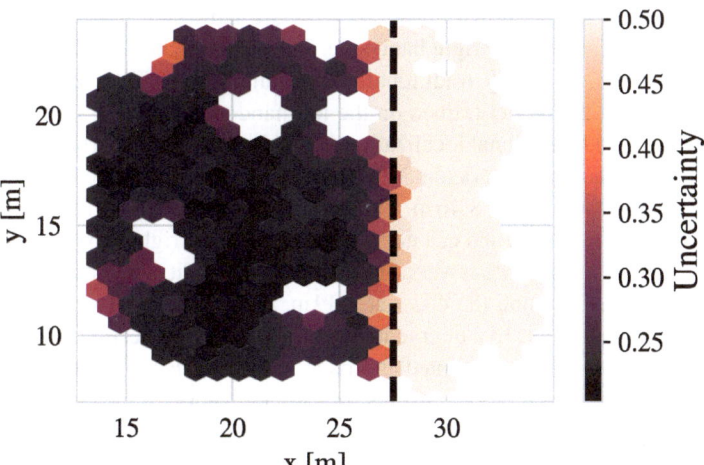

Fig. 10.8: Spatial distribution of the uncertainty of our probabilistic ensemble. The dashed black line indicates the transition from the training area (left) to the unseen area (right).

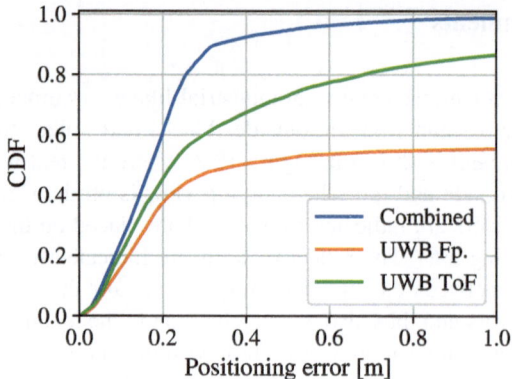

Fig. 10.9: CDFs of the errors for the fingerprint based model (UWB Fp.), the classical localization model (UWB ToF) and our proposed combined model (Combined).

positioning model very well. The data in the training area has a low uncertainty (black) compared to the samples out of the training data area (red). The decision boundary is very sharp, which means that an identification is feasible with the proposed approach.

To combine traditional localization methods with data-driven, fingerprint-based, methods, we use a Kalman filter. While the uncertainty is low, the positions of the model are fed to the Kalman filter. However, if the uncertainty rises above an uncertainty threshold, the distance estimations of only the three transceivers on the right hand side (grey), shown in Figure 10.7, are fed to the tracking filter as they have LOS in the area on the right hand side. The errors of the localization are shown in Figure 10.9, using only traditional positioning (UWB ToF), only data-driven positioning with the model trained on the left hand side (UWB Fp.) and the fusion of data-driven and traditional localization (Combined). It can clearly be seen that the data-driven method only works in 50% of the cases. Thus, the error is up to a MAE of of 2.72 m and a CE_{90} of 8.46 m. The model is only trained on the left hand side, see Figure 10.7 in orange, and can not extrapolate into the right hand side. Hence, the error on the right hand side is very high. If only the traditional positioning is used, the MAE is at 0.56 m and the CE_{90} at 1.39 m. NLOS causes errors in the distance estimations, which leads to a degradation of the positioning results. The best results can be achieved with the combined approach with a MAE of 0.21 m and a CE_{90} of 0.32 m. The data-driven model ensures robust localization in the cluttered left hand side area, while the traditional localization takes over on the right hand side. The evaluations have shown that a combination of both approaches is therefore useful to lower the effort for data recording and maintenance. To enable this, an explicit identification of the spatial limitation of the data-driven models is crucial to ensure reliable and robust positioning.

10.4 Conclusion and Outlook

In this work, we have successfully shown that AI can enhance the accuracy of radio localization systems in complex indoor environments. However, to enable a robust and reliable positioning an expensive life-cycle-management has to be employed. In an extensive ablation study, including 5 different scenarios, we evaluated that data-driven models trained on radio fingerprints cannot generalize to different environments and various environmental changes. While indoor environments often provide heterogeneous areas, with sections fully dominated by NLOS and more open areas with sufficient numbers of LOS connects. A combination of classical localization methods and data-driven methods is therefore useful to lower the expenses for maintenance and still achieve overall high localization accuracy.

We have shown how to combine traditional localization and data-driven localization based on radio fingerprints by uncertainty estimation. This allows us to identify the spatial limitations of the data-driven methods, which allows us to only employ the data-driven models in NLOS dominated areas, while classical positioning can be used elsewhere. This lowers the risk of out-dated radio fingerprints and the expenses for labeling and maintenance.

One potential avenue for future research involves the mitigation of the high cost associated with labeling extensive datasets. Addressing this challenge may entail exploring unsupervised or self-supervised learning methodologies, which hold promise in minimizing the dependency on labeled data points. Additionally, an ongoing inquiry pertains to the integration of fundamental physical principles, such as Maxwell's equations, into data-driven methodologies. Incorporating such knowledge has the potential to enhance the generalization abilities of these methodologies while concurrently diminishing the requisite volume of data.

Acknowledgements

We thank Maximilian Kasparek, Andreas Eidloth, Jan Niklas Bauer and Mohamed Soliman for implementing the proof-of-concept 5G uplink-TDOA positioning setup and its software-defined-radio-based processing pipeline.

References

1. L.I.N.K. application center of Fraunhofer IIS. https://www.iis.fraunhofer.de/de/profil/standorte/linkhalle.html.
2. T. Altstidl, S. Kram, O. Hermann, M. Stahlke, T. Feigl, and C. Mutschler. Accuracy-Aware Compression of Channel Impulse Responses Using Deep Learning. In *IEEE Intl. Conf. on Indoor Positioning and Indoor Navigation (IPIN)*, Lloret de Mar, Spain, Nov. 2021.
3. T. Brieger, N. L. Raichur, D. Jdidi, F. Ott, T. Feigl, J. R. van der Merwe, A. Rügamer, and W. Felber. Multimodal Learning for Reliable Interference Classification in GNSS Signals. In

Proc. of the Intl. Technical Meeting of the Satellite Division of the Institute of Navigation (ION GNSS+), pages 3210–3234, Denver, CO, Sept. 2022.

4. H. I. Fawaz, G. Forestier, J. Weber, L. Idoumghar, and P. A. Muller. "Deep learning for time series classification: a review". Data Min. Knowl. Discov., 33(4):917–963, 2019.

5. T. Feigl, E. Eberlein, S. Kram, and C. Mutschler. Robust ToA-Estimation Using Convolutional Neural Networks on Randomized Channel Models. In IEEE Intl. Conf. on Indoor Positioning and Indoor Navigation (IPIN), Lloret de Mar, Spain, Nov. 2021.

6. C. Gentner, T. Jost, W. Wang, S. Zhang, A. Dammann, and U.-C. Fiebig. Multipath assisted positioning with simultaneous localization and mapping. IEEE Transactions on Wireless Communications, 15(9):6104–6117, 2016.

7. D. Jdidi, T. Brieger, T. Feigl, D. C. Franco, J. R. van der Merwe, A. Rügamer, J. Seitz, and W. Felber. Unsupervised Disentanglement for Post-Identification of GNSS Interference in the Wild. In Proc. of the Intl. Technical Meeting of the Satellite Division of the Institute of Navigation (ION GNSS+), pages 1176–1208, Denver, CO, Sept. 2022.

8. A. Klaß, S. M. Lorenz, M. W. Lauer-Schmaltz, D. Rügamer, B. Bischl, C. Mutschler, and F. Ott. Uncertainty-aware Evaluation of Time-Series Classification for Online Handwriting Recognition with Domain Shift. In IJCAI-ECAI Intl. Workshop on Spatio-Temporal Reasoning and Learning (STRL), volume 3190, Vienna, Austria, July 2022.

9. S. Kram, C. Kraus, M. Stahlke, T. Feigl, J. Thielecke, and C. Mutschler. Delay Estimation in Dense Multipath Environments Using Time Series Segmentation. In IEEE Wireless Communications and Networking Conference (WCNC), Apr. 2022.

10. B. Lakshminarayanan, A. Pritzel, and C. Blundell. Simple and scalable predictive uncertainty estimation using deep ensembles. In Advances in neural information processing systems, volume 30, Long Beach, California, 2017.

11. A. Niitsoo, T. Edelhäußer, E. Eberlein, N. Hadaschik, and C. Mutschler. "A Deep Learning Approach to Position Estimation from Channel Impulse Responses". Sensors, 19(5):1064–1091, 2019.

12. F. Ott, T. Feigl, C. Löffler, and C. Mutschler. ViPR: Visual-Odometry-aided Pose Regression for 6DoF Camera Localization. In Proc. of the IEEE/CVF Intl. Conf. on Computer Vision and Pattern Recognition Workshops (CVPRW), pages 187–198, Seattle, WA, June 2020.

13. F. Ott, L. Heublein, D. Rügamer, B. Bischl, and C. Mutschler. Fusing Structure from Motion and Simulation-Augmented Pose Regression from Optical Flow for Challenging Indoor Environments. In arXiv preprint arXiv:2304.07250 [cs.CV], Apr. 2023.

14. F. Ott, N. L. Raichur, D. Rügamer, T. Feigl, H. Neumann, B. Bischl, and C. Mutschler. Benchmarking Visual-Inertial Deep Multimodal Fusion for Relative Pose Regression and Odometry-aided Absolute Pose Regression. In arXiv preprint arXiv:2208.00919 [cs.CV], Aug. 2022.

15. N. L. Raichur, T. Brieger, D. Jdidi, T. Feigl, J. R. van der Merwe, B. Ghimire, F. Ott, A. Rügamer, and W. Felber. Machine Learning-assisted GNSS Interference Monitoring Through Crowdsourcing. In Proc. of the Intl. Technical Meeting of the Satellite Division of the Institute of Navigation (ION GNSS+), pages 1151–1175, Denver, CO, Sept. 2022.

16. O. Ronneberger, P. Fischer, and T. Brox. U-net: Convolutional networks for biomedical image segmentation. In Medical Image Computing and Computer-Assisted Intervention–MICCAI 2015: 18th International Conference, Munich, Germany, October 5-9, 2015, Proceedings, Part III 18, pages 234–241. Springer, 2015.

17. M. Stahlke, T. Feigl, M. H. C. García, R. A. Stirling-Gallacher, J. Seitz, and C. Mutschler. Transfer learning to adapt 5g ai-based fingerprint localization across environments. In 2022 IEEE 95th Vehicular Technology Conference: (VTC2022-Spring), pages 1–5, 2022.

18. M. Stahlke, T. Feigl, S. Kram, B. M. Eskofier, and C. Mutschler. Uncertainty-based fingerprinting model selection for radio localization. In 2023 IEEE 13th International Conference on Indoor Positioning and Indoor Navigation (IPIN), pages 1–5, 2023.

19. M. Stahlke, S. Kram, C. Mutschler, and T. Mahr. Nlos detection using uwb channel impulse responses and convolutional neural networks. In 2020 International Conference on Localization and GNSS (ICL-GNSS), pages 1–6, 2020.

20. M. Stahlke, S. Kram, F. Ott, T. Feigl, and C. Mutschler. Estimating ToA Reliability with Variational Autoencoders. In *IEEE Sensors Journal*, volume 22(6), pages 5133–5140, Mar. 2022.
21. M. Stahlke, G. Yammine, T. Feigl, B. M. Eskofier, and C. Mutschler. Indoor Localization With Robust Global Channel Charting: A Time-Distance-Based Approach. In *IEEE Trans. on Machine Learning in Communications and Networking*, volume 1, pages 3–17, Mar. 2023.
22. J. R. van der Merwe, D. C. Franco, J. Hansen, T. Brieger, T. Feigl, F. Ott, D. Jdidi, A. Rügamer, and W. Felber. Low-Cost COTS GNSS Interference Monitoring, Detection, and Classification System. In *MDPI Sensors*, volume 23(7), 3452, Mar. 2023.
23. R. W. Wolcott and R. M. Eustice. Fast lidar localization using multiresolution gaussian mixture maps. In *2015 IEEE international conference on robotics and automation (ICRA)*, pages 2814–2821. IEEE, 2015.

Chapter 11
Comprehensible AI for Multimodal State Detection

Andreas Foltyn, Maximilian P. Oppelt

Abstract Affective computing enables computers to recognize and respond to human emotions and cognitive states using multimodal state recognition. This chapter explores cognitive load estimation through a machine learning life cycle, addressing data collection and preparation, modeling, and deployment challenges. Experiments demonstrate effective unimodal and fusion models for predicting cognitive load. Applications in gaming, healthcare, and driver monitoring have been discussed, providing valuable insights for researchers in this field.

Key words: affective computing, cognitive load, robustness, multimodal, fusion, biosignals, eye tracking

11.1 Introduction

Affective computing refers to the ability of computers to recognize, interpret, process, and respond to human emotions and cognitive states. It is an interdisciplinary field that combines computer science, psychology and neuroscience, and aims to automatically recognize and respond to human emotions and behavior patterns. An important component of affective computing is multimodal state recognition. This technology uses a variety of data sources, such as physiological measurements, facial expressions, and behavioral patterns, to determine a person's emotional and cognitive state. The use of multimodal state recognition has the potential to improve human-machine interaction and better understand human needs and preferences [10, 13].

These cognitive state detection systems can be utilized in various domains. One example is gaming where facial recognition and speech analysis can be used to de-

Fraunhofer Institute for Integrated Circuits IIS, Fraunhofer IIS, Erlangen, Germany

Corresponding author: Andreas Foltyn
e-mail: andreas.foltyn@iis.fraunhofer.de

tect the emotional state of players and adjust the game accordingly. If a player is stressed or frustrated during a game, the game can be automatically adjusted to calm the player down or help him overcome his frustration [3]. For healthcare applications the detection of stress helps during treatment and understanding of associated pathologies [1, 17]. Moreover, in the automotive industry, affective computing and multimodal state recognition are used in driver monitoring, e.g., study by Oppelt et al. [10] investigated the use of multimodal state recognition to detect cognitive load in drivers observing semi-autonomously driving vehicles.

Overall, affective computing and multimodal state recognition have the potential to improve human daily life and to optimize human-machine interaction. It is important though to ensure that the technology is used ethically and responsibly and that user privacy is preserved.

In this chapter, we aim to introduce a typical process for creating applications that predict cognitive load. Firstly, we discuss the study design, which involves collecting data from multiple sources, as well as a framework that can be used to systematically capture relevant high quality data. We show results from experiments of using models to predict cognitive load in well-established psychological cognitive load tests as well as in a laboratory real-world-motivated setting. By following this process, we hope to enable other researchers to create their own affective computing application.

11.1.1 Cognitive Load Estimation

Within the domain of affective computing and multimodal state detection, various psychological states and concepts can be detected. These include emotion recognition, stress detection, arousal, or valence detection, and cognitive load. This chapter however, focuses specifically on detecting cognitive load.

Cognitive load is a multifaceted concept with various definitions in the literature [8]. One common definition describes cognitive load as the amount of mental effort required by a learner to complete a task [12]. Factors such as task demand and subjective characteristics can contribute to cognitive load, and observable variables are used to assess it since it is not directly measurable.

To measure cognitive load, variables such as mental load, mental effort, and performance can be grouped into four categories of cognitive load measurement methods: subjective measures, performance measures, physiological measures, and behavioral measures [3]. Subjective measures involve self-reporting, such as questionnaires or rating scales, to assess cognitive load. Performance measures examine how well a task is completed, and the time taken to do so. Physiological measures include the activity of the heart measured through electrocardiographic recordings, eye tracking technology to measure pupil dilation or saccades and action units extracted from expressions using facial videos.

Since human psychology and emotional state are complex and multidimensional constructs, decision systems need to integrate multiple modalities to make accurate

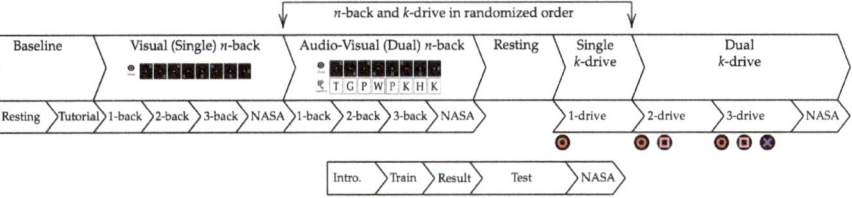

Fig. 11.1: Overview of the data collection setup.

predictions. By combining subjective, performance, physiological, and behavioral measures, decision systems can better understand and predict human behavior.

11.1.2 Challenges in Affective Computing

Building robust machine learning models in affective computing is not without its challenges. There are three key factors that pose unique challenges: The human factor, data collection, and modeling. Affective computing involves detecting human states, which can be expressed in many ways, such as behavior, facial expressions, physiological signals, or tone of voice. Unfortunately, the expression of these states is often highly subject-specific, depending on the task at hand. For example, facial expressions can be used for emotion recognition, but the expression is often person-dependent, e.g., some people might change their facial expressions to hide their inner state. Also, different internal states and factors can lead to the same expression in certain modalities, e.g., an increased heart rate could be due to stress or increased movement. Consequently, it is helpful to use several modalities in order to be able to make a more reliable statement. In our example, the two cases can be distinguished by measuring the increased heart rate by adding a motion sensor. Hence, there is a fundamental uncertainty in detecting human states, as we can only measure them using proxies.

Data collection poses a number of challenges. In this field, we work with people and their personal data. It is often necessary to comply with the legal framework as to what kind of data can be collected and how it can be used. Furthermore, the study design plays an important role in avoiding unwanted biases in the data. For example, the illumination of the face may be slightly different due to different stimuli on the monitor, such as pictures or games. This in turn can be seen in the video, leading to *spurious correlations*. In addition to the input data, the emotional or cognitive state annotation is also necessary to train models. However, obtaining the annotation is not trivial. First, we need to decide how to formulate the problem. It can be formulated as a binary, multi-class, multi-label or regression problem. Sometimes the problem can be formulated only as one type, but in other cases we need to make an informed decision. In addition, we have to decide whether a subjective or objective annotation should be used. In the case of subjective annotation, the participants of the study are asked which state they are in. However, the questionnaires are often not very reliable, because the person has to interpret and judge their state. For example, the more fine-grained the labeling, the harder it is to define their exact inner state. Alternatively, objective labels can be used. For example, you might assume that a particular stimulus induces a particular state and use the assignment to the stimuli as a label. However, this ignores the subjective perception of the stimuli.

Also in the modeling phase, we deal with various challenges. When models have been trained on the data, it should be evaluated which variables in particular are used for the decision. This may reveal biases that were not considered before. Examples could be skin color in emotion recognition, where certain ethnicities correlate with a skin type in the data set [5, 6]. It is also helpful to evaluate the robustness of models by testing them on different scenarios or domains. Often, because a dataset is randomly split, models are trained and tested on data that is fairly similar with respect to various factors. But if minor things change, such as the hardware of the sensors, the accuracy can decrease.

In summary, affective computing poses specific challenges that require careful consideration when building machine learning models. Addressing these challenges requires a deep understanding of the human factor, careful data collection, and modeling approaches that can generalize well across different data distributions.

11.2 Data Collection

In the following section, we present the data acquisition process of the data set *ADA*Base intended for the detection of cognitive overload [10, 18]. This unique multimodal data set encompasses a wide array of physiological measures, including electrocardiography (ECG), electrodermal activity (EDA), respiratory rate (RESP), photoplethysmography (PPG), video recordings, electromyography (EMG), eye-tracker (EYE) and skin temperature (TEMP). In total, the study involved 51 subjects who participated, resulting in an average recording duration of 155 minutes per subject. The primary objective of the study was to employ two distinct stimuli to deliberately induce cognitive overload.

The n-Back test is a standardised psychological test that places a heavy demand on a person's working memory. As shown in Figure 11.2, test subjects are shown a grid in which the position of a rectangle changes over time. The subject's task is to identify whether the current position of the rectangle matches the position n steps earlier. In the dual n-Back variant, the subject must simultaneously complete this memory task with an additional auditory signals. In the present study, both variants were performed with $n \in \{1, 2, 3\}$.

The k-Drive test is a semi-autonomous driving scenario in a simulator, as shown in Figure 11.3. While the car automatically drives a lap on the track, the participant has to react to certain events. The cognitive load was modulated by the number of

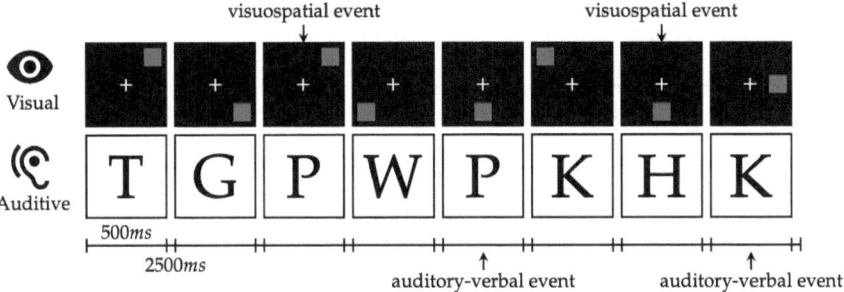

Fig. 11.2: Procedure for a Dual-n-Back test to induce cognitive overload. The aim is to respond when the current visual and auditory stimuli match those presented n steps earlier.

events and the additional task of creating a music playlist. The entire study design can be found in Figure 11.1.

11.2.1 Annotation

As already mentioned, the annotation for affective states can be implemented in a subjective or objective way. In this data set, both options were integrated. For a subjective annotation, a self-assessment with the NASA-TLX questionnaire was integrated. This is a standardised questionnaire that asks about various dimensions of stress. Furthermore, the performance and the allocation of the stimuli offer a possibility to implement objective annotations. Another possibility is to employ the various levels of our tasks as cognitive load classes and labels, categorizing them into two distinct groups based on both performance metrics and subjective feedback. The low cognitive load class encompasses baseline sessions and simpler task levels, while the high cognitive load class encompasses more complex task levels. This annotation strategy allows for a nuanced characterization of cognitive workload. For instance, in the n-back task, the 1-back level was categorized as low cognitive load (CL), reflecting its relatively simpler working memory demands. Conversely, the dual-task 3-back level was designated as high cognitive load, given its increased complexity requiring simultaneous management of driving tasks and higher working memory load. Similarly, in the k-drive scenario, baseline levels and simpler driving observations were classified as low CL, while more demanding tasks such as detecting high acceleration and deceleration events while controlling the music entertainment app in level 3 were classified as high cognitive load. Through carefully planning the experimentation schedule one should create a dataset that has a balanced class disbritubiton.

Table 11.1: This table shows the F1-Score and calibration for fusion methods based on XGBoost and FCN. All models are trained on *n*-Back.

Method	ECG	EDA	EMG	EYE	PPG	RESP	TEMP
XGBoost	0.68 ± 0.03	0.63 ± 0.05	0.56 ± 0.07	0.83 ± 0.04	0.64 ± 0.03	0.61 ± 0.03	0.68 ± 0.03
FCN	0.60 ± 0.09	0.59 ± 0.08	0.64 ± 0.03	0.85 ± 0.03	0.57 ± 0.06	0.62 ± 0.03	0.61 ± 0.16

11.2.2 Data Preprocessing

The recorded data is provided as time series, and we used a rolling window approach to extract segments from the data. This involved defining a specific window size and step size, allowing us to divide the entire signal into smaller segments. Additionally, modality-specific preprocessing steps were applied, such as outlier removal in the pupil size and detrending of the ECG signal. Subsequently, features were extracted from these segmented data points. For a detailed view of the exact preprocessing techniques and features used, please refer to [10].

Fig. 11.3: In the *k*-Drive test, participants react to specific events in an autonomous driving scene and complete a secondary task of playlist creation (taken from [10], licensed under CC-BY 4.0).

11.3 Modeling

In this section, we conduct a review of several experiments that have been carried out using the *ADA*Base data set. One focus is on understanding how the performance varies across different modalities. Additionally, we aim to identify significant differences, if any, between Deep Learning (DL) and classic Machine Learning (ML) approaches. Moreover, we investigate the potential benefits of fusion techniques in improving overall performance. Lastly, we carefully analyze how the models behave under a distribution shift, providing valuable insights into their robustness.

We discuss two models that were examined with the data in the following. XG-Boost [4] is a classic machine learning model utilizing manual extracted statistical and expert features. The Fully Convolutional Network (FCN) [20], on the other hand, is a 1D Convolutional Neural Network (CNN) that was introduced for processing time series data. This end-to-end network architecture does not rely on manual feature extraction and can be trained using the raw data as input. We also experimented with fusion methods. The first is late fusion, where the predictions of each modality are averaged. Early fusion, where the features of all modalities are concatenated, was also performed with XGBoost. In FCN, the features were linked in latent space before going to a final fully connected layer. All models were trained on the n-Back data with the objective annotation. For this purpose, the different difficulty levels of the test were clustered into two classes: low and high cognitive load.

11.3.1 In-Domain Evaluation

Table 11.1 shows the results of models trained and evaluated on n-Back. Across all modalities, the eye tracker achieves the best performance. There is no clear trend between the two methods as to which performs better. While the DL approach shows slightly better performance when used with the eye tracker, its performance with the other modalities is either worse or similar to XGBoost. A plausible explanation for this observation could be the difficulty in extracting generalizable features from some modalities, e.g., the ECG. Therefore, there is an opportunity to improve the representation learning using modality specific knowledge. This aspect is investigated in detail in Section 11.3.4 for the ECG. Table 11.3 displays the fusion results. For both methods, fusion does not lead to an improvement of the F1-score. Whether fusion can enhance the unimodal performance largely depends on the specific task and modalities involved. Previous studies on different domains and models have already demonstrated that incorporating more modalities does not necessarily lead to improved accuracy [19]. In addition, we report the calibration error to assess the reliability of the predictions. These scores are almost identical between the two models, but differ greatly between the fusion approaches. For early fusion it is significantly lower.

11.3.2 Cross-Domain Evaluation

As described in Section 11.1.2, it is important to evaluate the robustness of models [7]. In addition to the n-Back test, the k-Drive test was also carried out in the data collection. This is used here as an additional test set. Figure 11.2 shows models that were trained on n-Back but evaluated on k-Drive.

It is evident that the drift in domains leads to a change in model performance. The k-Drive scenario is characterized by increased movements and varying light

Table 11.2: This table shows the F1-Score for unimodal models trained and evaluated on n-Back.

Method	ECG	EDA	EMG	EYE	PPG	RESP	TEMP
XGBoost	0.83 ± 0.04	0.63 ± 0.03	0.65 ± 0.12	0.59 ± 0.03	0.70 ± 0.09	0.80 ± 0.06	0.77 ± 0.04
FCN	0.78 ± 0.07	0.67 ± 0.09	0.83 ± 0.04	0.73 ± 0.03	0.78 ± 0.06	0.76 ± 0.05	0.73 ± 0.19

conditions. Consequently, it is understandable that the F1-score of the eye tracker is lower compared to n-Back. However, we also observe that FCN exhibits relatively greater resilience to domain changes than XGBoost. One possible explanation for this could be that the expert features heavily aggregate information from the 60-second window, causing essential features that could aid in domain generalization to be lost. Interestingly, the other modalities show better performance than the in-distribution data. A potential reason for this could be that the altered task demands more activity from the participants. Participants move slightly more and may tend to engage in more movements during increased difficulty, as they need to switch their hands between two devices. This could explain corresponding changes in muscle activity, heart rate, skin temperature, and respiration.

Now, let us consider the question of achieving robust predictions across modalities. If we were to select a classifier solely based on the best unimodal in-distribution performance, it becomes apparent that we may encounter problems when the distribution changes. Even though the fusion approaches do not show improvement in the in-distribution data, it raises the question of whether we can achieve enhanced robustness by using fusion approaches. Table 11.3 presents the performance of the fusion models evaluated on the k-Drive data. We observe that all fusion approaches yield more robust results compared to relying solely on an eye tracker-based classifier. While XGBoost and FCN fusion models may produce similar outcomes on in-distribution data, the DL-based fusion methods prove to be substantially more robust.

11.3.3 Interpretability

Although the classification performance can give us a direction of the importance of individual modalities, it is important to look more closely at the features used and how they change between the two scenarios. This can give us an indication of the stability of individual features. For this we use the gain feature importance of XGBoost models trained with features from all modalities. This describes the mean improvement in loss that results from the use of a feature.

Figure 11.4a shows the feature importance of early fusion models trained with n-Back data. We see that pupil size is the most important feature across all modalities. However, models trained on drive data, as shown in Figure 11.4b, exhibit a different

Table 11.3: This table shows the F1-Score for unimodal models trained on n-Back and evaluated on k-Drive.

Domain	Method	Fusion Type	F1-Score ($\uparrow$)	Calibration Error ($\downarrow$)
n-Back	XGBoost	Late	0.82 ± 0.01	21.12 ± 1.94
		Early	0.83 ± 0.04	11.37 ± 2.17
	FCN	Late	0.80 ± 0.02	21.31 ± 1.85
		Feature	0.84 ± 0.02	11.05 ± 2.01
k-Drive	XGBoost	Late	0.74 ± 0.05	29.36 ± 2.13
		Early	0.65 ± 0.09	28.90 ± 4.72
	FCN	Late	0.80 ± 0.07	26.45 ± 2.79
		Feature	0.84 ± 0.05	20.83 ± 3.24

tendency. The eye tracker is still relatively important, but pupil-based features do not play a role. Instead, fixations are more important. The most important feature across all modalities is based on the slowly changing electrodermal activity the skin conductance level. It is noteworthy that features based on electrocardiographic measurements, such as heart rate and heart rate variability have also a high gain feature importance. This aligns with the observations of models trained using only a single modality, presented in Section 11.3.

11.3.4 Improving ECG Representation Learning

While classical machine learning algorithms can perform well with carefully crafted features and expert knowledge, DL models can often surpass their performance by automatically learning features from the data. This can be especially useful when working with complex data such as images, audio, or text. Additionally, DL models have shown great success in many applications and are constantly being improved upon. However, often a large amount of data is needed to learn generalizable features. Data is often hard to obtain and usually requires expert knowledge to annotate new recordings. Therefore, it is important to investigate how DL methods can be improved with modality-specific knowledge to learn robust features more efficiently. We look at this from two perspectives: data and architecture.

Data augmentation techniques can be used to artificially increase the size of the training data set by creating new, slightly modified versions of the existing data. This can help prevent overfitting and improve the generalization ability of the model. For this purpose domain-specific transformations were introduced, reflecting the variability and diversity of real-world data. This leads to improved performance of ECG DL models [14, 15]. The effects of different domain-specific augmentations are visualised in Figure 11.5. Typical artefacts that can be present in electrocardiographic

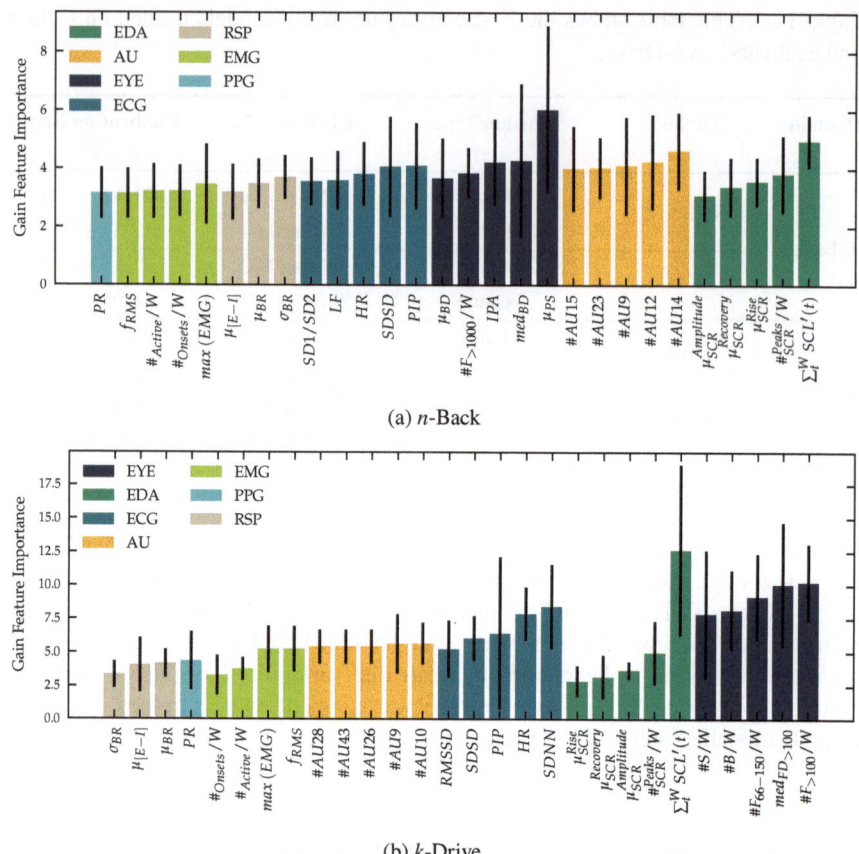

(a) n-Back

(b) k-Drive

Fig. 11.4: XGBoost feature importance for models trained on both scenarios.

recordings include muscle noise that interferes with the ECG electrode measurements and slowly varying wandering caused by sweat or respiration, shown in Figures 11.5c and 11.5a.

It is also possible to make specific changes to the architecture that can improve the performance of a particular modality. Since many important expert functions in the ECG are frequency-based, it is reasonable to add appropriate transformations to the architecture. For this a scattering transform to capture features on multiple time scales ECG recording was combined with a CNN. This architecture was investigated in various domains and was able to deliver good results in arrythmia recognition [11] compared to DL models for time series data.

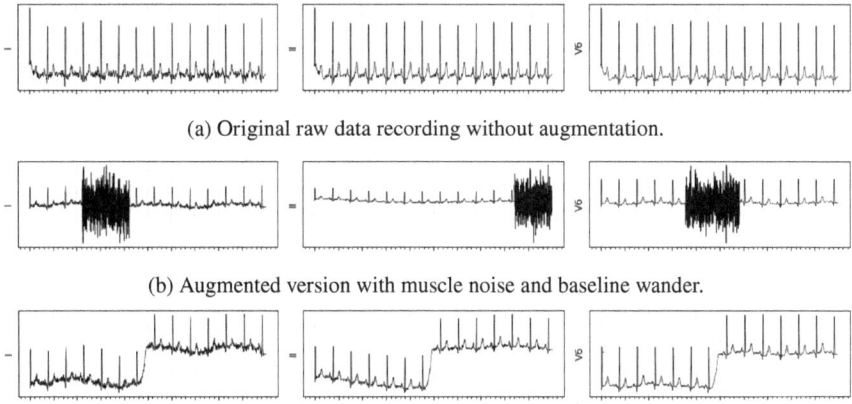

(a) Original raw data recording without augmentation.

(b) Augmented version with muscle noise and baseline wander.

(c) Augmented version with artificially added movement artefacts.

Fig. 11.5: Adding augmentations to the raw input data.

11.3.5 Deployment and Application

Detecting cognitive load can be achieved through various biosignals such as ECG, EEG, PPG, and respiration sensors, along with eye tracking data like pupil diameter, and facial expressions from videos. This approach works well in a laboratory setting, but to extend its application to real-world environments, wearables and integrated sensor technology are essential. Wearables can provide continuous monitoring of cognitive load, allowing for the collection of data outside of the laboratory and in real-world environments. The integration of multiple sensors can provide a more comprehensive understanding of cognitive load, making it easier to monitor and manage mental workload in both personal and professional settings. One example for integrating this kind of sensor technology in a day-to-day environment is Cardio-Textil, a wearable garment with integrated sensor technology [2]. Other examples might work only for some applications, such as the integration of webcam video data, specialized electrodermal sensors in game controllers or steering wheels in cars.

Models trained to detect cognitive load require a significant amount of computational resources, making them energy-intensive and not very efficient in terms of deployment. This issue can be addressed by employing certain strategies, such as the use of specialized hardware that can run inference on quantized neural networks. Quantized neural networks can optimize the energy consumption of time series processing models while maintaining their accuracy. Mueller et al. [9] introduce a hardware/software co-design methodology to create these hardware aware training routines efficently and well tested. Reiser et al. introduce a time-series specific – in their publication solely based on ECG recordings – framework to train, test and deploy quantized neural networks [16].

11.4 Conclusion and Outlook

This chapter has provided an insight into affective computing, in particular the estimation of cognitive load. To this end, a machine learning life cycle has been presented, from data collection through the modeling phase to deployment. For each of these steps, challenges are presented and examples of how to overcome them are given. In the data collection phase, data from two different scenarios were collected to build models for cognitive load estimation. In the experiments, the unimodal models based on the eye tracker were shown to perform best on in-distribution data. However, the unimodal models are sensitive to changes in the data distribution. This is where fusion models are comparatively robust. Furthermore, the importance of understanding the data and the models was presented in order to better understand biases and spurious correlations, among other things. We used feature importance to show that this can vary greatly between domains, and that some features are therefore not generalizable across domains. We also used the example of the ECG to discuss how the performance of DL models can be improved with modality-specific knowledge, such as special data augmentation. In future research, it is imperative to broaden the scope of data collection beyond current domains to ensure a well-rounded and diverse dataset. Additionally, expanding experimentation to encompass a wider range of scenarios, including safety-critical situations and entertainment contexts like video game play, will enhance the robustness and applicability of models. Exploring how models perform across different recording devices, such as wearable textiles for ECG recordings, is crucial for understanding their versatility and adaptability in real-world settings. Furthermore, assessing the robustness of classifiers should extend beyond evaluating predictive performance alone, incorporating analysis of uncertainty estimates to ensure the reliability and trustworthiness of machine learning models. Finally, investigating techniques for adapting models to various testing domains will be essential for their successful deployment and adoption in practical applications.

References

1. Bhargavi Mahesh, Weber Dominik, Garbas Jens, Foltyn Andreas, Maximilian P. Oppelt, Linda Becker, Nicolas Rohleder, and Nadine Lang. Setup for Multimodal Human Stress Dataset Collection. 2022.
2. G. Bramm, M. Hiller, C. Hofmann, S. Hristozov, M. P. Oppelt, N. Pfeiffer, M. Striegel, M. Struck, and D. Weber. CardioTEXTIL: Wearable for Monitoring and End-to-End Secure Distribution of ECGs. page 4, 2021.
3. F. Chen, J. Zhou, Y. Wang, K. Yu, S. Z. Arshad, A. Khawaji, and D. Conway. *Robust Multimodal Cognitive Load Measurement.* Human–Computer Interaction Series. Springer International Publishing, 2016.
4. T. Chen and C. Guestrin. XGBoost: A scalable tree boosting system. In *Proceedings of the 22nd ACM SIGKDD International Conference on Knowledge Discovery and Data Mining, KDD '16,* pages 785–794, New York, NY, USA, 2016. ACM.

5. J. Deuschel, B. Finzel, and I. Rieger. Uncovering the bias in facial expressions. *arXiv preprint arXiv:2011.11311*, 2020.
6. J. Deuschel, A. Foltyn, L. A. Adams, J. M. Vieregge, and U. Schmid. Benchmarking robustness to natural distribution shifts for facial analysis. In *NeurIPS 2021 Workshop on Distribution Shifts: Connecting Methods and Applications*, 2021.
7. A. Foltyn, J. Deuschel, N. R. Lang-Richter, N. Holzer, and M. P. Oppelt. Evaluating the robustness of multimodal task load estimation models. *Frontiers in Computer Science*, 6, 2024.
8. L. Longo, C. D. Wickens, G. Hancock, and P. A. Hancock. Human Mental Workload: A Survey and a Novel Inclusive Definition. *Frontiers in Psychology*, 13, 2022.
9. R. Müller, M. P. Oppelt, B. Kundu, B. R. A. Agashe, T. Thönes, E. Herzer, C. Schuhmann, S. Chakrabarty, C. Kroos, and L. Mateu. Hardware/Software Co-Design of an Automatically Generated Analog NN. In A. Orailoglu, M. Jung, and M. Reichenbach, editors, *Embedded Computer Systems: Architectures, Modeling, and Simulation*, volume 13227, pages 385–400. Springer International Publishing, 2022.
10. M. P. Oppelt, A. Foltyn, J. Deuschel, N. R. Lang, N. Holzer, B. M. Eskofier, and S. H. Yang. ADABase: A Multimodal Dataset for Cognitive Load Estimation. 23(1):340, 2022.
11. M. P. Oppelt, M. Riehl, F. P. Kemeth, and J. Steffan. Combining Scatter Transform and Deep Neural Networks for Multilabel Electrocardiogram Signal Classification. In *2020 Computing in Cardiology*, page 4. IEEE, 2020.
12. F. G. Paas, J. J. Van Merriënboer, and J. J. Adam. Measurement of cognitive load in instructional research. *Perceptual and motor skills*, 79(1):419–430, 1994.
13. R. W. Picard. *Affective Computing*. MIT press, 2000.
14. J. Qiu. Data Augmentation for Electrocardiogram Classification, Master's Thesis: Supervisor Oppelt, Maximilian P.; Nissen, Michael; Eskofier, Bjoern, 2021.
15. J. Qiu, M. P. Oppelt, M. Nissen, L. Anneken, K. Breininger, and B. Eskofier. Improving Deep Learning-based Cardiac Abnormality Detection in 12-Lead ECG with Data Augmentation. In *Annu Int Conf IEEE Eng Med Biol Soc. EMBC*, page 5, 2022.
16. D. Reiser, P. Reichel, S. Pechmann, M. Mallah, M. P. Oppelt, A. Hagelauer, M. Breiling, D. Fey, and M. Reichenbach. A Framework for Ultra Low-Power Hardware Accelerators Using NNs for Embedded Time Series Classification. 12(1):2, 2021.
17. B. Saha, L. Becker, J.-U. Garbas, M. P. Oppelt, A. Foltyn, S. Hettenkofer, N. Lang, M. Struck, N. Rohleder, and B. Mahesh. Investigation of Relation between Physiological Responses and Personality during Stress Recovery. In *2021 IEEE International Conference on Pervasive Computing and Communications Workshops and Other Affiliated Events (PerCom Workshops)*, pages 57–62. IEEE, 2021.
18. E. Stadtler. Bestimmung von Cognitive Load im autonomen Fahrsimulator durch Biosignalanalyse, Bachelor's Thesis: Supervisor Oppelt, Maximilian P.; Gradl, Stephan; Eskofier, Bjoern, 2021.
19. W. Wang, D. Tran, and M. Feiszli. What Makes Training Multi-Modal Classification Networks Hard?, Apr. 2020.
20. Z. Wang, W. Yan, and T. Oates. Time series classification from scratch with deep neural networks: A strong baseline. In *2017 International joint conference on neural networks (IJCNN)*, pages 1578–1585. IEEE, 2017.

Chapter 12
Robust and Adaptive AI for Digital Pathology

Michaela Benz[1], Petr Kuritcyn[1], Rosalie Kletzander[1], Volker Bruns[1]

Abstract The digitization of pathology opens up a wide field of applications that can be supported by AI-based analysis like the detection of tumors or a quantitative assessment of tissue composition. This contribution demonstrates possible ways on how to approach challenges in digital pathology like the robustness against data heterogeneity or the detection of out-of-distribution data. Moreover, the principle of prototypical few-shot models is explained, which can be adapted to new tasks with only a few labeled examples without any retraining of the underlying model parameters. In this chapter we show the suitability of a prototypical few-shot classification model for tumor detection in two different organs and a prototypical few-shot segmentation model for tumor composition analysis. Finally, a workflow for the creation of a dedicated AI model by only providing a few annotations within the MIKAIA® software of Fraunhofer IIS is presented.

Key words: few labels learning, data augmentation, digital pathology, prototypical few shot models

12.1 Introduction

The daily routine of a pathologist is defined by the examination of tissue samples from biopsies or surgeries. The processing of these tissue samples consists of steps such as cutting them into thin sections, staining them, and fixing them on glass slides. In traditional pathology, these tissue sections are then inspected directly by the pathologist under an optical microscope. In digital pathology, however, they are first digitized resulting in so-called whole-slide images (WSIs), for example by

[1]Fraunhofer Institute for Integrated Circuits IIS, Fraunhofer IIS, Erlangen, Germany

Corresponding author: Michaela Benz
e-mail: michaela.benz@iis.fraunhofer.de

229

C. Mutschler et al. (eds.), *Unlocking Artificial Intelligence*,
https://doi.org/10.1007/978-3-031-64832-8_12

capturing them using an automated microscopic scanner. The digitization opens up a wide range of possible applications for AI-based methods to support the pathologist in the subsequent assessment of the tissue sample, such as the detection of tumors or the identification and counting of specific cell types. However, these specific applications and data present a number of challenges.

The first major challenge is the heterogeneity of the data (Figure 12.1), which can arise, for example, from differences in sample preparation between different clinics, or from the use of microscopic scanners from different manufacturers. Likewise, the biological diversity in the appearance of tumors of even the same organ is high. This requires the AI models to be robust in the face of data heterogeneity. Moreover, it is important that an AI-model recognizes when it is presented with data it has not been trained with (out-of-distribution data). This might be artifacts like pen markings on the glass slide (Figure 12.1), or tissue types not contained in the training data.

In addition to the data heterogeneity, the enormous size of the digitized tissue sections – called whole slide images (WSIs) – poses a challenge, mainly regarding memory restrictions and computation time. With a typical resolution of 0.25 μm per pixel, a single WSI contains up to several billion pixels, each of which needs to be processed in the worst case scenario. Another challenge is that a very large amount of data (often labeled data) is needed for the training of the AI models. Creating an accurate labeled training set is very time-consuming and requires medically trained experts, which are often hardly available. Therefore, approaches relying on only few labeled data are preferable. Especially for pathological research with a wide range of different applications an approach that can be easily and quickly adapted to new tasks is desirable.

This contribution presents solutions to these challenges and validates their effectiveness in various experiments, demonstrated using the two applications i) tumor detection and ii) tumor characterization.

12.2 Applications: Tumor Detection and Tumor-Stroma Assessment

In the following sections, various solutions for robust and adaptable AI for two applications in digital pathology are presented. A more detailed description can be found in [13], [8] and [5].

The first application addressed here is the detection of tumorous tissue within a WSI. Delineating the tumor serves as prepossessing step for further analysis, for example, comparing the density of specific cells within and outside the tumor bounds. A classification approach is applied here in order to carry out an analysis of the tissue types present in the WSI.

The second application is a more detailed characterization of the tumor on a pixel level. Tumor tissue consists not only of viable tumor cells, but also of other benign tissue, called "stroma". The estimation of the tumor-stroma ratio is important for prognosis, since a higher ratio of non-tumorous tissue within a tumor correlates

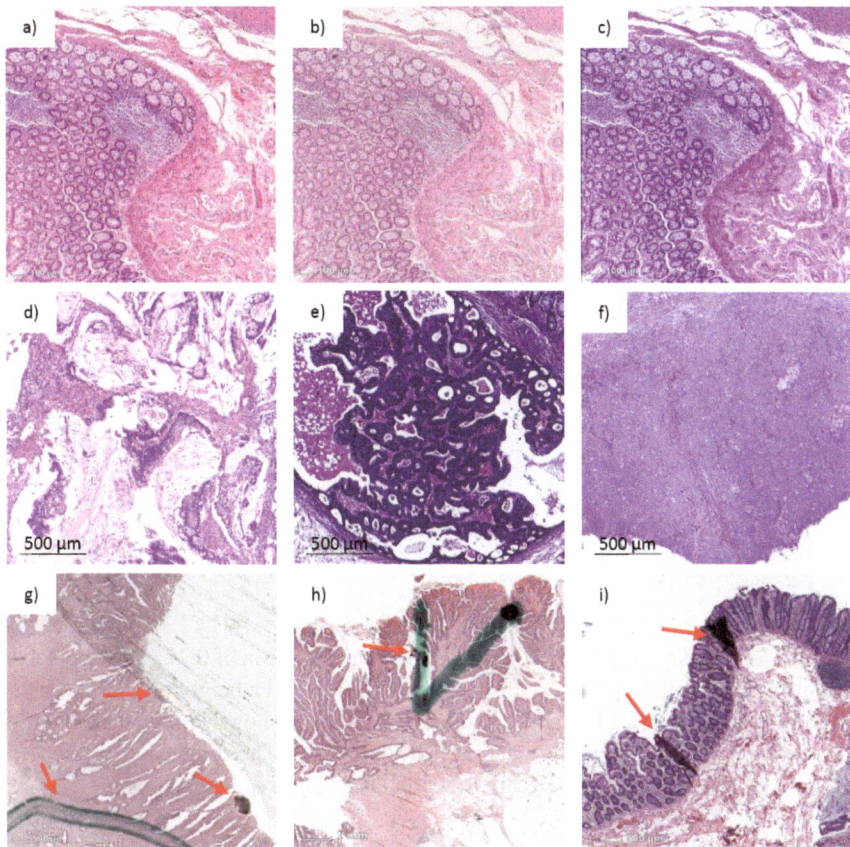

Fig. 12.1: Images a) to c) show part of the same tissue section digitized by different scanners, leading to visible color variance. An example for the biological diversity of tumors within the same entity is shown in images d) to f): all three examples are of adenocarcinoma in the colon. Different artifacts that can be present in WSIs are depicted in images g) to i) and marked with red arrows. Common artifacts are pen markings, tissue folds and dirt on the glass slide.

with a poorer outcome of the patient, i.e., with a higher risk of death. However, the assessment of this ratio by humans is prone to errors, and the ratios reported by multiple pathologists presented with the same case will oftentimes deviate significantly. Applying an AI-based model for the segmentation of the relevant tissue components supports a more reliable and objective assessment of the tumor-stroma ratio.

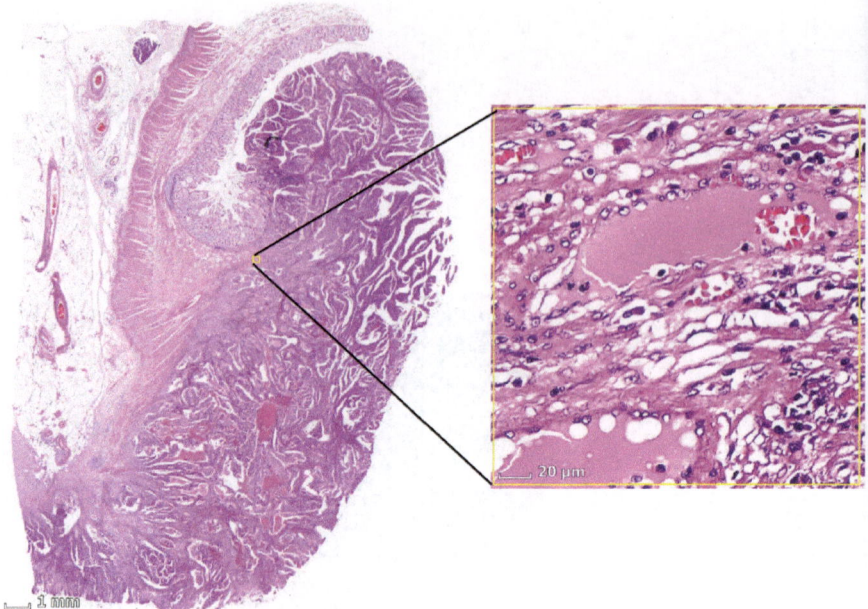

Fig. 12.2: The depicted WSI was scanned with a resolution of 0.22 µm per pixel and comprises 8.5 billion pixels, with a width of 81,000 and a height of 104,500 pixels. The section marked with a yellow frame in the WSI is shown enlarged on the right. The edge length of the image section is 1,000 pixels. The dimension of the tissue section on the glass slide is 18 mm in width and 23 mm in height.

12.2.1 Generation of Labeled Data Sets

The supervised training and quantitative evaluation of AI models requires labeled datasets. The type of labeling is dependent on the chosen approach: In a classification setting, the AI model only assigns one class label for each processed image tile, whereas the result of a segmentation model is a mask which assigns each pixel of the processed image tile to a class. Accordingly, the manually created ground truth is also required in the same way: Labeled image tiles for classification (typically squares with 200 to 300 pixels in width), and image tiles with corresponding masks for segmentation (typically larger squares with e.g. 512 pixels in width).

There are several options for creating a dataset consisting of image tiles and corresponding ground truth labels for the classification use case: One is to first divide the WSI into tiles and assign the class labels to each single tile afterwards. Another possibility is to first annotate tissue areas belonging to chosen classes within the WSI, and splitting the annotations into tiles afterwards, automatically assigning the annotation labels. The advantage of the second method is that the annotation of whole tissue regions in WSIs is much quicker than the labeling of single image tiles. One disadvantage is that larger annotations may contain tiles that are not representative of the specific tissue type, even though the entire annotated region clearly belongs to the respective tissue class. These not representative tiles are nonetheless added to the database with the label of the annotation (Figure 12.3). However, considering the total number of tiles, their influence is small.

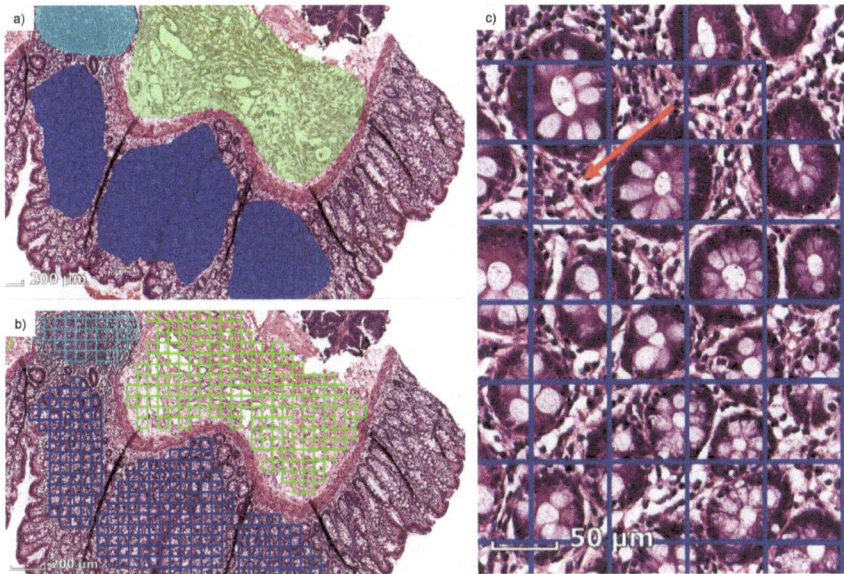

Fig. 12.3: Manually annotated tissue areas in a WSI a)). The colors of the annotations represent the assigned tissue classes: mucosa (blue), connective tissue (green), inflammation (turquoise). b) displays the image tiles that have a sufficient intersecting area with the annotations. The tiles are assigned to the tissue class of the corresponding annotation. Based on these few annotations shown in a), several hundreds labeled tiles are generated quickly. In c) the drawback of this approach is illustrated: sometimes tiles with unspecific image content are generated (highlighted by the red arrow).

Creating a pixel-precise ground truth data set to train and evaluate a segmentation model is even more time consuming, since a class label has to be assigned to each pixel of each image tile in the data set. One way to generate a larger number of annotated images serving as input to a segmentation model is to first annotate larger image tiles and then to randomly sampling patches from that annotated image, cf. Figure 12.3.

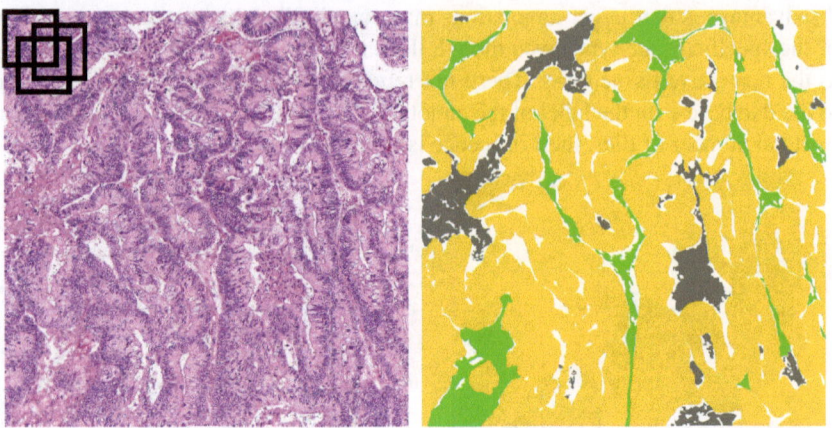

Fig. 12.4: Original image (left) and corresponding pixel precise ground truth mask (right). The assigned tissue labels are color-coded: tumor (yellow), stroma (green), necrosis (grey), background (white). The edge length of this image is 4,000 pixels, which corresponds to a length of approximately 900 μm. The required input size for our segmentation model is 512 x 512 pixel. Instead of tiling the original image with a grid of non-overlapping tiles of this size, the input image tiles are cropped at random positions (indicated by three exemplary squares with black outlines), thus generating a higher number of different input images.

12.2.2 Data Sets for Tumor Detection

Five different data sets were established for experiments conducted within the "tumor detection" application in order to show the robustness and adaptability of the created models. The primary data set was used mainly for training the model and for the baseline evaluation. In addition, multi-scanner and multi-center data sets were created to demonstrate the model robustness. To evaluate the capability to detect out-of-distribution data, a specific data set containing artifacts was generated. Finally, the adaptability to a new task based on only a few labeled examples was investigated on data originating from a different entity (urothelial cancer) than the primary data set (colon cancer).

12.2.2.1 Primary Data Set

The primary dataset was created based on 152 hematoxylin and eosin (H&E) stained colon sections. These were digitized with a resolution of 0.22 μm per pixel and manually annotated afterwards. The annotations consisted of seven tissue classes (Figure 12.5) and labeled image tiles with a size of 224 x 224 pixels were extracted from them as described in Section 12.2.1. The dataset was split up into three disjoint sets: A training set with 2,173,515 image tiles from 92 WSIs, a validation set with 719,010 image tiles from 30 WSIs, and a test set with 1,381,316 images tiles from 30 WSIs. The training set was used to train several deep neural network models, the validation set served to select the best-performing model, and the test set was employed for the evaluation of the model performance.

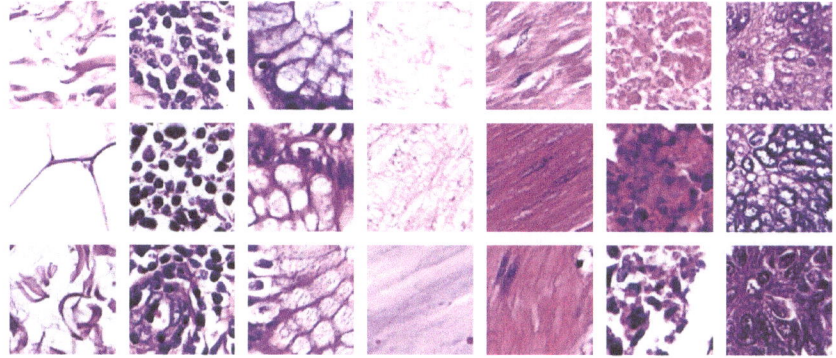

Fig. 12.5: In each column, examples for one tissue class are displayed. From left to right: connective combined with adipose tissue, inflammation, mucosa, mucus, muscle tissue, necrosis, and tumor cells.

12.2.2.2 Multi-Scanner Dataset

As mentioned in Section 19.1, it is crucial that an AI solution performs reliably, independent of the microscopic scanner used to digitize the tissue sections. As can be seen in Figure 12.1 (images a) to c)), there are immense differences in color appearances between different scanners. In order to evaluate the model robustness quantitatively, a multi-scanner data set was generated. Therefore, the 30 colon sections from the primary test set were also scanned with four different automated microscopic scanners and a manual digitization solution that uses a camera and a stitching software to create the WSI. Manual annotations were only performed on the primary WSIs. All additional WSIs were registered to their corresponding primary WSI so that the annotations could be transferred directly. Afterwards, labeled image tiles with a size of 224 x 224 pixel were generated based on the annotated WSIs for each scanner.

12.2.2.3 Multi-Center Dataset

Similar to the digitization process, the sample preparation, especially the staining process, might also introduce variances, e.g., in color. Therefore, a multi-center dataset with WSIs obtained from H&E stained colon section from different centers (clinics) was assembled. All WSIs were annotated manually, using the same seven tissue classes as in the primary data set. As for the other datasets, labeled image tiles with a size of 224 x 224 pixel were generated based on the annotated WSIs. In total, the multi-center data set comprises almost 900,000 image tiles derived from 110 WSIs.

12.2.2.4 Out-of-Distribution Data Set

In order to investigate whether out-of-distribution data can be detected with a suitable uncertainty measure, a small dedicated database was established. 15 H&E stained colon and lymph node sections were selected that contained artifacts as shown in Figure 12.1 (images g) to i)). On top of the seven tissue classes used in the primary dataset, three additional out-of-distribution classes were introduced: artifacts, debris and blood. Overall, this dataset comprises 14,001 image tiles assigned to one of the seven tissue classes (in-distribution data) and 11,388 out-of-distribution image tiles that are assigned either to the artifact (8,133), debris (1204), or blood (2051) class.

12.2.2.5 Urothelial Data Sets

With this dataset the adaptability of the classification model to the new entity of urothelial cancer using only a few annotations is evaluated. The distinctive characteristic of urothelial cancer is that it differentiates into a particularly large number of subtypes, some of which exhibit very different morphologies. Six different subtypes, combined into one "tumor" class, were included in this dataset. Accordingly, four "healthy" subclasses (e.g., connective tissue) were combined into one "healthy" class. Necrotic tissue, belonging to neither, was annotated as a separate class. For each subclass, three annotations in three different WSIs were included and set apart for the adaptation of the classification model. Further 37 WSIs were annotated, resulting in a test database with 13,963 image tiles of class tumor, 22,739 image tiles of class healthy and 2,126 image tiles of class necrosis.

12.2.3 Data Set for Tumor-Stroma Assessment

The tumor-stroma assessment application required a data set with a different type of ground truth, which was generated from 33 WSIs of the primary data set, all containing adenocarcinoma of various grades. Pixel-precise annotations were carried out within manually selected regions of interest (ROIs) with an average size of 1 mm x 1 mm in the WSIs (Figure 12.4) and each pixel was assigned to one of the six classes: tumor, stroma, mucus, necrosis, background and artifact. The artifact class was introduced for pixels that should be ignored during training and evaluation. The data set comprised three disjoint subsets: the training set with 29 ROIs from

23 WSIs (685,462,500 labeled pixels), the validation set with 6 RoIs from 4 WSIs (166,362,500 labeled pixels) and the test set with 9 ROIS from 6 WSIs (203,092,500 labeled pixels).

12.3 Prototypical Few-Shot Classification

For the task of tumor detection, a prototypical few-shot classification model [11] was chosen. The remarkable aspect of the prototypical model is that it can be adapted to a new task with only a few annotated examples and without retraining the underlying network weights [11]. The underlying neural network is very similar to a classical CNN approach: only the classification layers are removed and all layers that generate the feature vector are kept. The difference between classical neural networks and the prototypical network approach is how the classification itself is performed: In the classical approach, the classification starts from the feature vector via fully connected layers, often followed by a softmax layer. In the prototypical few-shot approach, the final layers that would perform the classification are omitted and the feature vector itself is used for computing the similarity of a query to so-called class prototypes and the class label of the nearest prototype is assigned to a query. Therefore, any CNN architecture such as e.g. Xception [2], ResNet [7] or EfficientNet [12] can be used as feature generator, or alternatively even a vision transformer [3].

The class prototypes are derived from a set of "supports", i.e., example images per class. These support images are propagated through the CNN backbone, resulting in support feature vectors. There are different methods for representing the class prototypes: the most straightforward way is to use the average over all support feature vectors as the prototype. A more sophisticated option is to apply a clustering algorithm on the support feature vectors and use the estimated cluster centers as prototypes. This is particularly advantageous if there are high variances in appearance within a single class.

The class prototypes, like the supports and the query images, are represented by their feature vectors in the so-called latent space. As a result, their similarity can be calculated by a distance metric, e.g., the squared euclidean norm. In order to classify a query, the class label of the most similar class prototype is assigned to the input query image.

A more detailed description of the prototypical few-shot approach, including the training procedure as well as investigations into the influence of the number of the support images per class and the number of prototypes per class, is given by Deuschel et al. [4].

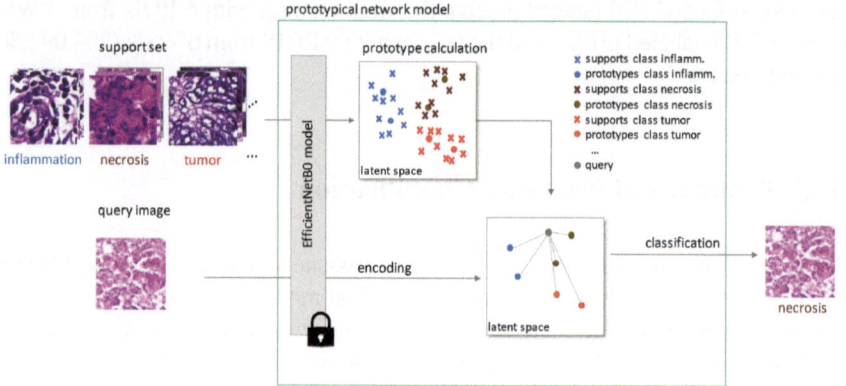

Fig. 12.6: Concept of classification with a prototypical few-shot model derived from [4]. Support images and a query image are propagated through an Efficient-NetB0 model (without classification layers), resulting in a feature vector for each of the input images. Based on the feature vectors of the support images for each class prototypes are calculated by applying e.g. a k-means clustering algorithm. The class label of the most similar prototype - in this case the one with the minimal squared euclidean distance - is assigned to the query image.

12.3.1 Robustness through Data Augmentation

One method of introducing more robustness into neural network models is to simulate the expected variance in the data by creating additional samples *de novo* during training, which is called data augmentation. This is particularly applicable in digital pathology, as the color changes in the data caused by using different scanners or staining protocols presents a major impact on the trained models and luckily is fairly easy to simulate. A further advantage of data augmentation is that it is performed only during training and thus the inference time remains unaffected, which is not the case with other methods that, for example, perform color normalization of the input image before the inference step.

Four different augmentations were applied randomly on the training data, each highlighting different possible variations in the input data. The first two augmentations randomly modified the hue and saturation components of the image colors, respectively. In order to access these components, the images were first transformed into the HSV color space to perform the modification, before being transformed back into the original RGB color space. Another augmentation was performed in a domain specific color space: the color space of the tissue staining. The tissue sections are stained with hematoxylin and eosin. In the so-called H&E color space, each of these components were separated into one channel each, plus a residual channel [1]. In order to simulate variations in the staining processes, the pixel values of both channels, the hematoxylin as well as the eosin channel, were altered randomly. Then, a transformation back into the RGB color space was performed. Finally, as a last augmentation, Gaussian blur was added to some of the input images. Each of these augmentations was applied to an image during training with a defined probability (25%). Examples of augmented images for these four different categories are depicted in Figure 12.2.

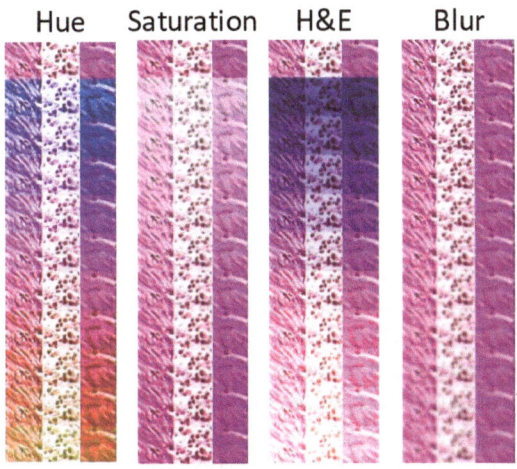

Fig. 12.7: Examples of images with four different augmentations applied: hue or saturation alteration of the image in the HSV color space, domain specific color space transformation ("H&E") with pixel value alteration of the resulting color channels, and Gaussian blurring. The first row shows the original images. The other rows show the corresponding augmented images obtained using different parameters, e.g., adding more or less Gaussian blur to the original image.

Other types of data augmentation are commonly used as well, such as scaling or contrast enhancement. In the presented application, however, applying such modifications like scaling the input images as an augmentation, did not significantly improve the robustness of the model. Investigations on the impact of further data augmentations and the robustness of different CNN network architectures can be found in [10] and [9]. The combination of the four above mentioned data augmentations showed not only to be very effective in our application but in fact, extremely necessary, as the classification accuracy without data augmentation proved to be very low on unseen scanner data, as the evaluation in the next section shows.

In order to find the most appropriate network architecture, several different models were compared, including Xception, ResNet50, EfficientNetB0, EfficientNetB3, EfficientB4, and QuickNet. In our case, the EfficientNetB0 architecture achieved the best trade-off between classification accuracy on the multi-scanner data set and inference time [9]. As a result, all evaluations reported in the following evaluation section were obtained with EfficientNetB0 models that were trained exclusively on WSIs scanned with the same automated microscopic scanner (primary data set) and applying hue, saturation, H&E and blur augmentations. The EfficientNetB0 was not only used for the prototypical few-shot model as a feature extractor, but also as classical CNN using fully connected-layers and a softmax layer for classification as a baseline comparison for the prototypical few-shot model.

12.3.1.1 Evaluation on the Multi-Scanner Data Set

Figure 12.8 shows the results of the prototypical few-shot and the classical CNN approach using an EfficientNetB0 architecture and a reference model (classical CNN with Xception architecture) that was trained without applying any data augmentation. For each approach – except for the reference without data augmentation – training was repeated three times using the training and validation set extracted from the primary data set. Afterwards, the respective models were applied to the test set of the primary data set ("original scanner") and to each set of the multi-scanner database and their results were averaged. In contrast to the reference trained without augmentation, the models trained with data augmentations showed robust performance on all automated scanners (scanners 1 to 4). The classification accuracy decreased only on the data obtained with a manual digitization process. Here, the scan quality is worse compared to the scan quality obtained in an automatized scanning process, and more stitching errors and out-of-focus regions are present. However, since manual digitization is fairly rare, no further investigations were carried out. To improve the performance on this data set, synthetic generation of images containing stitching errors could be investigated.

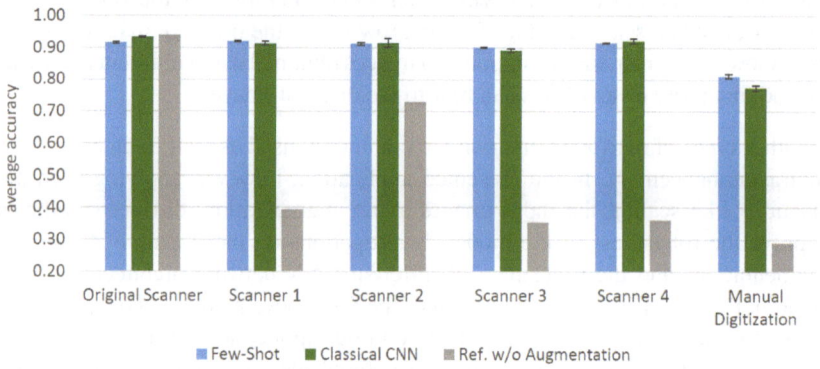

Fig. 12.8: Comparison of classification accuracy over data sets obtained with different scanners and a manual digitization process for the prototypical few-shot and the classical CNN approach (both trained using data augmentation), as well as a reference model (classical CNN with Xception architecture) trained without data augmentation.

The classical CNN approach achieved a higher classification accuracy on the original scanner data set than the prototypical few-shot approach (0.932 compared to

0.915). However, the average classification accuracy on the four automated scanners that were unseen during training is almost identical for both approaches (0.910 compared to 0.912). It is notable that the prototypical few-shot models performed very similarly on all five automated scanner data sets. In contrast, the reference model (classical CNN with Xception architecture) that was trained completely without applying data augmentations, obtained classification accuracy ranging between 0.29 and 0.73 on the multi-scanner data set. This strongly indicates the need for and the effectiveness of data augmentation.

12.3.1.2 Evaluation on the Multi-Center Data Set

For the evaluation on the multi-center data set, one model with EfficientNet B0 architecture was chosen from the three classical CNN and the three few-shot models, each. The decision was made based on which models performed best on the multi-scanner data set.

The results for both approaches on the multi-center data set, as well as for the primary test set, are denoted in Table 12.1. The overall accuracy for both approaches lies above 90% for all multi-center data set, demonstrating the models' robustness. Comparing the class specific metrics for the tumor class, a drop in recall compared to the primary test set is evident, whereas the precision lies almost in the same range as the precision on the primary test set. Since the recall is also involved in the calculation of the F1 score, these values are also lower than on the primary test set. In contrast to the multi-scanner data set, where the variance was caused only by the use of different scanners, this database has multiple sources of variance. First of all, these three data sets contain WSIs obtained from different tissue sections, unlike the multi-scanner data set, where the same 30 tissue sections were used for all scanners. As the tissue sections were stained at different sites, different staining protocols were applied, which lead to color variances. On top of that, different scanners were used at the different centers for the digitization. A closer look at the results of individual WSIs showed that a poor recall of the tumor class often occurred for WSIs containing tumor subtypes that were underrepresented in the primary training set. Therefore, the extension of the primary training set with WSIs containing these underrepresented subtypes would probably increase the performance of the models regarding the tumor recall on the multi-center data set.

Table 12.1: Overall accuracy as well as precision, recall and F1 score for the "tumor" class achieved with classical CNN and prototypical few-shot model on datasets from different centers and on the primary test dataset.

| | Few-Shot | | | Classical CNN | | | |
| | overall | Tumor | | | overall | Tumor | | |
	acc.	prec.	recall	F1	acc.	prec.	recall	F1
Center 1	0.94	**0.99**	0.73	0.84	**0.95**	0.97	**0.85**	**0.91**
Center 2	0.93	0.91	**0.91**	**0.91**	**0.95**	**0.93**	0.87	0.90
Center 3	**0.91**	0.96	**0.75**	**0.85**	0.91	**0.97**	0.70	0.82
Primary Set	0.92	**0.98**	0.94	**0.96**	**0.93**	0.97	**0.95**	**0.96**

12.3.2 Out-of-Distribution Detection

It is not only important that a classification model is robust w.r.t. variations in the input data, but also that a quantitative measure can be given of how reliable the predicted label for a given image tile is. This includes the recognition of so-called out-of-distribution inputs. As opposed to the images with which the respective model was trained (in-distribution data), out-of-distribution data refers to classes that were not part of the training data, for example images containing pen markings (Figure 12.1).

In the few-shot approach, the classification is performed based on the distance of the queries to the class prototypes. The class prototypes in turn are determined by their respective supports. The distribution of these supports can be considered as a multivariate normal distribution. Therefore, the Mahalanobis distance [6], which is calculated based on the covariance matrix of the support feature vectors and differences between the features of the query and the corresponding class prototype, can be utilized for estimating the reliability that the query image belongs to this cluster.

This approach was evaluated on the out-of-distribution data set with a prototypical few-shot model using an EfficientNetB0 architecture that generates a 32-dimensional feature vector. The model was trained on the primary training set. The classified image tiles of the out-of-distribution data set can be grouped into three categories: in-distribution image tiles that are classified correctly, in-distribution image tiles with wrong classification, and out-of-distribution image tiles. For each image tile the Mahalanobis distance from its feature vector to the nearest class prototype was calculated. Then all image tiles with a Mahalanobis distance above a defined threshold were rejected. The ratio of rejected images tiles to the total amount of images tiles for each of the categories are depicted in Figure 12.9 for varying thresholds. For instance, a threshold of 125 leads to the rejection of only 1.2% of correctly classified tiles, but 23.5% of incorrectly classified tiles and 41.4% of all out-of-distribution tiles. Thus the Mahalanobis distance seems to be a promising measure for the reliability of the model predictions. One further advantage is that the inference time is barely increased by computing the Mahalanobis distance, which is an important aspect considering the enormous size of the WSIs.

12.3.3 Adaptation to Urothelial Tumor Detection

The design of the prototypical few-shot model with its explicit representation of classes by prototypes in the latent space allows an adaptation to a new classification task without retraining the weights of the CNN (here an EfficientNetB0) that is used for embedding the images in latent space. It is sufficient to generate new prototypes that represent the classes of the new task. Therefore, support images of each new class are required. The model was initially trained on the primary data set that was derived from colon tissue sections containing adenocarcinoma. The new task

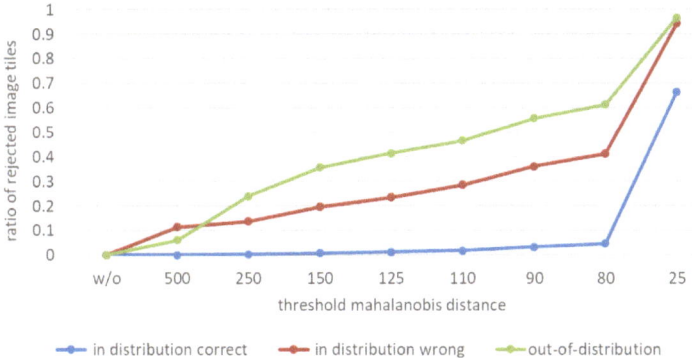

Fig. 12.9: All images tiles with a Mahalanobis distance greater than the respective threshold to the nearest class prototype were rejected. Three different groups were considered: image tiles belonging to the out-of-distribution classes, image tiles belonging to the in-distribution classes that were classified correctly and image tiles belonging to the in-distribution classes with wrong classifications. The ratio of rejected image tiles to the total amount of tiles for each category is depicted in dependence of the applied threshold for the Mahalanobis distance.

is to distinguish the three classes tumor, healthy and necrosis in WSIs containing urothelial tumor. As described in Section 12.2.2.5, 33 regions were annotated (three for each tumor subtype, as well as three for each healthy subclass) in 18 WSIs. The resulting labeled image tiles were taken as supports to calculate prototypes for each class. Two different methods were compared in [8]: a) using a k-means clustering (with k=6) on the set of all supports of each class and b) calculating one prototype per annotation by averaging all supports derived from this annotation. The latter method is particularly suited to support an interactive process in which the user also annotates previously misclassified regions. Calculating the prototype per annotation ensures that each annotation is represented by a prototype, which is not the case when using k-means clustering.

The results of the evaluation on a disjoint test set are shown in Table 12.2. Both models achieved similar results and were able to recognize all three classes well, as shown by the high recall values. These results demonstrate that the prototypical few-shot model was successfully adapted to the new task.

12.3.4 Interactive AI Authoring with MIKAIA®

The entire workflow of adapting the prototypical few-shot model and applying it to WSIs was realized in the MIKAIA® software (Figure 12.10) of Fraunhofer IIS. Without any knowledge about deep neural networks users can create their own model

Table 12.2: Classification results of the adapted prototypical few-shot model on the urothelial test set with two different prototype calculation methods: A) using k-means clustering on the set of all supports of each class and B) one prototype per annotation calculated as average over all supports derived from this annotation.

prototype method	overall accuracy	avg. precision	avg. recall	avg. F1
A (k-means)	0.936	0.898	0.930	0.912
B (per annotation)	0.929	0.908	0.919	0.913

by providing annotations for the classes to be distinguished interactively. Hereby, the model can be adapted iteratively. As soon as at least one annotation is available for each class, the new prototypes can be calculated. Depending on the size of the provided annotations this process only takes a couple of seconds. Afterwards, the model can be applied. If the result is not sufficient, further annotations may be carried out and added to the prototype calculation.

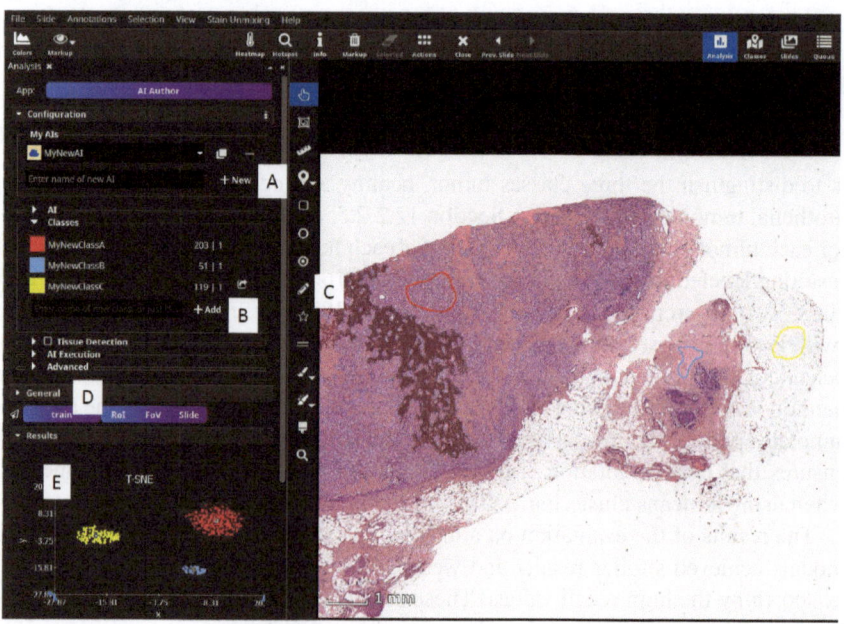

Fig. 12.10: User interface of the "AI Author" module of the MIKAIA® software of Fraunhofer IIS. To create a new AI the user specifies a name and clicks on the button "New" (A). Then, classes can be named and added (B). Afterwards, annotations for each class need to be provided (C). Clicking "Train" triggers the calculation of the new prototypes (D). Hereby, all image tiles within the annotations are taken as support images. A two-dimensional projection of the latent space is displayed in (E). The supports of each class are color-coded and form separate clusters in latent space.

Performing a tissue analysis on a WSI with a deep neural network requires first of all a tiling step into smaller image tiles as the input sizes of the network are significantly smaller (e.g., 224 x 224 pixels) compared to the size of a WSI (e.g., 80,000 x 100,000 pixel). Tiling into non-overlapping image tiles and generating a color-coded overlay according to the predicted label results in a checker board visualization. A smoother overlay can be obtained by using overlapping tiles, hence increasing the computation time due to the increased number of tiles to be classified. Another option is to introduce a preprocessing step that divides the WSI in *superpixels* by grouping visually similar adjacent pixels. Afterwards, image tiles within the superpixels are classified and their classification labels are combined into a superpixel label. Thereby, it is sufficient to classify only a random subset of images tiles contained in the superpixel which leads to a reduction of the computation time [13]. Figure 12.11 shows the classification results obtained with the prototypical few-shot model that was adapted based on the three annotations displayed in Figure 12.10 for both tiling strategies.

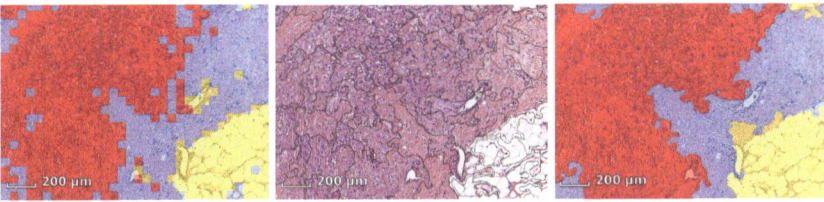

Fig. 12.11: Color-coded classification results for two different tiling strategies are shown that were obtained with the adapted prototypical few-shot model based on the annotations shown in Figure 12.10. Tiling into non-overlapping image tiles yields a checkered classification overlay (left). A prior segmentation into superpixels (middle) yields smoother results (right).

12.4 Prototypical Few-Shot Segmentation

Deep neural network segmentation models are chosen for tasks where a pixel-precise analysis is required. The basic principle of the prototypical few-shot approach can be applied here as well as shown in Figure 12.12, ensuring an easy adaptability to new tasks. Instead of single classification labels per support image, labeled masks are required in this case. Similar to the classification approach, the input images are propagated through a CNN architecture with the difference that the result for one image is not a single feature vector but a spatially arranged map of feature vectors. This feature map has smaller dimensions than the input image. For each support image the feature map is fused with the accordingly downsampled annotation mask. Afterwards, masked average pooling is applied, resulting in one support feature vector representing the average of all pixels belonging to this class in this support image. The subsequent calculation of prototypes is identical to the classification case. Query images are also propagated through the network. For all pixels in their feature map the similarity to the class prototypes is determined and the label of the most similar one is assigned to this pixel. Finally, an upsampling step is performed resulting in a segmentation mask in the original resolution of the input image. Further details can be found in [5].

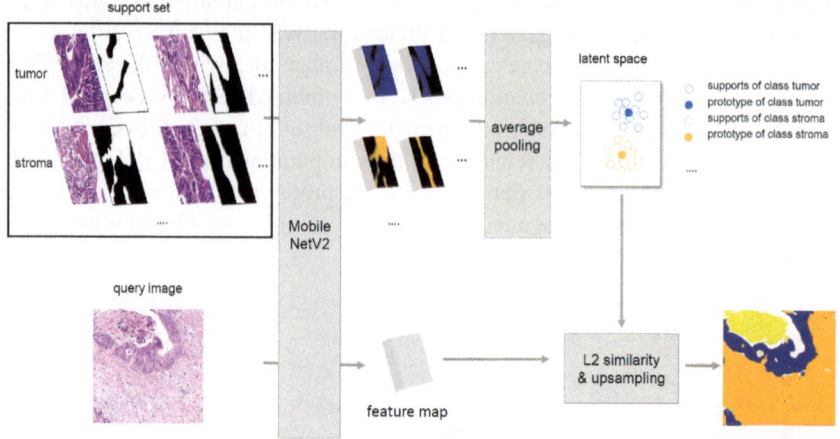

Fig. 12.12: Concept of segmentation with a prototypical few-shot model derived from [5], licensed under CC-BY 4.0. Support images are propagated through a MobileNetV2, resulting in a downsampled feature map which is fused with the annotation mask. Afterwards, masked average pooling is applied to obtain one support feature vector per support image. Class prototypes in latent space are calculated by, e.g., applying k-means clustering to all support feature vectors of the respective class. Query images are propagated through the MobileNetV2 as well. For each feature vector in the query feature map the most similar prototype is determined and its label is assigned to the corresponding pixel in the feature map.

12.4.1 Tumor-Stroma Assessment

On the data set described in Section 12.2.3 a prototypical few-shot segmentation model with a MobileNetV2 architecture was trained. A detailed description of the training parameters is provided in [5]. Table 12.3 shows the results obtained on the corresponding test set. Especially, the classes tumor and stroma that are important for an accurate tumor-stroma assessment are reliably recognized. Confusions between the other three classes are less relevant as they do not affect the tumor stroma ratio. This opens up new opportunities for the tumor-stroma characterization like the complete segmentation of the tumor and detection of regions with high and low stroma density. Currently, manual assessment is restricted to determining the ratio of the tumor and stroma area within small representative region of interest within the tumor due to time constraints.

Table 12.3: Results of pixel-wise evaluation on tumor-stroma test set.

Class	precision	recall	F1-score
tumor	0.917	0.925	0.921
stroma	0.881	0.910	0.895
necrosis	0.631	0.682	0.655
mucus	0.848	0.525	0.638
background	0.727	0.710	0.718

12.5 Conclusion and Outlook

Prototypical few-shot approaches are easily adapted to new tasks based on only a few annotated examples without retraining the underlying network weights. They can be designed as classification as well as segmentation model and are suited for the use in digital pathology applications, as shown in this contribution for the example of tumor detection and tumor-stroma assessment. An important aspect when developing AI solutions for digital pathology is to ensure their robustness against the data heterogeneity in "real-world" scenarios. Domain-specific data augmentations are an effective way to cope with variance that are introduced by differences in staining procedures or by digitization with scanners from different manufactures. This was demonstrated for the tumor detection task on multi-scanner and multi-center data. Moreover the biological diversity of, e.g., tumor subtypes, has to be reflected in the training data. Missing or underrepresented sub types might lead to a poorer detection rate which probably caused the lower recall values for the tumor class on the multi-center data. Furthermore, a desirable property is that the AI model recognizes when it is confronted with out-of-distribution data that it has not encountered during training. These might include yet unseen tumor subtypes or artifacts. Therefore, the model should not only predict a result but also provide an estimate how reliable the prediction is. In case of prototypical few-shot models, a promising reliability measure can be derived from the Mahalanobis distance between the query and the corresponding prototype in latent space. Its effectiveness in reducing incorrectly classified and out-of-distribution image tiles was shown for the tumor detection application. To make the developed AI approaches usable for pathologists, the workflow to customize and generate an own AI was realized in a software with graphical user interface. The introduction of AI solutions to support pathologists in the assessment of digitized tissue sections opens up new possibilities in the research of new prognostic factors like the tumor-stroma ratio: instead of only evaluating individual small hotspots, analyses can be performed on the entire tumor area that were not feasible manually in terms of time. Also, the results become more objective, since especially the estimation of areas is very subjective and high variations between different pathologists occur.

In addition to analyzing the composition of different tissue types, which can be performed with the proposed few-shot models, a quantitative analysis of the cell types and their distributions provides further insights into the tumor biology. Therefore, future work will focus on developing a reliable cell detection and classification and on combining it with the described prototypical few-shot approaches. General classification and segmentation of a WSI into different tissue types and sub-structures with prototypical networks will permit subsequent informed cell classification analyses. The detection of cell types and densities in relation to the segmented substructures will provide the pathologist with an even more comprehensive and quantitative characterization, which will facilitate more specific diagnosis required for personalized medicine..

Acknowledgements

We want to thank our clinical partners for their support: Carol I. Geppert, Markus Eckstein, Arndt Hartmann (Institute of Pathology, University Hospital Erlangen), Katja Evert, Felix Keil, Matthias Evert (Institute of Pathology, University of Regensburg), Paul K. Ziegler, Katrin Bankov and Peter Wild (Dr. Senckenberg Institute of Pathology, University Hospital Frankfurt).

The results shown here are in part based upon data generated by the TCGA Research Network: https://www.cancer.gov/tcga.

References

1. B. E. Bejnordi, G. Litjens, and et al. Stain specific standardization of whole-slide histopathological images. *IEEE Trans Med Imaging*, 35(2):404–415, 2016.
2. F. Chollet. Xception: Deep learning with depthwise separable convolutions, 2017.
3. L. Deininger, B. Stimpel, A. Yuce, S. Abbasi-Sureshjani, S. Schönenberger, P. Ocampo, K. Korski, and F. Gaire. A comparative study between vision transformers and cnns in digital pathology, 2022.
4. J. Deuschel, D. Firmbach, C. I. Geppert, M. Eckstein, A. Hartmann, V. Bruns, P. Kuritcyn, J. Dexl, D. Hartmann, D. Perrin, T. Wittenberg, and M. Benz. Multi-prototype few-shot learning in histopathology. In *2021 IEEE/CVF International Conference on Computer Vision Workshops (ICCVW)*, pages 620–628, 2021.
5. D. Firmbach, M. Benz, P. Kuritcyn, V. Bruns, C. Lang-Schwarz, F. A. Stuebs, S. Merkel, L.-S. Leikauf, A.-L. Braunschweig, A. Oldenburger, L. Gloßner, N. Abele, C. Eck, C. Matek, A. Hartmann, and C. I. Geppert. Tumor-stroma ratio in colorectal cancer - comparison between human estimation and automated assessment. *Cancers*, 15(10), 2023.
6. H. Ghorbani. Mahalanobis distance and its application for detecting multivariate outliers. *Ser. Math. Inform.*, 34(3):583–595, 2019.
7. K. He, X. Zhang, S. Ren, and J. Sun. Deep residual learning for image recognition, 2015.

8. R. Kletzander, P. Kuritcyn, V. Bruns, M. Eckstein, C. Geppert, A. Hartmann, and M. Benz. Domain transfer in histopathology using multi-protonets with interactive prototype adaptation. *Current Directions in Biomedical Engineering*, 9(1):491–494, 2023.

9. P. Kuritcyn, M. Benz, J. Dexl, V. Bruns, A. Hartmann, and C. Geppert. Comparison of CNN models on a multi-scanner database in colon cancer histology. In *Medical Imaging with Deep Learning*, 2021.

10. P. Kuritcyn, C. I. Geppert, M. Eckstein, A. Hartmann, T. Wittenberg, J. Dexl, S. Baghdadlian, D. Hartmann, D. Perrin, V. Bruns, and M. Benz. Robust slide cartography in colon cancer histology. In C. Palm, T. M. Deserno, H. Handels, A. Maier, K. Maier-Hein, and T. Tolxdorff, editors, *Bildverarbeitung für die Medizin 2021*, pages 229–234, Wiesbaden, 2021. Springer Fachmedien Wiesbaden.

11. J. Snell, K. Swersky, and R. S. Zemel. Prototypical networks for few-shot learning, 2017.

12. M. Tan and Q. V. Le. Efficientnet: Rethinking model scaling for convolutional neural networks, 2020.

13. F. Wilm, M. Benz, V. Bruns, S. Baghdadlian, J. Dexl, D. Hartmann, P. Kuritcyn, M. Weidenfeller, T. Wittenberg, S. Merkel, A. Hartmann, M. Eckstein, and C. I. Geppert. Fast whole-slide cartography in colon cancer histology using superpixels and CNN classification. *Journal of Medical Imaging*, 9(02), mar 2022.

Chapter 13
Safe and Reliable AI for Autonomous Systems

Axel Plinge[1], Georgios Kontes[1], Sebastian Rietsch[1], Christopher Mutschler[1]

Abstract A big challenge in implementing autonomous systems using reinforcement learning in a way to be used in the real world is to make them dependable, i.e., explainable and reliable. Automated and autonomous driving poses one of the biggest challenges to the development of artificial intelligence (AI), as it is technically demanding to solve the tasks involved in order to make a car act autonomously in real-world situations. However, unless autonomous systems become truly safe and dependable, they cannot be deployed in a real-world setup. Operating autonomous vehicles not only efficiently but also safely and reliably is even more challenging. This chapter explains several unique and innovative methods to illustrate dependable reinforcement learning in autonomous vehicles.

Key words: reinforcement learning, autonomous systems, dependable artificial intelligence, autonomous vehicles

13.1 Introduction

Autonomous agents determine – independently of human supervision – how a given system should operate in response to observed environmental parameters. They comprise actuators for interacting with the environment, sensors for perceiving it, and components for the aggregation, analysis, and interpretation of data, as well as situation assessment and planning future actions. Applications include, among others, robotic movement [14], the control of autonomous vehicles [12] and drones in industrial use cases [15], regulation of chemical processes [13] or hydraulic pumps

Fraunhofer Institute for Integrated Circuits IIS, Fraunhofer IIS, Nuremberg, Germany

Corresponding author: Axel Plinge
e-mail: axel.plinge@iis.fraunhofer.de

© The Author(s) 2024
C. Mutschler et al. (eds.), *Unlocking Artificial Intelligence*,
https://doi.org/10.1007/978-3-031-64832-8_13

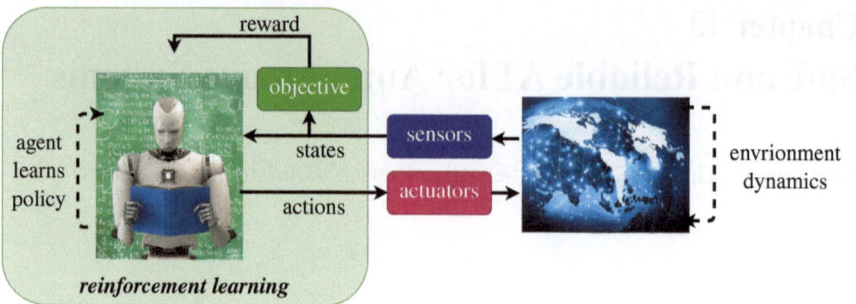

Fig. 13.1: An autonomous system with reinforcement learning. Images (C) phonla-maiphoto - Fotolia.com, Victoria - Fotolia.com

in industrial plants [10], control of intelligent buildings [46], and renewable energy production through wind turbines [32].

13.1.1 Reinforcement Learning

The complexity of real-world applications paves the way for the development and adoption of *learning* autonomous systems. Here, the agent progressively learns to improve on a given task through interaction with the environment. For the learning component of the agent, we adopt the reinforcement learning (RL) paradigm.

RL is a branch of machine learning that addresses sequential decision problems under uncertainty. Therefore, it is more suited for systems that operate autonomously and affect the environment with actuators compared to supervised and unsupervised learning algorithms that are able to tackle one-step decision problems.

Here, the agent interacts with the environment, and through trial-and-error, progressively derives rules that define how actions or decisions are selected. This process is guided by a reward signal which is available to the agent after every time step – with the final goal to learn a function (called *policy*) that selects actions that maximize the future expected total reward gathered by the agent. The generality of the reward definition allows for optimizing the behavior of the system towards multiple desired objectives, such as safety and effectiveness.

RL has shown to reach super human-level performance across a wide range of tasks. Unfortunately, even though generalization in supervised learning is generally mature and well-established, those insights do not generally transfer well to the reward-based learning paradigm. It is generally understood that standard RL tends to be brittle and unfit to address unknown situations or cope even with subtle deviations from their training conditions (even to the random seeds of the (simulated) environments [16]), seriously restricting their relevance in real-world applications [21]. The problem further intensifies due to novel RL algorithms being commonly evaluated based on training performance inside singleton environments like Atari or ALE [5], slowing a deeper understanding of such algorithmic qualities. We address these shortcomings by adopting dependable RL approaches and best practices. To achieve this, we make the connection between dependability and algorithmic transparency – a crucial factor in the design of autonomous systems, as highlighted in the concept discussed by Lipton [26]. Here, we ensure that the inputs, internal mechanisms, and outputs of the decision-making process are comprehensible, thereby enabling humans to gain insight into the decision-making process. In our approach, we achieve transparency through five aspects:

Understandable observations and states: Prefer hand-engineered or learned features over raw sensor measurements for algorithmic transparency as they carry more semantic meaning (e.g.,[35]).

Semantic actions: Semantic, discrete action spaces are favored over direct low-level control for better understanding. The behavioral planner employs a few discrete, semantic actions executed through low-level controllers like model predictive control (MPC) (e.g., see [40, 28, 33, 24]).

Explicit safety components: While approaches like reward shaping (see Chapter 3) have been effective, they rely on an implicit trade-off between safety and performance. We strive to make this trade-off explicit, enabling separate learning, analysis, and verification of the risk estimator (e.g., see [38]).

Interpretable learning framework: For example, value-based RL methods like deep Q-networks (DQNs) are easier to interpret compared to policy-gradient methods like proximal policy optimization (PPO), cf. chapter 3. DQN estimates future rewards for state-action pairs, providing a clear theoretical definition that can be numerically approximated. Policy gradients optimize the policy directly, which is less intuitive to explain (e.g., see [17, 23, 37, 38] and the references therein).

13.1.2 Reinforcement Learning for Autonomous Driving

As dependable reinforcement learning is best applied in a design framework geared towards safety, we develop a dependable driving assistant in order to demonstrate our technology. Autonomous driving provides a unique context for research, as it requires learning correct behavior over variations of road layouts and distributions of traffic situations, including individual driver personalities and hard-to-predict traffic incidents.

The typical processing pipeline of autonomous vehicles is made up of perception, behavioral planning, motion planning, and actuators, as illustrated in Figure 13.2. The perception module outputs an abstract representation of the environment. Behavioral planning selects a high-level action, e.g., changing lanes to avoid an obstacle. Motion planning prescribes the path of this motion, whereas actuators implement the motion along that path.

Recently, several algorithmic flavors of RL have been utilized to address various sub-problems in the autonomous driving domain [20]. In this direction, there have

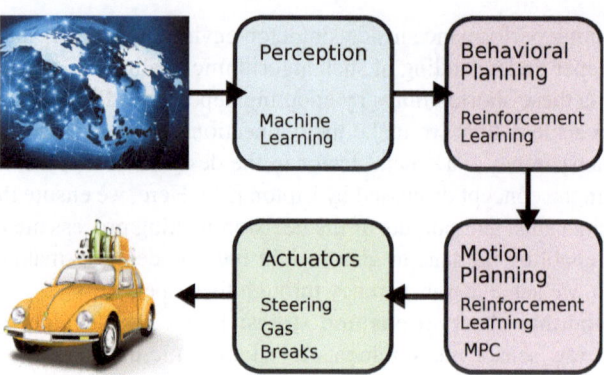

Fig. 13.2: Autonomous vehicle pipeline: The perception is typically handled by supervised machine learning (e.g., computer vision), while behavioral planning is a key application for RL. The motion planning can be handled by RL or model predictive control (MPC) before the actuators bring the action to the road. Images (C) Victoria - Fotolia.com, fotokalle - Fotolia.com

been notable successful real-world applications (e.g. [7, 3]) that leverage imitation learning algorithms [30] training on available data from vehicles, usually combined with some form of data augmentation to address the distribution shift of the deployment environment. Other, smaller-scale studies utilize an available high-fidelity simulator and simulation-to-reality transfer approaches [31] or even learn a driving policy online [18]. A central concern in these approaches is the extent to which we can *guarantee* safety [41] for all possible situations that can occur in the real world.

Unless autonomous systems become truly safe and dependable, they cannot be deployed in a real-world setup. Operating autonomous vehicles not only efficiently but also safely and reliably is even more challenging. There are many unique tasks that need to be solved reliably for safe and efficient performance. Among them, behavioral planning is one of the hardest challenges [25].

We focus on the realization of behavioral planning by reinforcement learning. Safe and efficient behavior are the key guiding principles for autonomous vehicles. Manually designed rule-based systems need to act very conservatively to ensure a safe operation. This limits their applicability to real-world systems. On the other hand, more advanced behaviors, i.e., policies learned through means of reinforcement learning (RL), suffer from non-interpretability as they are usually expressed by deep neural networks that are hard to explain. Even worse, there are no formal safety guarantees for their operation.

In Figure 13.3, we see the overall pipeline of our proposed approach to behavioral planning and its dependable implementation. For each traffic scenario, our training framework Driver Dojo [34] generates multiple challenging environments. These are used to train a safe reinforcement learning policy with, e.g., our SafeDQN algorithm [38]. The safe policy is used to train a safe decision tree using our SafeVIPER algorithm [37]. These trees can also be validated analytically in case there is a usable

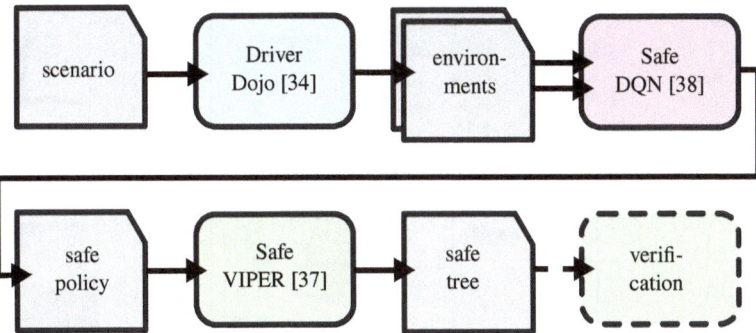

Fig. 13.3: Our proposed pipeline for dependable RL in autonomous driving.

analytic formulation of the scenario. In the following sections, each of our methods is explained and evaluated in more detail.

13.2 Generating Environments with Driver Dojo

To address RL generalizability and robustness, we require a framework that is able to emulate a highly-variable driving environment that allows for training and testing agent behavior in many distinct scenarios. To compare the trade-off between high-level actions and direct control inputs or different observation spaces against each other, it must also be easily configurable and provide a broad set of ready-to-use features. For this, we introduce a sophisticated, application-focused environment benchmark for autonomous driving called Driver Dojo built on top of the simulation of urban mobility (SUMO) [27] traffic simulator.

Our environment framework offers a broad suite of features: i) fully randomized street networks for intersections, roundabouts, and multiple highway driving tasks; ii) fine-grained rule over traffic initialization and randomization including a sampling-based method for physical and behavioral of non-ego driver attributes on a per-vehicle basis; iii) a direct and semantic action space, multiple vehicle dynamics models, and a catalog of composable observation methods; iv) co-simulation to Carla [11] for realistic observations from different types of sensors; v) an underlying code-base that is modular and performant and allows for fine-grained composition of environment designs. Furthermore, Driver Dojo has a simple workflow to create and deploy static scenario definitions on par with most simulation benchmarks. The code for the benchmark is available at https://github.com/seawee1/driver-dojo.

In our benchmark, we mainly concern ourselves with i.i.d. generalization [21], where training and evaluation scenarios are drawn from the same underlying distribution. However, as we show in our experiments, in-distribution generalization in an autonomous driving application already poses highly challenging problems.

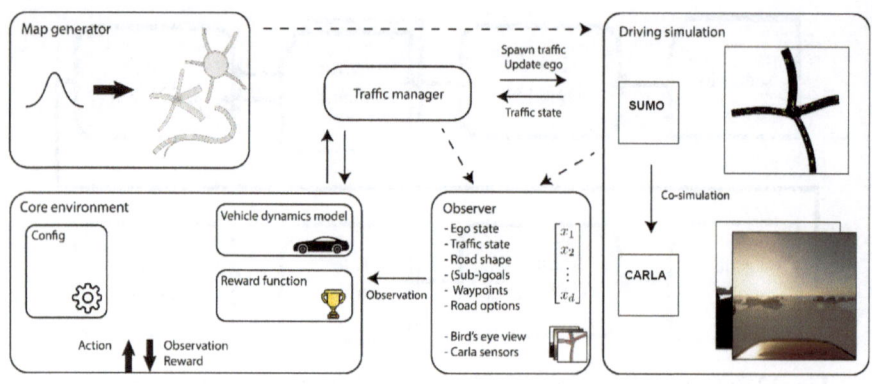

Fig. 13.4: Overview of Driver Dojo.

We note that efforts towards adopting real-world traffic datasets aim to address a similar problem [47, 22] like Driver Dojo, though ultimately suffer from the shortcoming that prerecorded traffic trajectories deprive the user of the ability to dynamically respond to movements of the controlled ego vehicle. We argue that the issue has to be alleviated first for datasets to become an acceptable alternative to simulation benchmarks.

13.2.1 Method

In Driver Dojo [34], we employ SUMO as the core engine and use its traffic model to simulate traffic and its interactions with the ego vehicle. It is a microscopic traffic simulator that is under active development for over 20 years and, thus, very mature and offers a pallet of features and a versatile ecosystem of range applications, supporting in many aspects for building a naturalistic driving benchmark for autonomous agents. The driving simulator Carla [11] offers official SUMO co-simulation support in its recent versions. Despite SUMO's lack of a native 3D engine, the integration of Carla allows us to provide diverse sensor readings through an extensive library of implemented sensor models. We take advantage of the vehicle dynamics library of CommonRoad [2] to simulate ego vehicle motion. It offers a range of dynamic models of varying complexity and physical parameter sets of three distinct real-world passenger cars. The architecture of Driver Dojo is visualized Figure 13.4.

Actions Our implementation allows granting the agent *direct* access to the car's throttle, brake pedal, and steering wheel or commanding it through high-level *semantic* actions. While control actions have an immediate and direct impact on the vehicle, semantic actions affect the vehicle many steps into the future, abstracting the underlying control complexity and, in other words, decoupling the behavioral execution problem from the planning problem, allowing to solve them independently. In its current form, the semantic action space is realized through a low-level Stanley control that tracks the path along a selected lane under a specified velocity, both of which the agent can modify through its actions.

To ensure maximum flexibility, action spaces are parameterized to allow for discretization of the continuous control action space or selection from various discrete space configurations. Additionally, we offer an interface for outside motion planners through passing a list of waypoints to the agent vehicle, enabling combined methods of classical motion planning and RL.

To reduce the computational complexity in experimental settings where accurate physical modeling is not the main concern, Driver Dojo offers the TargetPositionSpeed (TPS) vehicle model, which simply interpolates the ego position and orientation towards the next waypoint. In all other cases, we adopt the kinematic single-track, single-track drift or multi-body dynamics vehicle models offered by the CommonRoad [2] suite, along with the accompanying physical parameters for a Ford Escort, BMW320i, and a VW Vanagon.

Observations. It is common practice to either provide raw perceptions in the form of sensor readings to the agent or to let it perceive the environment in the form of driving affordances [8, 1], which is a more condensed representation of the road situation inside a meaningful and compact feature space. Whereas the latter eases the learning task, reward-based feature learning on raw perceptions could be an important angle for solving the generalization problem, as convolutional neural networks have proven immensely powerful in learning expressive feature representations in other problem domains. Driver Dojo enables both paradigms by implementing a multitude of ready-to-use ground-truth observations extracted from the world state inside SUMO and by using Carla to generate raw perception sensor signals.

Environment generation. We randomize on both the street network level and the traffic level through different methods. Our core benchmark scenarios include roundabouts, intersections and high-ways. Driver Dojo provides programmatic scenario definitions for maximum flexibility in creating highly diverse scenarios. We translate between conventional map formats, including OpenDRIVE, through netconvert, which is part of the SUMO software ecosystem. For the roundabout and intersection scenarios, we define distributions over the number of incoming lanes, their angle in relation to the structural center, their lengths, as well as the distance of their connection point with the junction. For roundabouts in particular, we added probabilistic deformations as real-world roundabouts are often not perfectly round.

Traffic behavior. SUMO's microscopic traffic simulation modularizes traffic behavior through separate car-following, lane-change, and junction models. Commonly, such models are fixed for one traffic simulation, but different model parameters can be assigned to groups of vehicles. These span across attributes that define, for instance, pushiness or willingness for strategic lane changes, but also safety-related factors like targeted time headway, driving imperfections or malicious overlooking when crossing a junction. To make our benchmark as challenging as possible, we define distributions over 34 such parameters of interest, from which we sample a fixed set of parameter constellations, and randomly assign them to non-egos entering

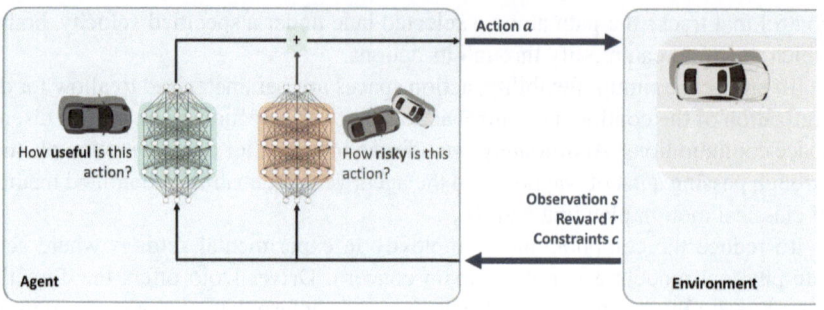

Fig. 13.5: Safe DQN approach: Two separate networks are learning safe and efficient behavior in parallel. One optimizes for utility, one for safety.

the scenario. These also include physical properties such as vehicle dimensions or acceleration and deceleration profiles.

13.3 Training safe Policies with SafeDQN

In order to ensure the dependability of reinforcement learning behavior planning, [38] presents a novel approach that splits its knowledge of safe behavior into two components. The idea starts from Q-Learning, a fundamental algorithm in RL that iteratively updates Q-values using the Bellman optimality equation. It is an off-policy method aiming to approach the optimal policy. To address large or continuous state spaces, a non-linear function approximator like a neural network (NN) can represent the Q-function. Its parameters can be learned through back-propagation by minimizing the Bellman error of the Q-function. The DQN algorithm [29] builds upon this concept and incorporates additional techniques to enhance learning stability (see Chapter 3).

13.3.1 Method

In our approach, we train two neural networks in parallel, as illustrated in Figure 13.5. The first network is used as an approximator of the optimal state-action value (Q-value) function, meaning the maximum expected utility or cumulative future return of an action conditioned on a state, in the same manner as in tabular Q-Learning or Deep Q-Networks. For our safety-oriented approach, we add the second network to approximate the optimal state-action risk function, meaning the minimum expected cumulative cost of an action conditioned on a state. This allows us to make an explicit risk-utility trade-off by subtracting the risk estimate from the utility estimate and choosing the action that maximizes the combined risk-utility signal. To enable

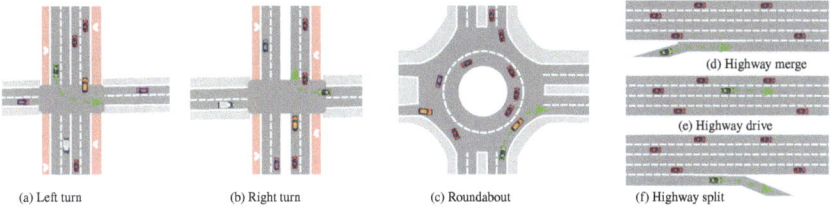

Fig. 13.6: Driver Dojo generated maps for the different scenarios.

the user to manually specify a desired threshold safety, we automatically learn the linear weighting factor alongside both network weights during training.

Further, enhanced interpretability is achieved through the disentangled estimates, enabling separate analysis and evaluation of risk and utility, thereby facilitating better understanding and reasoning.

Furthermore, we assessed a variant of SafeDQN with an alternative objective function, referred to as SafeDQN-Alt. In this modified objective, the action from which to bootstrap the next Q-value is determined from the Lagrangian combination $Utility(s, a)+\lambda Risk(s, a)$. The rationale behind this adjusted objective is to optimize both estimators by incorporating information from each other. Specifically, this change in objective aims to mitigate overestimation in states where the best action according to utility carries a high risk and should be avoided. This concept bears resemblance to Double DQN [45], which employs the best action from the target network to alleviate overestimation bias. CDQN [17] also incorporates constraints, albeit in a more stringent manner.

13.3.2 Evaluation

Through evaluation of SafeDQN in diverse traffic scenarios, we demonstrate its ability to generate effective strategies while significantly outperforming baseline methods in terms of safety. Thus, the goal for the agent is to successfully complete each scenario along a predefined route without a collision. In Figure 13.6, we see the target environments generated with Driver Dojo [34]. We included three typical highway tasks (merge, drive, split) and inner-city scenarios (left/right turns and a roundabout). The street networks of scenarios Right turn, Left turn, and Roundabout are part of the *Town03* map from the CARLA simulator [11], which were integrated utilizing SUMO's tooling scripts.

The other algorithms considered in our evaluation include DQN with reward shaping, PPO with reward shaping, and RCPO+. RCPO+ is similar to RCPO [43] but utilizes two critics, akin to the approach employed in SafeDQN and CPPO [42], to separately estimate reward and safety.

In Figure 13.7, the results are shown as the average return per episode (top) and the cumulative number of constraint violations (bottom) of SafeDQN along with the

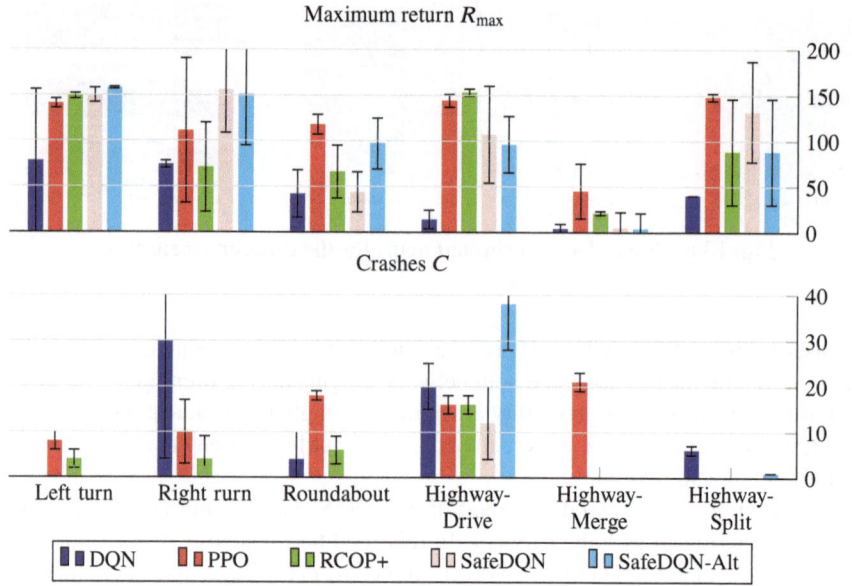

Fig. 13.7: Evaluation results for SafeDQN against baseline algorithms in the scenarios.

baselines after training. Overall, our method operates much safer than all baselines in five out of six scenarios. SafeDQN encounters difficulty in quickly discovering a secure policy during Highway-Drive scenarios. Unlike other scenarios, Highway-Drive exhibits a distinct characteristic where a random policy exhibits competitive performance. Typically, a random policy operates at speeds lower than the average traffic speed, and the randomness in lane changes can be more easily adjusted for by the traffic's lane-changing model.

13.4 Extracting tree policies with SafeVIPER

While most work on neural network interpretability [39, 6] fails to explain policy networks, previous work by Bastani et al. [4] uses imitation learning on the policy to extract decision trees that are both interpretable and easy to comprehend through manual analysis. Moreover, the decision trees can be formulated as sets of logical clauses of the input because the policy is piece-wise linear. Through formulating system clauses, the logical rules can be used to formally prove correctness [4].

The method introduced in [37] provides a pipeline that builds on this technique to create policies that are both safe and interpretable. The pipeline trains a non-interpretable RL agent for safe behavior, modifying existing reward structures and training techniques. The pipeline combines (Deep) RL advancements with safety and interpretability requirements. It extracts rules resembling the policy through a decision tree, which matches the performance and safety of the DNN agent while being easily interpretable. Additionally, the pipeline employs formal verification to ensure safety for linearized system dynamics. This approach extends beyond overtaking maneuvers, encompassing risk-sensitive driving scenarios like roundabout entry, exit, and intersection navigation. Notably, our work represents the first application of VIPER [4] and its verification concepts in an autonomous driving context.

13.4.1 Training the Policy

The SafeVIPER algorithm we introduced [37] works as follows: Decision trees are trained form the learned RL policy as student policies. We begin by initializing several elements, including a dataset of transitions denoted as $\mathcal{D}$, a dataset for critical transitions referred to as $\mathcal{K}$, the initial policy for aggregating experiences represented as $\hat{\pi}_0$, and the set of safe students denoted as Π_s. Subsequently, we repeatedly collect experience from the current policy $\hat{\pi}_i$ and incorporate this experience into our dataset $\mathcal{D}$. We identify critical trajectories, which are trajectories where $\hat{\pi}_i$ results in a violation of constraints, and include them in $\mathcal{K}$. Then, we train a new student policy using a dataset that comprises both $\mathcal{K}$ and a subset of $\mathcal{D}$. If the trained student policy successfully passes an evaluation test without violating constraints, it is added to Π_s. Eventually, after training a total of N students, we evaluate all policies within Π_s. We select and return the policy that attains the highest reward, taking into account a substantial penalty for constraint violations.

13.4.2 Verification of Decision Trees

The *verification* step formally proves the safety of the agent policy. Both the environment and the trained decision trees are formulated into sets of mathematical constraints. Then, we show that, when these constraints hold, a crash is formally impossible. This verifies the safety of the trained policies for linearized dynamics. We use a linearized lane change scenario as a running example.

Our verification approach is based on work by Bastani et al. [4]. The general idea is to translate the system dynamics, the agent policy, and the definition of a catastrophic episode into sets of constraints. Let S be the set of clauses that describe our system, $\mathcal{P}$ the set of clauses that describe the agent policy, and C the set of clauses that describe catastrophic situations, i.e., crashes. Formally, we prove that the following holds: $S \wedge \mathcal{P} \Longrightarrow \neg C$ In other words, if both the system dynamics and the policy hold, there is no solution that satisfies the crash conditions. As a consequence, such a system is deemed *safe*.

To prove this, we use the satisfiability modulo theories (SMT) solver Z3 [9] to find solutions to the negation, $\neg(S \wedge \mathcal{P} \Longrightarrow \neg C) \equiv \cdots \equiv S \wedge \mathcal{P} \wedge C$. If there are no

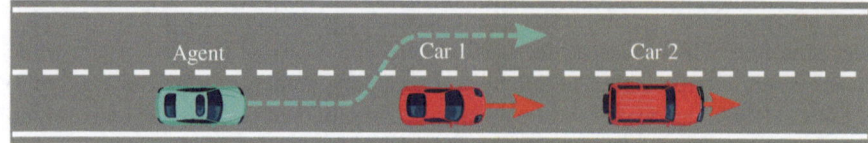

Fig. 13.8: Experimental scenario. The agent follows two cars on the right lane of a two-way highway. Car 1 is faster than Car 2, and will overtake. The agent has to decide when and how to overtake.

assignments that satisfy these equations, the first assumption must hold, and thus the system and the agent together are safe.

13.4.3 Evaluation

Autonomous vehicles frequently encounter overtaking scenarios on various road types and traffic conditions. The goal is to optimize speed and efficiency while ensuring safe maneuvers. Striking a balance between caution and performance is essential, as overly conservative behavior hinders efficiency. Meeting high safety standards is crucial for real-world deployment, achieved through rigorous verification methods. The objective is to minimize risk when making decisions. To show the viability of our approach, we therefore choose an overtaking maneuver.

For our scenario shown in Figure 13.8, we place three cars on the right lane of a highway. The foremost car (Car 2) drives slower than the middle car (Car 1). The agent's car is the fastest but behind the two. The longitudinal decisions of non-ego cars are made by standard intelligent driver model [44] rules, while lane change decisions are made according to the minimizing overall braking induced by lane change (MOBIL) model [19]. The car controlled by the agent faces a challenge in overtaking both vehicles simultaneously due to the possibility of the second vehicle initiating its own overtaking maneuver. Consequently, the agent must predict the expected response of the second vehicle and make a decision to either overtake both cars confidently or wait until the second car has completed its maneuver. Every episode begins with a random setup of the distances and velocities between the vehicles and lasts for a fixed duration of 40 time steps or stops in the event of a collision.

To evaluate the trained policies, we report the final longitudinal position of the agent car as a good score for the agent performance and the probability of crashes in 20 000 time steps of simulation. Additionally, we evaluate the policies in a *randomized* test environment. In this environment, the non-vehicle cars randomly change their speed. This makes safe operation harder for the agent.

The performance and probability of constraint violations of our training method and the baseline are presented in Figure 13.9. When compared to the unmodified baseline, our training method exhibits significantly safer behavior in terms of crash

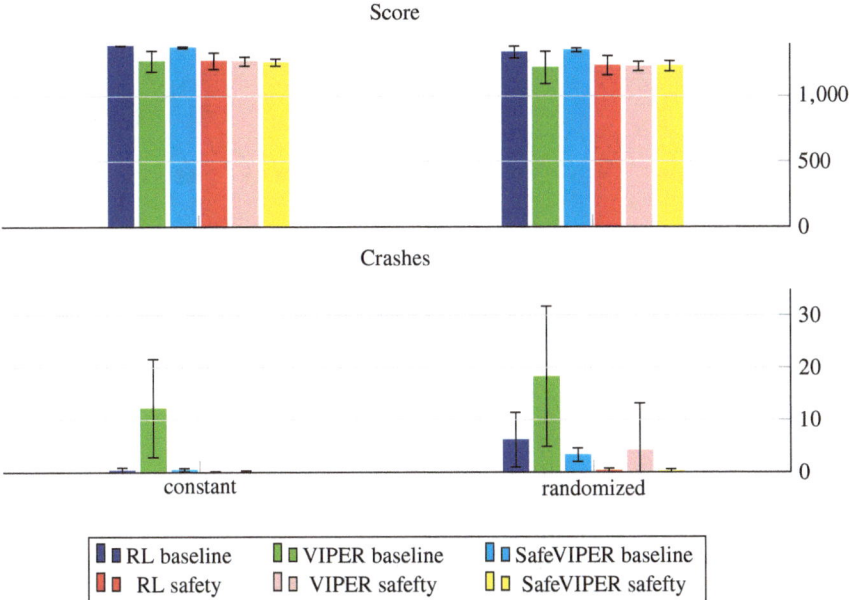

Fig. 13.9: Evaluation performance and crash frequencies for the *baseline* and *safety* training methods with mean and standard deviation across five replications. All policies are evaluated on a (*constant*) environment, and on a version where the other cars change their speed randomly (*randomized*). Trees are extracted both using the original VIPER algorithm [4], and with our modifications (SafeVIPER).

incidents, with a trade-off of less than 10% in performance. This holds also for the randomized movement of other cars, where the improvement is even more substantial. These findings demonstrate that our training method greatly enhances the system's resilience to changes in the environment. In individual replications, our method achieved a score of up to 1358 in the constant environment, without any constraint violations, and maintained perfect safety during the randomized evaluation.

Figure 13.10 shows a visualization of a tree extracted from a *safety* teacher policy using SafeVIPER. Its small size and clear decision criteria make the tree highly interpretable. Each decision can be understood easily by manually retracing the decisions made by the tree. This gives the opportunity to inspect the generated tree in detail, and to figure out if certain decisions are not optimal. We found that many decisions and subtrees permit a hierarchical interpretation: Tree nodes at the top of the tree establish a certain situation (e.g., slow non-agent-vehicles), while the sub-trees below these nodes perform detailed controlling tasks for, or reactions to, these situations.

In Figure 13.11 a detailed look at a part of the full tree is shown. Grey `subtree` nodes show parts of the tree that were cut away. This subtree splits between the relative speed and the lane distance of the two vehicles in our scenario. `closest`

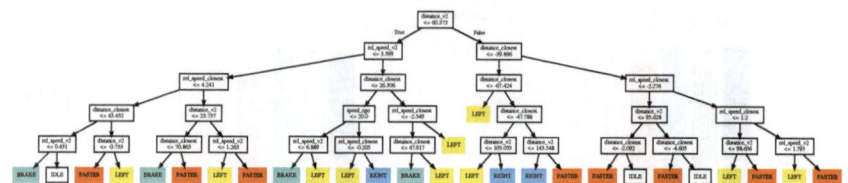

Fig. 13.10: Policy tree extracted from a PPO policy trained with the *safety* training modifications. Each node represents at the decision. To compute the output of the decision tree, start at the top node, and evaluate each node, stepping left or right, until a leaf node is reached. Leaf nodes are colored according to the action: **Change lane to the right**, change lane to the left, accelerate, brake, and **idle** (no dedicated action to take).
Image taken from [37] (C) IEEE 2021

denotes the vehicle with the smaller absolute lane distance to our vehicle, v2 is the vehicle further away. Here the decisions are chosen as follows: Nodes A and B check for the relative velocity of the two non-agent vehicles. Thus, if C is activated, we know that both vehicles are slower or only barely faster than the agent is. C splits on the distance to closest, going into the unlabeled subtree on the left if closest is either behind the ego vehicle or less than 43 m in front of it. E uses the distance to v2: if it is behind us, the agent accelerates, if it is in front of us, the agent changes lanes to the left. In our scenario, this is sensible: If closest is the slower vehicle and v2 (the faster one) is behind us, we can assume that we already changed the lane to overtake closest. Thus, with our lane clear, we can accelerate. If v2 is the slower vehicle, and both vehicles are at most around 4 m/s bit faster than we are (recall nodes A and B), we can change our lane to indicate that we will overtake, and expect the other vehicles to cooperate. On the other labeled subtree, nodes F and G seem to perform a distance control task.

13.5 Conclusion and Outlook

In our work, we developed several interconnected methods to make artificial intelligence (AI) explainable and reliable in the application of reinforcement learning (RL) to autonomous driving, thus showcasing dependable AI is attainable.

We introduced Driver Dojo [34] as a benchmark for prototyping, training, and evaluating agents across a wide range of scenario variations. It allows to generate a wide range of different driving scenarios under extended randomization. Apart from pre-implemented scenarios, we elaborated on how Driver Dojo allows for fast and efficient prototyping of new training environments and solutions.

SafeDQN [38] combines explicit risk estimation and Lagrangian learning to discover the best solutions for constrained Markov decision processes (CMDPs). Unlike

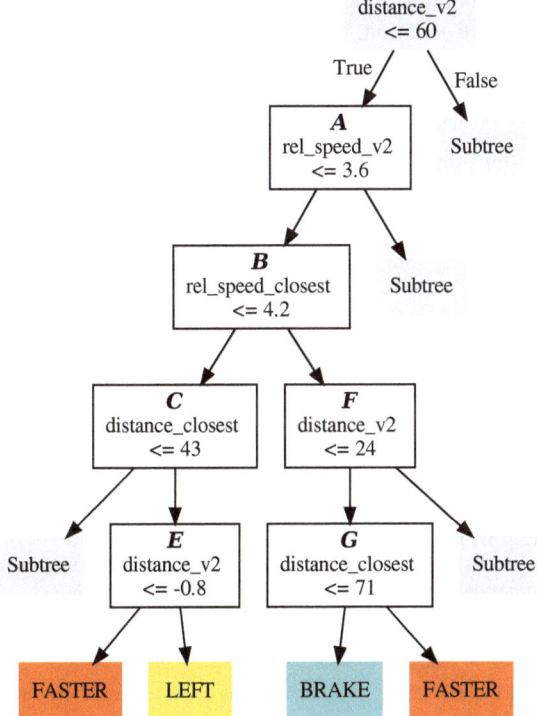

Fig. 13.11: Subtree of the full extracted tree.   Image from [37] (C) IEEE 2021

other methods, SafeDQN does not require manual tuning of hyperparameters, particularly for the reward function. It surpasses baseline approaches in terms of safety and average return across various traffic scenarios. Furthermore, SafeDQN's separate risk and utility estimators offer independent interpretation, bootstrapping, and training capabilities. The insights by the interpretability underscore the importance of evaluating safety components independently.

Through the combination of prior work on safe RL and interpretability in [37], we constructed a pipeline for the creation of interpretable, safe policies. To the best of our knowledge, we were the first to propose the adoption of VIPER [4] to tasks in autonomous vehicles with a verification of the safety of the generated trees. We presented SafeVIPER, adapting VIPER for extraction in CMDPs. Our policies achieve perfect safety in the environment they are trained in, and can be verified to be provably safe in linearized environments. They also generalize robustly to more demanding evaluations. Furthermore, the policies are small decision trees, which makes them highly interpretable.

We have shown approaches to introduce explainability and reliability in several instances. However, AI is being deployed more and more in various applications and devices that operate in a connected and cooperative environment. This is also important in scenarios involving autonomous mobility and traffic, where multiple agents must collaborate to find a shared solution. Since cooperation and communication among agents play a crucial role in these situations, approaches that focus solely on single-agent solutions often do not meet the desired outcomes. Instead, multi-agent reinforcement learning (MARL) addresses multi-agent problems and aims to discover policies that assist multiple vehicles in achieving both their individual and collective objective [36]. Thus, future work should incorporate the explainable and reliable aspects of dependable AI into MARL.

References

1. T. Agarwal, H. Arora, and J. Schneider. Affordance-based Reinforcement Learning for Urban Driving. *arXiv preprint arXiv:2101.05970*, 2021.
2. M. Althoff, M. Koschi, and S. Manzinger. Commonroad: Composable benchmarks for motion planning on roads. In *IEEE Intelligent Vehicles Symp.*, pages 719–726, 2017.
3. M. Bansal, A. Krizhevsky, and A. Ogale. Chauffeurnet: Learning to drive by imitating the best and synthesizing the worst. *arXiv preprint arXiv:1812.03079*, 2018.
4. O. Bastani, Y. Pu, and A. Solar-Lezama. Verifiable reinforcement learning via policy extraction. In *Advances in neural information processing systems*, pages 2494–2504, 2018.
5. M. G. Bellemare, Y. Naddaf, J. Veness, and M. Bowling. The arcade learning environment: An evaluation platform for general agents. *Journal of Artificial Intelligence Research*, 47:253–279, 2013.
6. A. Binder, G. Montavon, S. Lapuschkin, K. Müller, and W. Samek. Layer-wise relevance propagation for neural networks with local renormalization layers. In *Artificial Neural Networks and Machine Learning (ICANN)*, pages 63–71, Barcelona, Spain, 2016.
7. M. Bojarski, D. Del Testa, D. Dworakowski, B. Firner, B. Flepp, P. Goyal, L. D. Jackel, M. Monfort, U. Muller, J. Zhang, et al. End to end learning for self-driving cars. *arXiv preprint arXiv:1604.07316*, 2016.
8. C. Chen, A. Seff, A. Kornhauser, and J. Xiao. DeepDriving: Learning affordance for direct perception in autonomous driving. In *IEEE Int. Conf. on Computer Vision*, pages 2722–2730, Dec. 2015.
9. L. M. de Moura and N. Bjørner. Z3: an efficient SMT solver. In C. R. Ramakrishnan and J. Rehof, editors, *TACAS 2008*, volume 4963 of *Lecture Notes in Computer Science*, pages 337–340. Springer, 2008.
10. J. Deng, S. Sierla, J. Sun, and V. Vyatkin. Reinforcement learning for industrial process control: A case study in flatness control in steel industry. *Comput. Ind.*, 143:103748, 2022.
11. A. Dosovitskiy, G. Ros, F. Codevilla, A. Lopez, and V. Koltun. Carla: An open urban driving simulator. In *Conference on robot learning*, pages 1–16, 2017.
12. A. R. Fayjie, S. Hossain, D. Oualid, and D. J. Lee. Driverless car: Autonomous driving using deep reinforcement learning in urban environment. *2018 15th International Conference on Ubiquitous Robots (UR)*, pages 896–901, 2018.
13. Q. Göttl, Y. Tönges, D. G. Grimm, and J. Burger. Automated flowsheet synthesis using hierarchical reinforcement learning: Proof of concept. *Chemie Ingenieur Technik*, 2021.
14. S. S. Gu, E. Holly, T. P. Lillicrap, and S. Levine. Deep reinforcement learning for robotic manipulation with asynchronous off-policy updates. *2017 IEEE International Conference on Robotics and Automation (ICRA)*, pages 3389–3396, 2016.
15. L. He, N. Aouf, and B. Song. Explainable deep reinforcement learning for uav autonomous path planning. *Aerospace Science and Technology*, 118:107052, 2021.
16. P. Henderson, R. Islam, P. Bachman, J. Pineau, D. Precup, and D. Meger. Deep reinforcement learning that matters. In *Proceedings of the AAAI conference on artificial intelligence*, volume 32, 2018.

17. G. Kalweit, M. Huegle, M. Werling, and J. Boedecker. Deep constrained q-learning. *arXiv preprint arXiv:2003.09398*, 2020.
18. A. Kendall, J. Hawke, D. Janz, P. Mazur, D. Reda, J.-M. Allen, V.-D. Lam, A. Bewley, and A. Shah. Learning to drive in a day. In *IEEE Int. Conf. Robotics and Automation (ICRA)*, pages 8248–8254, 2019.
19. A. Kesting, M. Treiber, and D. Helbing. General lane-changing model mobil for car-following models. *Transportation Research Record*, 1999(1):86–94, 2007.
20. B. R. Kiran, I. Sobh, V. Talpaert, P. Mannion, A. A. Al Sallab, S. Yogamani, and P. Pérez. Deep reinforcement learning for autonomous driving: A survey. *IEEE Trans. Intell. Transportation Systems*, 2021.
21. R. Kirk, A. Zhang, E. Grefenstette, and T. Rocktäschel. A survey of generalisation in deep reinforcement learning. *arXiv preprint arXiv:2111.09794*, 2021.
22. R. Krajewski, J. Bock, L. Kloeker, and L. Eckstein. The highd dataset: A drone dataset of naturalistic vehicle trajectories on german highways for validation of highly automated driving systems. In *IEEE Int. Conf. Intell. Transportation Systems (ITSC)*, pages 2118–2125. IEEE, 2018.
23. H. Krasowski, X. Wang, and M. Althoff. Safe reinforcement learning for autonomous lane changing using set-based prediction. In *IEEE Int. Conf. on Intelligent Transportation Systems*, 2020.
24. D. Landgraf, A. Völz, G. Kontes, C. Mutschler, and K. Graichen. Hierarchical learning for model predictive collision avoidance. *IFAC-PapersOnLine*, 55(20):355–360, 2022.
25. E. Leurent. *Safe and Efficient Reinforcement Learning for Behavioural Planning in Autonomous Driving*. PhD thesis, University of Lille, France, 2020.
26. Z. C. Lipton. The mythos of model interpretability. *Commun. ACM*, 61(10):36–43, 2018.
27. P. A. Lopez, M. Behrisch, L. Bieker-Walz, J. Erdmann, Y.-P. Flötteröd, R. Hilbrich, L. Lücken, J. Rummel, P. Wagner, and E. Wießner. Microscopic traffic simulation using sumo. In *IEEE Int. Conf. Intell. Transportation Systems (ITSC)*, 2018.
28. R. McAllister, Y. Gal, A. Kendall, M. Van Der Wilk, A. Shah, R. Cipolla, and A. Weller. Concrete problems for autonomous vehicle safety: Advantages of bayesian deep learning. In *Int. Joint Conference on Artificial Intelligence*, Aug. 2017.
29. V. Mnih, K. Kavukcuoglu, D. Silver, A. A. Rusu, J. Veness, M. G. Bellemare, A. Graves, M. Riedmiller, A. K. Fidjeland, G. Ostrovski, et al. Human-level control through deep reinforcement learning. *Nature*, 518(7540):529–533, 2015.
30. T. Osa, J. Pajarinen, G. Neumann, J. A. Bagnell, P. Abbeel, J. Peters, et al. An algorithmic perspective on imitation learning. *Foundations and Trends® in Robotics*, 7(1-2):1–179, 2018.
31. B. Osiński, A. Jakubowski, P. Zięcina, P. Miłoś, C. Galias, S. Homoceanu, and H. Michalewski. Simulation-based reinforcement learning for real-world autonomous driving. In *IEEE Int. Conf. Robotics and Automation (ICRA)*, pages 6411–6418, 2020.
32. V. R. Padullaparthi, S. Nagarathinam, A. Vasan, V. P. Menon, and D. Sudarsanam. Falcon-farm level control for wind turbines using multi-agent deep reinforcement learning. *Renewable Energy*, 2022.
33. N. Rhinehart, J. He, C. Packer, M. A. Wright, R. McAllister, J. E. Gonzalez, and S. Levine. Contingencies from observations: Tractable contingency planning with learned behavior models. *arXiv preprint arXiv:2104.10558*, 2021.
34. S. Rietsch, S.-Y. Huang, G. Kontes, A. Plinge, and C. Mutschler. Driver dojo: A benchmark for generalizable reinforcement learning for autonomous driving, 2022.
35. A. Sauer, N. Savinov, and A. Geiger. Conditional affordance learning for driving in urban environments. In *Conference on Robot Learning*, pages 237–252, 2018.
36. L. M. Schmidt, J. Brosig, A. Plinge, B. M. Eskofier, and C. Mutschler. An introduction to multi-agent reinforcement learning and review of its application to autonomous mobility. pages 1342–1349, 2022.
37. L. M. Schmidt, G. D. Kontes, A. Plinge, and C. Mutschler. Can you trust your autonomous car? interpretable and verifiably safe reinforcement learning. In *IEEE Intelligent Vehicles Symposium*, pages 171–178, Nagoya, Japan, July 2021.

38. L. M. Schmidt, S. Rietsch, B. M. Eskofier, A. Plinge, and C. Mutschler. How to learn from risk: Explicit risk-utility reinforcement learning for efficient and safe driving strategies. pages 1913–1920, 2022.
39. R. R. Selvaraju, M. Cogswell, A. Das, R. Vedantam, D. Parikh, and D. Batra. Grad-cam: Visual explanations from deep networks via gradient-based localization. In *Proc. IEEE Intl. Conf. Computer Vision*, pages 618–626, 2017.
40. S. Shalev-Shwartz, S. Shammah, and A. Shashua. Safe, multi-agent, reinforcement learning for autonomous driving. *arXiv preprint arXiv:1610.03295*, 2016.
41. S. Shalev-Shwartz, S. Shammah, and A. Shashua. On a formal model of safe and scalable self-driving cars. *arXiv preprint arXiv:1708.06374*, 2017.
42. A. Stooke, J. Achiam, and P. Abbeel. Responsive safety in reinforcement learning by PID lagrangian methods. In *Int. Conf. on Machine Learning (ICML)*, volume 119 of *Proceedings of Machine Learning Research*, pages 9133–9143. PMLR, 2020.
43. C. Tessler, D. J. Mankowitz, and S. Mannor. Reward constrained policy optimization. In *Int. Conf. Learning Representations, (ICLR)*, 2019.
44. M. Treiber, A. Hennecke, and D. Helbing. Congested traffic states in empirical observations and microscopic simulations. *Physical Review E*, 62(2):1805–1824, Aug 2000.
45. H. van Hasselt, A. Guez, and D. Silver. Deep reinforcement learning with double q-learning. In *AAAI Conf. Artificial Intelligence*, pages 2094–2100, Phoenix, AZ, USA, 2016.
46. T. Wei, Y. Wang, and Q. Zhu. Deep reinforcement learning for building hvac control. *2017 54th ACM/EDAC/IEEE Design Automation Conference (DAC)*, pages 1–6, 2017.
47. W. Zhan, L. Sun, D. Wang, H. Shi, A. Clausse, M. Naumann, J. Kummerle, H. Konigshof, C. Stiller, A. de La Fortelle, et al. Interaction dataset: An international, adversarial and cooperative motion dataset in interactive driving scenarios with semantic maps. *arXiv preprint arXiv:1910.03088*, 2019.

Chapter 14
AI for Stability Optimization in Low Voltage Direct Current Microgrids

Georg Roeder[1], Raffael Schwanninger[2], Peter Wienzek[1], Moritz Kerscher[1], Bernd Wunder[1], Martin Schellenberger[1]

Abstract Low voltage direct current (LVDC) is an enabling technology to foster a sustainable resilient energy supply. LVDC microgrids comprising energy generators, storage systems, and loads work as independently controlled units in connection with common alternating current networks. Precise digitized control applying intelligent power converters enables new AI-based approaches for DC microgrid layout and operation. In this work, a new method involving connected machine learning and optimization is established together with a novel measurement system, which enables the measurement and improvement of microgrid stability. The application is successfully validated by experimental assessment on a testbed with a four-terminal DC network operating at a voltage of 380 V_{DC} and the advantages of the AI-based approach are demonstrated.

Key words: LVDC microgrid, stability, digital twin, random forest, optimization, PRBS measurement

14.1 Introduction

The demand for climate-friendly, resource-efficient, and resilient energy supply requires an increasing share of renewable energy sources and implementation of efficient energy distribution systems. For more than a century, large-scale energy supply has been realized by alternating current (AC) power grids [4], [5]. Through transformation to high voltage, AC enables cost-effective transport of large amounts of

---

[1]Fraunhofer Institute for Integrated Systems and Device Technology IISB, Erlangen, Germany
[2]Friedrich-Alexander-Universität Erlangen-Nürnberg, Institute for Power Electronics, Nuremberg, Germany

Corresponding author: Georg Roeder
e-mail: `georg.roeder@iisb.fraunhofer.de`

© The Author(s) 2024
C. Mutschler et al. (eds.), *Unlocking Artificial Intelligence*,
https://doi.org/10.1007/978-3-031-64832-8_14

energy over long distances. At the same time, the grid frequency can be kept stable during changing power supply and load requirements due to the power generators inertia. Today, the growing usage of sustainable energy sources, storage systems, and consumers has led to an increasing relevance of Low Voltage Direct Current (LVDC) microgrids with a supply system voltage up to 1500 V [10], where these components are efficiently and directly connected and operate as a subsystem of the AC grid [19]. The interconnection of DC sources, storage systems, and consumers avoids unnecessary AC/DC conversions, enables a cost-efficient design of the sub-grid [30], and precise digitized control due to the application of intelligent power converters and AI methods [20]. With the increasing complexity of LVDC networks due to the growing number of components, stabilization of grid control is a challenge. New measurement systems are required to non-intrusively measure grid stability involving many grid components. At the same time, artificial intelligence techniques should be applied in the stability assessment to handle the complex and time-consuming analysis and provide means for parameter adjustment to optimize grid stability during operation. A new solution involving AI for stability optimization in LVDC microgrids is described in the following sections.

14.2 Low Voltage DC Microgrids

14.2.1 Control of Low Voltage DC Microgrids

A typical topology of LVDC microgrids is the bus topology [10], [19], whereby DC grids can also be implemented in other configurations such as radial, multi-terminal, ring-bus, ladder, and zonal topologies [19]. Figure 14.1 shows an example four-terminal DC microgrid network with a bus topology as a simplified part of a larger DC grid, which is realized in a laboratory testbed and was investigated further in this work.

For the control of multiple source based DC microgrids, hierarchical control is frequently employed to maintain DC bus voltage, control load sharing, maintain power quality, and to increase independence between the control levels [10], [19]. The primary control level addresses tasks such as current and voltage regulation and power coupling. The secondary control level addresses voltage maintenance and improvement of power quality, whereas the tertiary, regulatory level focuses energy management, system optimization, and economic distribution [10], [19]. The DC microgrid shown in Figure 14.1 is realized applying a frequently employed decentralized control scheme for power sharing and performance and voltage regulation [10], [19], [30], [20] at the secondary control level due to its inherent reliability in case of failures [10], [19], [30] and the avoidance of centralized control of local converters and digital communication links. Primary control is realized by using a current-mode droop control scheme [10], [30], [20] with the objective to maintain the bus voltage at 380 V_{DC}. Figure 14.2 shows the arrangement of the droop controller and the and inner control loop of a DC/DC converter for maintaining the DC grid

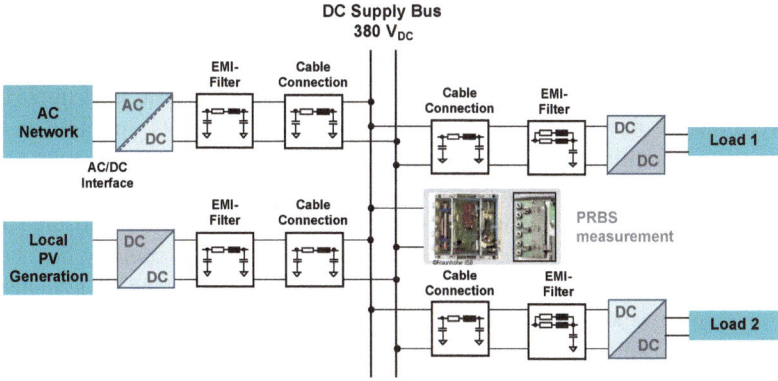

Fig. 14.1: Example four-terminal DC microgrid network with a bus topology. The source and the load branches contain the converters including control elements, electromagnetic interference (EMI) filters, and the cable connection. Additionally, the pseudorandom binary sequence (PRBS) measurement for stability determination is indicated, which is conducted at the DC supply bus.

at a given bus voltage. The droop curves adjust the power sharing between sources and loads, whereas the current flow controller and power electronics maintain stable coupling to the network depending on the adjustments of the droop controller. By using a PID controller, deviations between the the target value I_{set} generated by the droop curve and the actual value I are compensated.

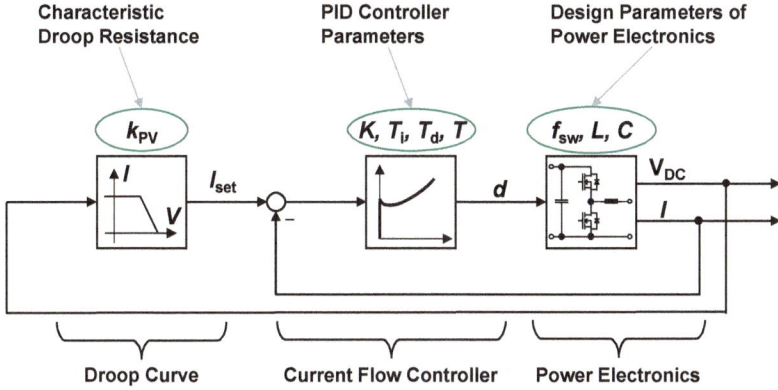

Fig. 14.2: Droop controlled power electronics system for maintaining the DC grid at a given bus voltage.

14.2.2 Stability of Low Voltage DC Microgrids

The monitoring and optimization of the stability in DC microgrid is of importance to avoid reverse power flows in the distributed generation units. Instabilities arise due to the generation of oscillations, which are caused by switching operations that can generate high-frequency alternating currents due to low inertia characteristics of the microgrid and supply-demand uncertainties [24]. Hence, even if the source and load subsystems are stable themselves, the overall system stability may be degraded [22]. A review of stability criteria for DC power distribution systems and control techniques for stability improvement are given in [22], [17]. Stability may be distinguished in small-signal stability, voltage stability, and transient stability [17]. Small-signal stability is the system's ability to maintain stable under low amplitude disturbances occurring at various frequencies. Voltage stability describes the ability to maintain the steady-state value of the voltage in the normal and abnormal operating conditions. Transient stability denotes the system's ability to maintain a stable state under severe transient disturbances. Within this work, small-signal and voltage stability are addressed as relevant topics in regular microgrid operation and the gain and phase margin criterion was used for stability assessment, which relates to the minor-loop gain criteria (MLGC) [22]. The concept of small-signal stability assumes, that a system can be subjected to perturbation analysis, where the leading terms dominate and enable modeling as a linear time invariant (LTI) system.

The stability of DC grids can be measured by means of systems theory and control engineering. For the investigation of stability, the DC grid is typically divided into two systems, usually the first subsystem consisting of all sources and the second consisting of all loads (Figure 14.3). However, a division can also be carried out for a spatially resolved analysis to the converter and network side in the branches of the network. The figure depicts the impedances of the source Z_s and the load Z_l as well as the voltages V_s, V_l and currents I_s, I_l to calculate these impedances, which are required to derive the stability from small-signal analysis or the PRBS measurement. The PRBS measurement and calculation of these impedances are explained in more detail in Section 14.4.1.

The transfer function G_{sl} of two individually stable source and load subsystems as shown in Figure 14.3 is given as:

$$G_{sl} = \frac{V_s}{V_l} = G_s G_l \frac{Z_l}{Z_l + Z_s} = G_s G_l \frac{1}{1 + T_{MLG}}, \tag{14.1}$$

where T_{MLG} is the minor loop gain, which is given by:

$$T_{MLG} = \frac{Z_s}{Z_l} = T_{bus}, \tag{14.2}$$

and G_s, G_l are the transfer functions of the source and load system. According to control theory, the interconnected system is stable if the Nyquist contour of T_{MLG} does not encircle the $(-1|0)$ point in the complex plane [22]. The gain margin and phase margin criterion enables that in certain frequency ranges the source impedance

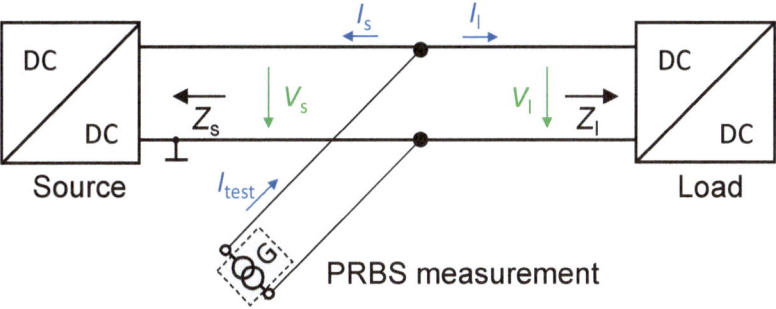

Fig. 14.3: Division of the DC microgrid into two systems, usually the sources and the loads. The impedances of the source Z_s and the load Z_l as well as the voltages V_s, V_l and currents I_s, I_l, which are required to calculate these impedances, are indicated. These parameters are required to derive the stability from small-signal analysis or the PRBS measurement.

amplitude $|Z_s|$ may be larger than the load amplitude $|Z_l|$ but ensures margins such that the Nyquist criterion is satisfied [22] and avoids the excitation of the system. At a phase angle of $-180°$, which occurs in the resonance point of a system, the gain margin ensures that the amplitude of $|T_{bus}(j\omega)|$ is sufficiently damped, i.e., is sufficiently distant from the point $(-1|0)$. The gain margin GM is defined as [31]:

$$GM = \frac{1}{|T_{bus}(j\omega)|}.$$ (14.3)

In the Bode plot, where the amplitude is depicted in decibels, the gain margin can be determined as:

$$GM \, dB = 20 \, log_{10} GM = 0 \, dB - 20 \, log_{10} |T_{bus}(j\omega)|.$$ (14.4)

The phase margin describes whether the phase angle is sufficiently distant to the resonance angle at $-180°$. The phase margin is defined as:

$$PM = arg \, T_{bus}(j\omega) - (-180°).$$ (14.5)

A system is then considered to be stable if both, the gain margin GM and the phase margin PM, exceed the threshold values GM_c, PM_c characterizing a sufficient damping and distance to the resonance angle:

$$GM \geqslant GM_c \wedge PM \geqslant PM_c.$$ (14.6)

Typical values for GM_c and PM_c are 6 dB and 45° but may be varied depending on the application. The MLGC is a sufficient stability criterion, i.e., the system is stable if the criteria are met but otherwise the system is not necessarily unstable. If the system is unstable according to MLGC, typically the operating point is retained until the margins get close to 0. In this case, the system undergoes resonance excitation and the operating point can no longer be maintained. This effect will be discussed further in Section 14.4.2.

14.3 AI-based Stability Optimization for Low Voltage DC Microgrids

14.3.1 Overview

LVDC microgrid layout for stability is typically based on empirical knowledge and is carried out by experts with a high level of expertise in network design, control technology, and stability analysis. Moreover, until now, there have been no sufficiently compact, interaction-free and fast measurement methods available for routine monitoring and optimization of grid stability. Through the use of the pseudorandom binary sequence (PRBS) measurement technology [18] in conjunction with artificial intelligence methods, a solution is created to automatically analyze and evaluate the stability in DC grids and to determine stable settings from the multitude of parameters influencing grid stability. This is achieved by optimization of software adjustable parameters, e.g., changing the droop control characteristics. The methods developed are intended for use in regular operation of DC grids and to support experts in the DC grid design. The approach comprises several steps related to machine learning, optimization, and implementation, which are depicted in Figure 14.4. The approach is described further in the following sections.

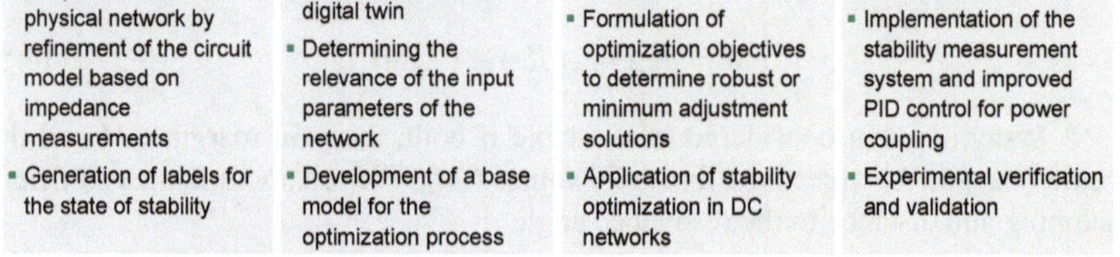

Digital Network Twin	LVDC Microgrid Surrogate Model	Stability Optimization	Implementation and Assessment
• Automated calculation of small-signal DC network impedance and stability assessment for many input parameter settings • Adaptation to the physical network by refinement of the circuit model based on impedance measurements • Generation of labels for the state of stability	• Modeling the relationship between grid parameters and stability using random classification forests based on the labels derived in the digital twin • Determining the relevance of the input parameters of the network • Development of a base model for the optimization process	• Reformulation of the random forest decision process as mixed-integer and possibly non-linear program (MINLP) • Formulation of optimization objectives to determine robust or minimum adjustment solutions • Application of stability optimization in DC networks	• Stability optimization for a four participant DC network, which is available for experimental assessment in a testbed • Implementation of the stability measurement system and improved PID control for power coupling • Experimental verification and validation

Fig. 14.4: Approach for optimization of the grid stability.

14.3.2 Digital Network Twin and Generation of Labels to Describe the Stability State

Electrical networks such as DC microgrids comprise numerous subsystems, components, and devices such that network analysis requires electronic circuit simulation. To obtain the small-signal stability behavior of the example microgrid as depicted in Figure 14.1 for multiple possible input parameter settings, a digital network twin for automated calculation of the small-signal DC network impedance and the impedance-based stability evaluation according to the MLGC was realized [23]. The digital twin is based on a grid circuit model capable to determine the static and dynamic behavior of the entire grid and to assess the stability for different operating points. The digital twin was realized using the circuit simulator LTspice [2]. In this model, the power converters are linearized around the operating point with their frequency response sufficiently below the converter switching frequency [20]. In the digital twin, relevant software-controlled factors such as the droop curve at the source converters and the cut-off frequency of the output filter at the DC/DC converter can be varied, which may also be changed for control purposes during network operation. Furthermore, important parameters of the grid design may be investigated, e.g., the output capacitance and equivalent series resistance of the source converters, the wire lengths from the AC source, the PV system, and of the load groups to the bus node. The calculations are conducted varying predefined network parameters as described above and for multiple load value combinations at Load 1 and Load 2. Parameter variation is conducted by Latin hypercube sampling [14], [28]. A small-signal frequency sweep in the range of 10 Hz to 100 kHz is conducted to calculate T_{bus} according to (14.2), assess the gain and phase margin according to (14.4) and (14.5), and to provide a label that the grid is stable if both the gain and the phase margin are larger than the threshold values as given in (14.6). Otherwise, the network state is labeled as unstable. As a result, a table comprising input vectors, which optimally fill the input parameter space and the respective labels indicating the stability state is available for further use in surrogate modeling by random forests.

14.3.3 LVDC Microgrid Surrogate Model Applying Random Forests

To describe the relation of the inputs and labels obtained in the digital twin, surrogate modeling with random forests was employed [1]. Random forests were selected as a machine learning algorithm in preparation of the stability optimization task since tree and random forest classifiers may either be trained in a globally optimal way using mixed-integer programming methods [6], [3] or the decisions of trained classifiers may suit as base models in optimization (see [12] and references therein). Random classification forests are an ensemble learning method that invoke the construction of a multitude of decision trees [8], [13], which averages predictions over the individual trees with the objective to reduce the generalization error on unknown samples or parameter input [7], [15], [16], [25] (Figure 14.5). Furthermore, they enable the analysis of variable importance of the input on the target parameter [7], i.e., they provide information to the engineer, which parameters are the most relevant for obtaining stability.

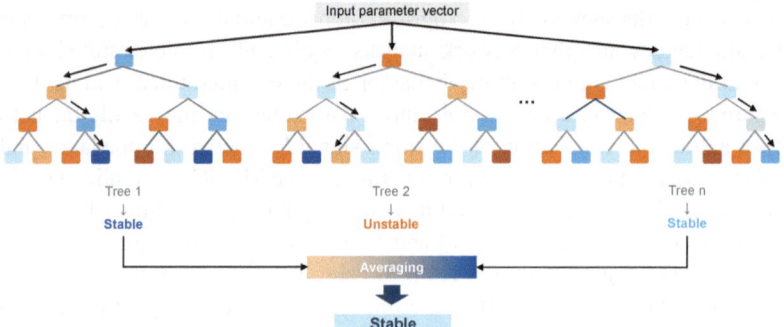

Fig. 14.5: Schematic representation of ensemble learning with random forests. The colors and hue indicate the probability of a stable or unstable class.

For random forest modeling the data set was split into a training and an independent test data set with a ratio of 70:30. The hyperparameters of the random forest models were optimized using a grid search strategy with a 10-fold stratified cross-validation [21]. The potential imbalance of the stable and unstable classes was taken into account [23], [9], [26] by adjusting the loss function L to maximize the average area under the receiver operating characteristics curve A_{ROC} and the precision-recall curve A_{PR}:

$$L = max[\frac{1}{2}(A_{ROC} + A_{PR})]. \tag{14.7}$$

The adjustment of L improves the balanced accuracy score and reduces false positives, i.e., to predict unstable states as stable. The random forests were calculated for a variety of input parameter variations applied in the digital twin typically varying all possible parameters within specified parameter ranges or varying the parameters k_{AC}, k_{PV}, which adjust the droop characteristic during regular operation. The load values were not considered as parameters in the random forest as the predictions should be valid over the complete range of load settings. As an example case with a variation of two parameters to be adjusted, Figure 14.6 shows the map of the random forest probabilities to obtain a stable state when varying k_{AC}, k_{PV} and both loads between 0 W and 10000 W. If adjustment for more parameters is required, the additional respective cross sections in the parameter space may be visualized [12]. The labels for the random forest modelling were generated with the digital twin for 24981 different parameter settings with a resulting imbalance ratio of the positive to the negative class labels of $n_{stable}/n_{unstable} = 1.26$ [9]. The random forest model was generated with the hyperparameter optimization procedures as described above with the load values not included. The minimum samples per leaf node were set to $n_{min} = 11$ to obtain reasonably large leaf node sizes in order to estimate the decision probability. For the optimized model, a balanced accuracy of 78% and a precision of 87% was obtained selecting 500 trees, considering the square root of the maximum number of features for the best split, and adjusting the class weights for the stable class as $w_{stable} = 0.35$ and for the unstable class as $w_{unstable} = 0.65$ from the hyperparameter input space.

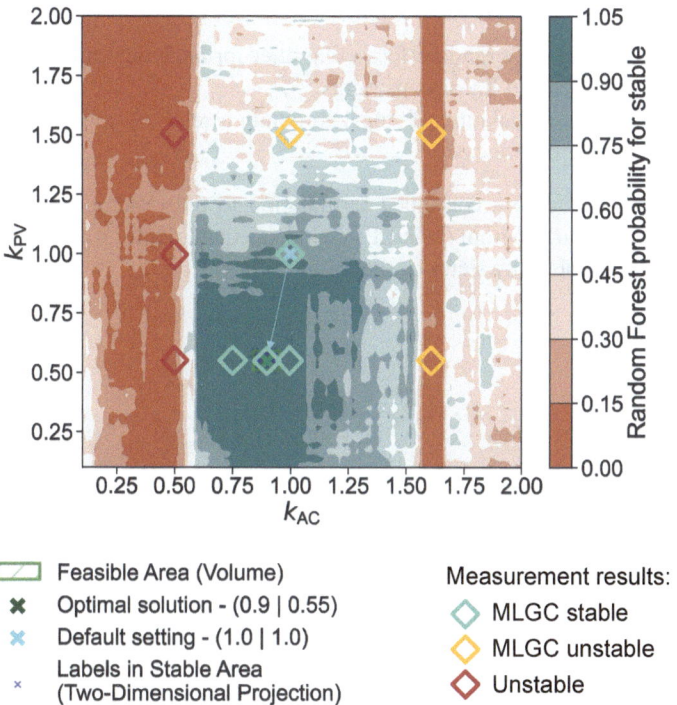

Fig. 14.6: Map of the random forest probabilities to obtain a stable state when varying the characteristic parameters k_{AC}, k_{PV} of the converter droop curves and loads between 0 W and 10000 W. The results are shown for an adapted digital twin model of the four-terminal DC laboratory testbed. Additionally, the figure shows the result of the optimization to find the largest area around the new setpoint, where a larger stable region can be obtained by changing the free optimization parameters the k_{AC}, k_{PV} from (1.0|1.0) to (0.9|0.55). The results from measurements in the testbed are indicated as squares. Green squares indicate MLGC stable systems, orange squares indicate MLGC violating systems that can be measured and red squares indicate unstable systems not able to find stable operating points.

14.3.4 Stability Optimization Applying Decision Trees

For the optimization of the parameter setting of the DC network, a novel optimization approach was proposed, which exploits the relations of the input parameters and stability labels and enables the identification of improved settings in a classification problem. The approach is described in detail in [12] and foundations of optimization and the integration of decision trees are provided in Chapter 7 of this book, so here, only the major concepts are depicted. The approach for optimization over the random forests classification comprises the following steps:

1. Reformulation of the random forest classifier for analytical modeling by mixed-integer variables and linear constraints,
2. refinement the formulation of step 1 to obtain a relevant optimization objective for the considered use case.

These steps lead to a continued formulation to the general mixed-integer and possibly non-linear program (MINLP) with decision variables x and their non-linear relation with the stability states of the following form:

$$\min \quad c^T x \tag{14.8}$$

$$s.t. \quad Dx \leq d \tag{14.9}$$

$$x \in \mathbb{Z}^{p-q} \times \mathbb{R}^q \tag{14.10}$$

$$x \in \mathcal{F}. \tag{14.11}$$

Step 1 provides a formulation of Eq. (14.11) to describe the decisions in an existing random forest model. The inclusion of different optimization objectives in step 2 provides additions and modifications to Eq. (14.8) and the constraints of Eqs. (14.9) and (14.11). Here, several relevant use cases were implemented with two different solvers [11], [29]: Finding the closest solution from a starting point, finding the minimum number of parameters to be adjusted, finding solutions with a margin distance to compensate for parameter fluctuations, and finding solutions, with the largest stable area either as the largest volume of the polytope or the largest volume of an inscribed sphere to maintain equal distances to the boundaries. Figure 14.6 shows the result of the optimization finding the largest stable volume of the polytope, which is a rectangle for two dimensions. It is seen that the adjustment of the software-controllable parameters k_{AC}, k_{PV} leads to an improved setting from the default starting point at $(1.0|1.0)$ to $(0.9|0.55)$, i.e., the stability may be adjusted during regular grid operation.

14.4 Implementation and Assessment

14.4.1 Measurement of Grid Stability

As shown in the previous sections, the use of impedance-based criteria in DC microgrids allows for stability assessments of an otherwise unknown grid. If the individual components of the grid are partially known, the impedance can also be used to refine the models and therefore increase the model accuracy, even in not directly measured operating points. The prerequisite for either approach is the online impedance measurement. In this work a pseudo-random binary sequence (PRBS) is employed as the excitation signal for the impedance measurement. The measurement principle is depicted in Figure 14.3. A broad band test current I_{test} is injected on the bus bar between the sources and loads. I_{test} is then split up into one part directed towards the source I_s and a part flowing into the load I_l. Together with the voltages V_s and V_l, the source and load impedances Z_s and Z_l can be calculated according to eq. (14.12). Z_s and Z_l are then used to determine T_{bus} as shown in Section 14.2.2.

$$Z_s = \frac{V_s}{I_s} \quad Z_l = \frac{V_l}{I_l} \tag{14.12}$$

The benefit of the PRBS-based impedance measurement is that the coupling system can be built very compact (Figure 14.8) and can operate minimally invasive between measurement sequences. In this work, the test signal is coupled through the periodic switching of a 100 Ω power resistor in and out of the grid using a power MOSFET. Other measurement systems based on serially coupled sine wave excitation require coupling transformers, which become very large at high DC currents and change the physical properties of the system due to their non negligible inductance. An exemplary result of a Bode plot displaying $|Z_s(j\omega)|$ and $arg\,Z_s(j\omega)$ of a PRBS impedance measurement of an individual converter in comparison to a reference measurement using a serial coupling is shown in Figure 14.7.

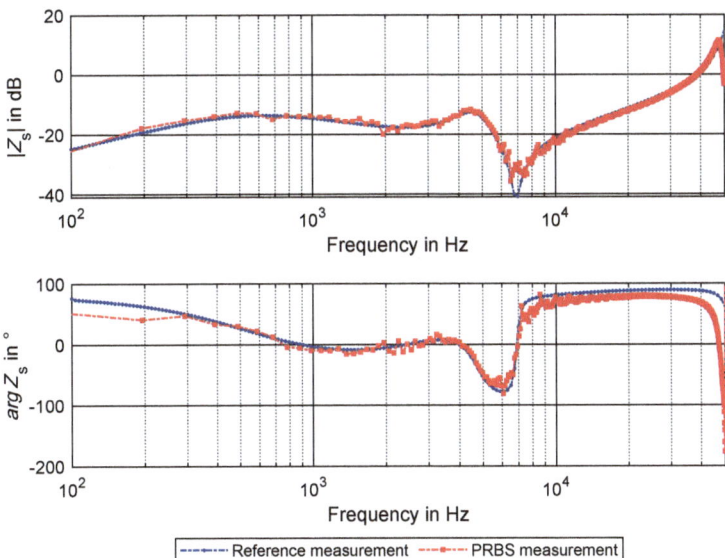

Fig. 14.7: Bode plot determined for a single converter with PRBS compared to a highly accurate reference measurement. The PRBS measurement shows comparable results, while increasing measurement speed at reduced volume and weight.

Aside from very low impedances, the PRBS measurement shows comparable results to the reference measurement. It is important to note, that the simultaneous measurement of 2047 frequencies with PRBS was carried out within 112 ms, while each frequency point with the reference measurement took 330 ms for the shown resolution. Using the reference measurement system for the same frequencies would therefore lead to a 6031 fold increase of measurement time. Also, the volume and weight of the PRBS impedance measurement system is significantly lower compared to the reference measurement system, while being able to carry five times the DC current.

14.4.2 Experimental Validation

The four-terminal DC microgrid network with a bus topology together with the PRBS measurement (Figure 14.1) were realized as demonstrator on a testbed operating at 380 V_{DC} to prove the feasibility of the approach in an experimental validation. Figure 14.8 shows the demonstrator of the four-terminal network with the integrated PRBS measurement system.

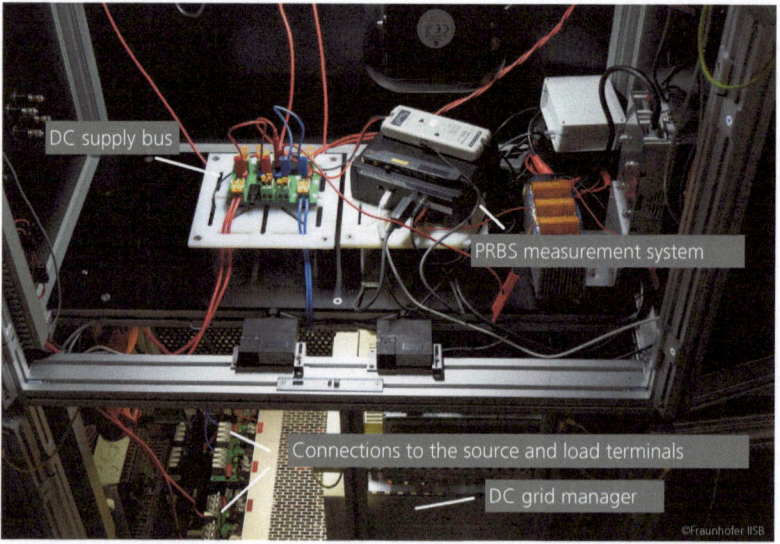

Fig. 14.8: Demonstrator of the four-terminal network with the integrated PRBS measurement system.

For validation of the overall approach, the map showing the random forest probabilities (Figure 14.6) was verified using online PRBS impedance measurement and applying the MLGC from Eq. (14.6). The map projects a multitude of power values for both loads, whereas in the dedicated validation experiments on the testbed, the load dependencies were evaluated at dedicated values. Hence, the grid was measured

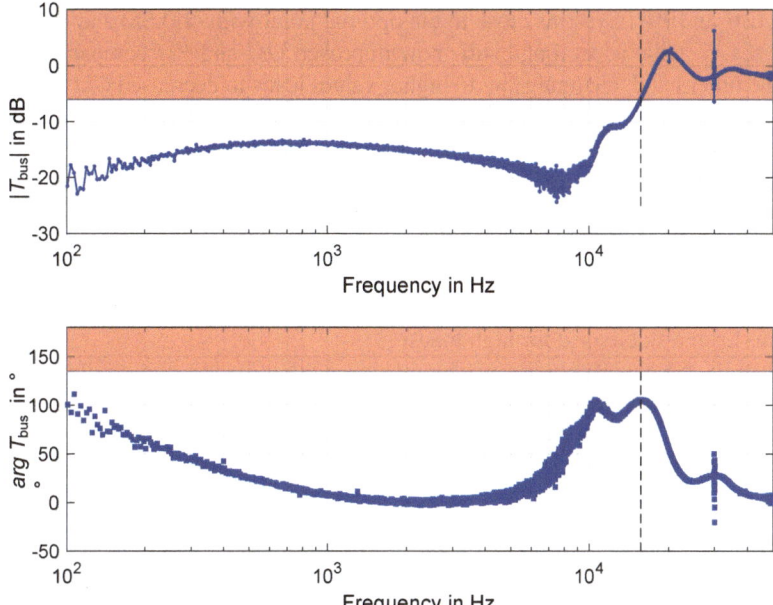

Fig. 14.9: Bode plot determined from a PRBS measurement for a stable operation point with k_{AC}, k_{PV} at (0.9|0.55).

at various k_{AC} and k_{PV} values at low loads (800 W), medium loads (6 kW), and high loads (12 kW). The resulting stability states are indicated as squares in Figure 14.6. To illustrate the approach for determining the stability state, Figure 14.9 shows the Bode plot displaying $|T_{bus}(j\omega)|$ and $arg\,T_{bus}(j\omega)$ of a PRBS measurement recorded at a stable operating point with k_{AC}, k_{PV} at (0.9|0.55). The red areas indicate the regions where the gain or phase requirements are exceeded. It can be seen, that while the gain requirement is violated from 18 kHz to 23 kHz, the phase margins at both corner frequencies, i.e., $GM = 0$ dB with 90° and 155° and for $GM = 6$ dB with 100° are sufficient for classification as a stable system.

For interpretation of summarized stability results in Figure 14.6 it has to be considered that a system that is unstable according to MLGC can have different characteristics (see Section 14.2.2). If a system that is stable according to MLGC, the operating point is retained, as well as if the limit values are exceeded. However, if the remaining stability and phase margins get close to 0, the system undergoes resonance excitation and the operating point can no longer be maintained. This leads to a deteriorated impedance measurement, whereas otherwise the determination of the impedance and the calculation of the gain and phase margins is possible. These states are indicated by green, orange, and red squares in Figure 14.6, where green indicates an MLGC stable, orange an MLGC unstable and red a system with no stable operating point. Starting at the default parameters (1|1), the system is already MLGC stable. Changing the free parameters to (0.9|0.55) leads to an increase in both *GM* and *PM*. Systems close to the optimal point with (1|0.55) and (0.75|0.55) are MLGC stable as well and still show improved *GM* and *PM* compared to (1|1). On the other hand, changing k_{PV} to higher values leads to decreased *GM* and *PM*. If the thresholds of GM_c, PM_c are set to 6 dB and 45°, the system has to be considered unstable for (1|1.5) at medium loads. Increasing k_{AC} to higher values i.e. (1.6|0.55) and (1.6|1.5) also leads to MLGC unstable systems. Here, especially low and medium loads lead to unfavourable *GM* and *PM*. Reducing k_{AC} to lower than 0.5 can lead to generally unstable systems. For additionally low k_{PV} like for (0.5|0.55), the system can not maintain the desired grid voltage for medium and high loads. Higher k_{PV} as in (0.5|1.5) will be MLGC unstable for low and medium loads, but will also not be able to form a stable grid for high loads.

Overall, the experimental verification showed that the measured points are fully in agreement with the predicted stability states from the digital twin and the random forest model and that the optimization algorithm finds a stable operating region for the DC microgrid.

14.5 Conclusion and Outlook

We developed, implemented, and validated a novel approach for applying artificial intelligence for stability optimization in LVDC microgrids. The developed solution addresses the increasing usage and complexity of LVDC microgrids, which are used to support environmental-friendly and resilient energy supply in connection with the common AC networks. The LVDC microgrids realize efficient energy generation and sharing by connection of DC power sources, loads, and storage systems with a more material- and cost-effective design of the network and avoiding unnecessary AC/DC conversions. The application of digitized converters enables intelligent and fast control and opens the possibility to solve specific problems in DC networks through AI support, such as the design for grid stability or maintenance of grid stability during operation.

To address a flexible, widely applicable solution that can be transferred to other application areas, e.g., for DC on-board networks in automotive applications, the approach was divided into four development areas:

First, by automating an electronic circuit simulator for small-signal analysis and upgrading with an integrated, configurable possibility for functional evaluation, a flexible digital twin of DC networks can be realized. The digital twin automatically calculates the small-signal stability for a variety of input parameterizations of the network and generates labels for the description of the stability state as stable or unstable. The layout of the grid is carried out in the simulator by the design expert as in the regular workflow and can be flexibly adapted to different components, devices, and topologies. For the automated stability evaluation, the minor-loop gain criterion (MLGC) was used in this work, but further stability criteria can be implemented.

In a second step, based on the generation of labels for the stability state, a surrogate classification model is created based on random forests. These establish a model relationship between the input parameters of the DC grid and the state of stability and, by evaluating the variable influences, give the domain expert an overview of which parameters mainly influence stability. The random forest model can be easily adapted to new models from the digital twin, since in particular class imbalances, which can result from different network configurations, are taken into account during model creation. Moreover, the random forest models prepare the stability optimization step from a classification task.

A new method for optimization from classification trees was developed as a third step that exploits the models from the trained random forest classifiers. The optimization method determines optimized stability parameters of the network, for example for finding favorable design parameters or by adapting software parameters such as parameters of the droop characteristics during operation. Different target functions can be used for optimization, for example, to determine a stable operating point with the largest possible stable surrounding area. The steps for stability optimization were first carried out on a digital twin of a four-terminal network.

As a fourth implementation, verification, and validation step, this four-terminal network was realized on an experimental testbed as a demonstrator. The new PRBS method for online impedance and stability measurement was improved and enhanced for minimal invasive measurement at 380 V_{DC}, including data acquisition, power coupling, and service functions. The impedance measurement was used to characterize relevant components of the testbed and to refine the circuit model in the digital twin with subsequent processing of the following modeling and optimization steps. Validation experiments were able to show that with the AI-based modeling and optimization steps, the adjustment of the stability during operation by modifying the droop characteristic curves is possible.

Overall, the application demonstrated the feasibility and advantage of AI-supported power electronics for DC networks as envisioned in the research field of cognitive power electronics [27]. The work and results provided foundations for further research topics. These are the extension of the developed methods for LVDC microgrids, where characteristic parameters of participants are not accessible, and data-based assessment of control stability for large-signal analysis and stability optimization in combination with the presented droop control approach. Evolving research fields also include fault detection and remaining useful lifetime assessment in DC grids, e.g., for industrial and domestic energy systems as well as on device and component level.

References

1. R. Alizadeh, J. Allen, and F. Mistree. Managing computational complexity using surrogate models: a critical review. *Research in Engineering Design*, 31(3):275–298, 07 2020.
2. Analog Devices, Inc. LTspice, version XVII, 2021. Accessed: 2023-07-28.
3. D. Bertsimas and J. Dunn. Optimal classification trees. *Machine Learning*, 106:1039–1082, 2017.

4. T. J. Blalock. The rotary era, part 1: Early ac-to-dc power conversion [history]. *IEEE Power and Energy Magazine*, 11(5):82–92, 2013.
5. T. J. Blalock. The rotary era, part 2: ac-to-dc power conversion, continued [history]. *IEEE Power and Energy Magazine*, 11(6):96–105, 2013.
6. A. Bonfietti, M. Lombardi, and M. Milano. Embedding decision trees and random forests in constraint programming. In L. Michel, editor, *Integration of AI and OR Techniques in Constraint Programming*, pages 74–90, Cham, 2015. Springer International Publishing.
7. L. Breiman. Machine learning, volume 45, number 1 - springerlink. *Machine Learning*, 45:5–32, 10 2001.
8. L. Breiman, J. Friedman, C. J. Stone, and R. Olshen. *Classification and Regression Trees*. Imprint Routledge, 1984.
9. A. Fernández, S. García, M. Galar, R. C. Prati, B. Krawczyk, and F. Herrera. *Performance Measures*, pages 47–61. Springer International Publishing, Cham, 2018.
10. F. Gao, R. Kang, J. Cao, and T. Yang. Primary and secondary control in dc microgrids: a review. *Journal of Modern Power Systems and Clean Energy*, 7(2):227–242, 2019.
11. Gurobi Optimization, LLC. Gurobi Optimizer Reference Manual. https://www.gurobi.com, 2023. Accessed: 2023-07-28.
12. D. Gutina, A. Bärmann, G. Roeder, M. Schellenberger, and F. Liers. Optimization over decision trees: a case study for the design of stable direct-current electricity networks. *Optimization and Engineering*, pages 1–41, 02 2023.
13. T. Hastie, R. Tibshirani, and J. Friedman. *The elements of statistical learning: data mining, inference and prediction*. Springer, New York, NY, 12 edition, 2017.
14. C. D. Lin and L. Kang. A general construction for space-filling latin hypercubes. *Statistica Sinica*, 26(2):675–690, 2016.
15. W.-Y. Loh. Classification and regression trees. *WIREs Data Mining and Knowledge Discovery*, 1(1):14–23, 2011.
16. W.-Y. Loh. Fifty years of classification and regression trees. *International Statistical Review*, 82, 06 2014.
17. S. Mehta and P. Basak. A comprehensive review on control techniques for stability improvement in microgrids. *International Transactions on Electrical Energy Systems*, 31(4):e12822, 2021.
18. B. Miao, R. Zane, and D. Maksimovic. System identification of power converters with digital control through cross-correlation methods. *IEEE Transactions on Power Electronics*, 20(5):1093–1099, 2005.
19. B. Modu, M. P. Abdullah, M. A. Sanusi, and M. F. Hamza. Dc-based microgrid: Topologies, control schemes, and implementations. *Alexandria Engineering Journal*, 70:61–92, 2023.
20. L. Ott, Y. Han, B. Wunder, J. Kaiser, F. Fersterra, M. Schulz, and M. März. An advanced voltage droop control concept for grid-tied and autonomous dc microgrids. In *2015 IEEE International Telecommunications Energy Conference (INTELEC)*, pages 1–6, 2015.
21. F. Pedregosa, G. Varoquaux, A. Gramfort, V. Michel, B. Thirion, O. Grisel, M. Blondel, P. Prettenhofer, R. Weiss, V. Dubourg, J. Vanderplas, A. Passos, D. Cournapeau, M. Brucher, M. Perrot, and E. Duchesnay. Scikit-learn: Machine learning in Python. *Journal of Machine Learning Research*, 12:2825–2830, 2011.
22. A. Riccobono and E. Santi. Comprehensive review of stability criteria for dc power distribution systems. *IEEE Transactions on Industry Applications*, 50(5):3525–3535, 2014.
23. G. Roeder, L. Ott, A. Meier, B. Wunder, P. Wienzek, A. Baermann, F. Liers, and M. Schellenberger. Analysis and improvement of lvdc-grid stability using circuit simulation and machine learning - a case study. In *NEIS 2021; Conference on Sustainable Energy Supply and Energy Storage Systems*, pages 1–7, 2021.
24. R. Sabzehgar. A review of ac/dc microgrid-developments, technologies, and challenges. In *2015 IEEE Green Energy and Systems Conference (IGESC)*, pages 11–17, 2015.
25. O. Sagi and L. Rokach. Ensemble learning: A survey. *WIREs Data Mining and Knowledge Discovery*, 8(4):e1249, 2018.
26. T. Saito and M. Rehmsmeier. The precision-recall plot is more informative than the roc plot when evaluating binary classifiers on imbalanced datasets. *PLOS ONE*, 10(3):1–21, 03 2015.

27. M. Schellenberger, V. Lorentz, and B. Eckardt. Cognitive power electronics – an enabler for smart systems. In *PCIM Europe 2022; International Exhibition and Conference for Power Electronics, Intelligent Motion, Renewable Energy and Energy Management*, pages 1–5, 2022.
28. B. Tang. Orthogonal array-based latin hypercubes. *Journal of the American Statistical Association*, 88(424):1392–1397, 1993.
29. T. A. M. Toffolo and H. G. Santos. Python mip (mixed-integer linear programming) tools, version 1.13. https://pypi.org/project/mip/, 2020. Accessed: 2023-07-28.
30. B. Wunder, L. Ott, J. Kaiser, Y. Han, F. Fersterra, and M. März. Overview of different topologies and control strategies for dc micro grids. In *2015 IEEE First International Conference on DC Microgrids (ICDCM)*, pages 349–354, 2015.
31. K. Åström and R. Murray. *Feedback Systems: An Introduction for Scientists and Engineers*. Princeton University Press, 01, 2008.

Chapter 15
Self-Optimization in Adaptive Logistics Networks

Julius Mehringer, Ursula Neumann, Friedrich Wagner, Christopher Scholl

Abstract The spectrum of applications for AI is extremely broad, ranging from support for complex decisions and data-driven corporate strategies to the automation of everyday processes. In logistics networks, a plurality of decisions are to be made on a daily basis. Typically, those decisions are comprised of a combination of forecasting and optimization, forming the area of *prescriptive analytics*. In this chapter, we present two use cases for arriving at optimal decisions: the case of prescribing cost-optimal order policies for the stocking of spare parts, and the case of mixing raw materials to final products with varying raw material quality.

Key words: prescriptive analytics, bayesian modeling, Gompertz growth model, life cycle modeling, non-linear optimization, robust optimization.

15.1 Introduction

For manufacturing companies, being able to supply spare parts to markets along the product life cycle is a vital aspect of company success, contributing up to 70% to the revenue of service business [4]. For durable goods, spare part demand occurs not only during the phase of serial production but a considerable proportion of total demand occurs after production has ceased, called the *End of Production* (EOP) date. In order to satisfy this demand, five strategies exist [3]:

- reproduction,
- use of compatible parts,
- remanufacturing (internal & external),
- reuse,

Fraunhofer Institute for Integrated Circuits IIS, Fraunhofer IIS, Erlangen, Germany

Corresponding author: Julius Mehringer
e-mail: julius.mehringer@iis.fraunhofer.de

• (final) stocking.

All of the five strategies are characterized by certain drawbacks: remanufacturing and reuse require maintenance of a logistics network and respective processes. Reproduction leads to considerably high production costs. The use of compatible parts requires sufficient quality and quantity of the respective parts, which can be challenging to find. The main drawback of a (final) stocking strategy is the difficulty of quantifying the final order quantity, called the *All-Time Buy* [25]. This is especially difficult when the EOP date lies early in the product's life cycle (as discussed in detail in Section 15.3). If a confident forecast of the All-Time Buy is possible though, the final stocking approach is preferable over other approaches. Thus, quantifying the All-Time Buy is considered a central element of differentiation in the competition, contributing to a company's economic success [11].

In the following section, we first present a brief overview of relevant and recent literature regarding the All-Time Buy. We then describe the prediction problem formally in Section 15.3, and present an approach to forecasting the All-Time Buy quantity in Section 15.4. In Section 15.5 we describe a robust optimization approach, which can deliver a cost-optimal ordering policy based on All-Time Buy predictions. This policy balances out possible surplus stock and possible out of stock situations, both of which are costly.

15.2 A Brief Overview of Relevant Literature on Predicting the All-Time Buy Quantity

Since the All-Time Buy is a quantity of key interest for after sales departments in industries such as automotive, consumer durable goods, industrial equipment, and electronics, there is a variety of forecasting approaches in order to tackle this challenge. In [13], a dynamic inventory policy is developed, allowing for adjustments as more and more observations occur in the life cycle of a product. A key requirement for their methodology is the availability of a peak of sales, which is only given reliably if a company is in charge of the whole supply chain; in modern supply chain structures, this requirement can rarely be reliably met. The authors of [5] assume a very simple, yet in practice valid calculation of the final order quantity by modeling an exponential decay with rate 0.7 after EOP. We found this approach to be well established in the industry as a benchmark, due to its simplicity. In [6], the authors develop a forecasting model for individual spare parts, assuming the availability of spare part failure rates, replacement rates, and sales data. Due to the assumptions about data availability, the forecasting model lacks practical implementation possibilities, since individual and reliable failure rates and precise sales data are oftentimes scarce. Although [29] describes and formalizes main aspects of spare part demand forecasting, and develops an appealing Bayesian hierarchical modeling approach, their modeling approach only allows a single step prediction. As we will see in Section 15.3, this is unsatisfactory for the prediction of a All-Time Buy quantity.

More recent literature suggests that the relevant patterns are observed in the demand of earlier products as well as partially in the early phase of a spare part life cycle. As [10], [23] and [12] propose, clustering methods are able to identify typical demand patterns in a historical data set. Those patterns are characterized by differing shapes of life cycle curves, e.g., rather flat versus concentrated demand, or the main proportion of total demand occurring before or after EOP. Depending on the available data, different clustering algorithms seem feasible for this pattern recognition task; the Partitioning Around Medoids (PAM) Algorithm [20] or the k-modes Algorithm [8] are promising candidates, as typically numeric as well as categorical features are available. Then, in a subsequent step, a single spare part for which a forecast has to be conducted, can be assigned to a cluster with a classification algorithm. Finally, for each cluster, the median or mean demand at each time step can be used as a typical demand curve, as proposed independently both by [23] and [12]. Determining forecasts for the All-Time Buy quantity hence consists of finding the respective typical demand curve and scaling this curve in such a way that the last observed spare part demand matches the scaled demand prediction. While building on core ideas from this stream of literature, we present a more robust approach to modeling in chapter 15.4 than the approached described in previous literature.

15.3 Predicting the All-Time Buy

In general, quantitative forecasting is the craft of finding patterns and extrapolating those patterns into the future. Along a product's life cycle, spare part management can be divided in the three main phases 1) *initial procurement phase*, 2) *the normal operation phase* and 3) the *End-of-Life phase* [7]. In this chapter, we address one of the challenges in phase 3, i.e., the End-of-Life phase. We're focusing on "[the necessity] to set a final order on spare parts according to the demand patterns at the end of the product life cycle (known as an 'all-time buy' or 'lasttime buy')" [7]. Thus, with respect to the product life cycle, the manufacturer has to decide at EOP how much demand is likely to occur during the total length of the final phase, see Figure 15.1.

While the literature suggests that the All-Time Buy is a single quantity of interest, we have learned from industry that the problem appears to be more nuanced. In some supply chain settings, it is possible to split the All-Time Buy quantity into a number of orders, each of which is a fraction of the All-Time Buy quantity. By doing so, a company can significantly improve ordering and storage cost. However, a prediction model then should not only predict the All-Time Buy quantity, but also quantify the speed at which demand likely occurs. We address this by modeling the shape of a life cycle with a growth model, as described in detail in Section 15.4.

Usually, in a forecasting setting, a sufficiently long time series of observed values $y_1, \ldots, y_T$ is presented and the forecasting task is to predict the next H values of the

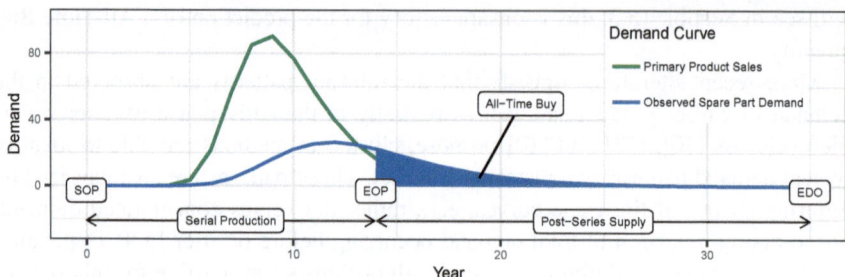

Fig. 15.1: Determination of the *All-Time Buy* quantity, embedded in the product life cycle. Primary Product Sales predate and influence observed spare part demand. The shaded area represents the All-Time Buy quantity. *SOP* and *EDO* denote the *Start Of Production* and the *End of Delivery Obligation* respectively.

series $y_{T+1}, \ldots, y_{T+H}$. For the All-Time Buy quantity, we typically observe the time series until the EOP date in a product's life cycle $y_{T \hat{=} EOP}$. Then, H is the prediction horizon.

Importantly, T and H can be of variable lengths per spare part i, denoting an EOP date early or late in a spare part's life cycle. This variability leads to a gradual decrease of difficulty for the prediction task with an increasing number of observations, see Figure 15.2. In general, for life cycles of spare parts, EOP is a few time steps before the peak of it's demand, leading to a high variability of predictions based on observed historical demand only as observations increase [17]. Additionally, a forecaster typically does not know about the position of the EOP date in the life cycle at the time of forecasting.

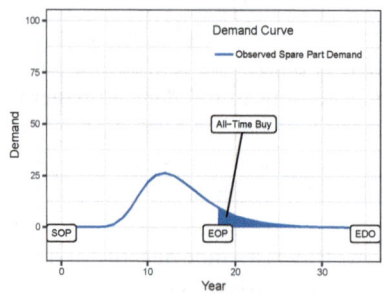
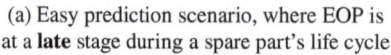
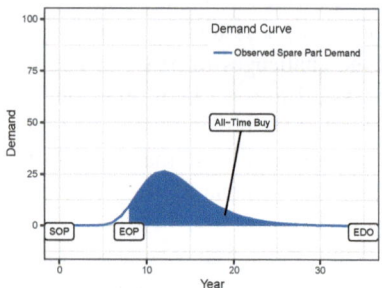

(a) Easy prediction scenario, where EOP is at a **late** stage during a spare part's life cycle

(b) Hard prediction scenario, where EOP is at an **early** stage during a spare part's life cycle

Fig. 15.2: Easy vs. hard scenario; the exact position of EOP in a product's life cycle is generally unknown to the forecaster.

As noted by [29], the "input to the forecasting process consists not only of previous demands for a particular part, but also of observed demands for other parts – even parts that were recently withdrawn from the manufacturing process." Due to the possibly very short observation length of T, we thus need a model that is able to form predictions mainly from a collection of supporting predecessors, but can take into account all recent observed demands in T of our target spare part i.

For our modeling purposes, we translate the distinctive, observed life cycle curve to its cumulative form, see figure Figure 15.3 for an illustration. Then, y_t is the cumulative demand at t and the All-Time Buy is the sum of all predictions during H:

$$A = y_{T+H}.$$ (15.1)

15.4 A Probabilistic Hierarchical Growth Curve model

Since the shape of a spare part's demand process follows a life cycle curve, we propose a Bayesian hierarchical growth model for the prediction of the All-Time Buy quantity. Modeling growth is typically conducted with sigmoid functions, describing growth as starting slowly and increasing over time before reaching an upper asymptote.

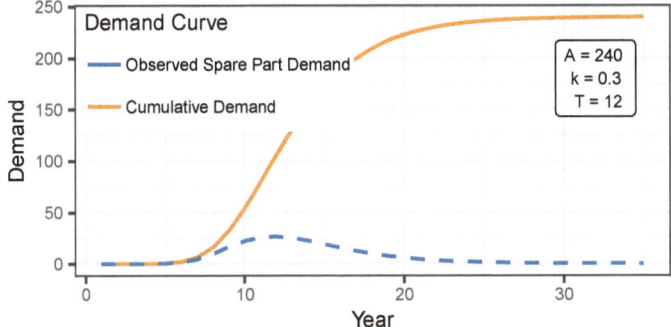

Fig. 15.3: Cumulative demand modeled with the Gompertz function.

A widely used growth model is the Gompertz model. We follow the convention of [26] and formulate the Gompertz growth model as a *Type I* model:

$$y_t = A \cdot e^{-e^{-k(t-T)}}$$ (15.2)

In Figure 15.3, we show the values of the cumulative demand over time, with the respective parameters of the Gompertz function. This formulation is appealing, since it allows an interpretation of the parameters. Here, y_t is the value as a function of time, in our case the cumulative demand at time t. A is the upper asymptote (reflecting the All-Time Buy, see Equation (15.1)), k is a growth-rate parameter that affects the slope and T is the time at inflection. This time point is of special relevance for our use case, since it is the point where maximum demand occurs. T is coined a location parameter, since it shifts the curve horizontally; the parameters A and k determine the shape of the curve and are thus shape parameters. Inferring proper parameters of the growth model from the few observations of a single spare part only proves to be difficult. As a remedy, we propose a hierarchical modeling approach that allow us to use a set of reference spare part demand curves in order obtain parameters of our target demand curve.

Hierarchical modeling assumes the existence of groups in a data set. By modeling the growth curves hierarchically, we're able to relate the parameters of our target growth curve on the parameters of (almost) complete life cycle curves from reference spare parts identified e.g. by similarity on a spare parts master data. Formally, we thus assume the model parameters as being sampled from a population distribution of parameters, i.e., we use information about the population to improve estimated of individual items by regularizing toward the population mean [15].

We specify the model's generative process as

$$
\begin{aligned}
y_{ijt} &\sim logN\left(f\left(t;\mu_{\Theta_i}\right),\sigma^2\right) & \mu_{k_j} &\sim N\left(\mu_{k_l},1\right) \\
& & \tau^2_{k_j} &= 1 \\
\mu_{\Theta_i} &= \begin{bmatrix} \mu_{A_i} \\ \mu_{k_i} \\ \mu_{T_i} \end{bmatrix} & \mu_{T_j} &\sim N\left(\mu_{T_l},\tau^2_{T_l}\right) \\
& & \tau^2_{T_j} &\sim N^+\left(0,0.25\right) \\
\mu_{A_i} &\sim logN\left(\mu_{A_j},\tau^2_{A_j}\right) & \mu_{A_l} &\sim N\left(5,1\right) \\
\mu_{k_i} &\sim N\left(\mu_{k_j},\tau^2_{k_j}\right) & \tau^2_{A_l} &\sim N^+\left(0,0.25\right) \\
\mu_{T_i} &\sim N\left(\mu_{T_j},\tau^2_{T_j}\right) & \mu_{k_l} &\sim N\left(0,0.01\right) \\
\mu_{A_j} &\sim logN\left(\mu_{A_l},\tau^2_{A_l}\right) & \mu_{T_l} &\sim N\left(0,25\right) \\
\tau^2_{A_j} &\sim N^+\left(0,0.25\right) & \tau^2_{T_l} &\sim N^+\left(0,0.25\right) \\
& & \sigma^2 &\sim N^+\left(0,1\right),
\end{aligned}
\tag{15.3}
$$

with f being the Gompertz function from Equation (15.2). The model's priors are chosen as uninformative, but allowing for all assumed possible values of y.

For the target time series y, we specify i as the item's index, j as the respective group membership, and t as the time. The respective Bayesian network for this generative process is shown in Figure 15.4. This non-centered parametrization does not model the parameters of individual items independently, but jointly by assuming common hyper parameters from the latent population, μ_{Θ_i}.

We then can use the draws generated by a sampler (Hamiltonian Monte Carlo [1] in our case) and generate predictions for y_h, given all the observations from the target time series y_t and observations from other, similar and complete life cycle curves:

$$
p\left(y_h|y_t\right) = \int p\left(y_h|\mu_\Theta\right) p\left(\mu_\Theta|y_t\right) d\mu_\Theta.
$$

The essential part is the likelihood $p(y_h|\mu_\Theta)$ as a model of the values to be predicted, conditioned on the parameter vector μ_Θ [19]. Since the parameters of the data generating process μ_Θ are computationally intractable, we can't analytically derive the posterior $p\left(\mu_\Theta|y_t\right)$ and thus have to revert to a sampling-based approach in order

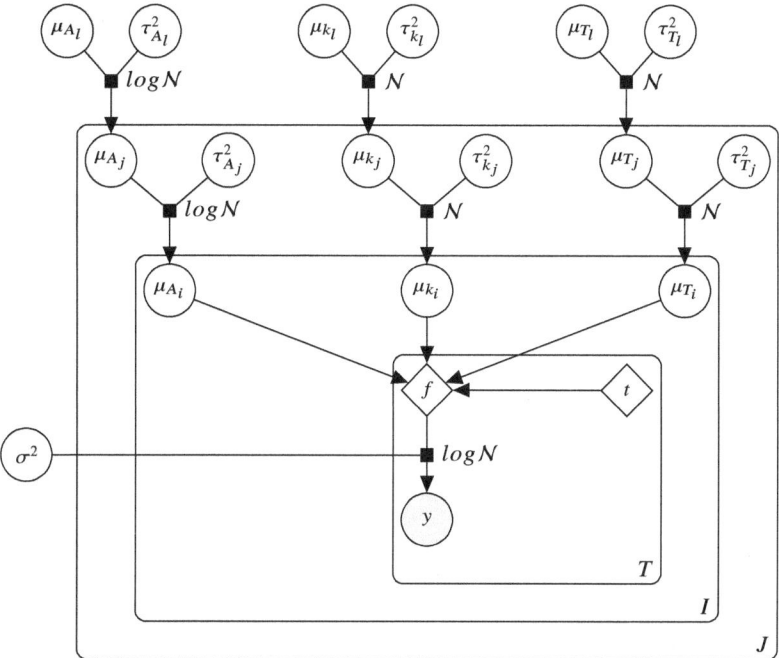

Fig. 15.4: Bayesian network for the assumed data generating process. Each y_i is the result of the value of the Gompertz function f of item i at time t. The Gompertz function f has the latent parameters $\theta = \left[\mu_{A_i}, \mu_{k_i}, \mu_{T_i}\right]$, them being modeled as inherit from the respective hyper parameters μ_{A_j}, μ_{k_j} and μ_{T_j} with respective τs for all groups J. The μs are drawn from global hyper priors, according to the data generating process specified in (15.3).

to sample from the posterior predictive distribution $p(y_h|y_t)$ [9]. For $r = 1, \ldots, R$, the *HMC sampler* [1] draws samples $\mu_{\Theta}^{(r)} \sim p(\mu_{\Theta}|y_t)$ from the posterior [14], which allow sampling from the posterior predictive distribution $y_h^{(r)} \sim p(y_h|y_t)$ through this strategy:

$$y_h^{(r)} \sim log\mathcal{N}\left(f\left(h; \mu_{\Theta}^{(r)}\right), \sigma^{2(r)}\right)$$

with f being the Gompertz function from Equation (15.2).

In short, we obtain R samples $y_h^{(1)}, \ldots, y_h^{(R)}$ from our posterior predictive distribution (see Figure (15.4)) for each time step h and plug those samples in a robust optimization procedure, guaranteeing a cost-optimal ordering policy. The formulation of this optimization model will be described in the following section.

15.5 Determining the Optimal Order Policy

Given a predicted demand of spare parts, the question for a cost-optimal production or order policy arises naturally. Depending on the specific application scenario, different types of costs may arise. We focus on the most common ones, namely storage costs as well as order costs per order and per part. A cost-optimal order policy satisfies the demands while simultaneously minimizing the total costs. Optimization problems with the goal of determining optimal order or production policies are generally known as *lot sizing problems*. More specifically, the problem at hand belongs to the subclass of *single-item uncapacitated lot sizing problems* (SIULSP). For a comprehensive survey on lot sizing problems, we refer to [16, 27].

Lot sizing is a representative example for optimization problems arising in production and logistics. Such problems are often successfully tackled by *Mixed-Integer Programming* (MIP) techniques. There, the problem of interest is modeled via a linear objective together with linear constraints over integer and continuous variables. The success of MIP methods is based on decades of developing sophisticated solution methods for such linear models, typically employing *branch-and-cut* algorithms.

As a starting point, we restate the MIP model for SIULSPs from [24]. Here, the time horizon for the production plan is sliced into T many time intervals of equal length (e.g., weeks or months). For each time interval $t \in \{1, \ldots, T\}$, an integer variable $x_t \in \mathbb{Z}_0^+$ models the quantity of produced parts. A variable $s_t \in \mathbb{Z}_0^+$ models the stock at time interval $t \in \{1, \ldots, T + 1\}$, while a binary variable $z_t \in \{0, 1\}$ indicates whether parts are produced in time interval $t \in \{1, \ldots, T\}$ or not. The basic model assumes storage costs c_{storage} per part as well as fixed production costs c_{fix} per time interval with production, independent of the produced amount. Furthermore, the demand $d_t \in \mathbb{Z}_0^+$ is assumed to be known for each time interval. In the context of this work, the demand is derived from the forecast, see Eq. (15.2). Finally, a sufficiently large constant M is needed, where $M \gg \sum_t d_t$. With these variables and parameters the SIULSP can be model by

$$\min_{x} \quad \sum_{t=1}^{T} (c_{\text{fix}} \cdot z_t + c_{\text{storage}} \cdot s_t) \tag{15.4a}$$

$$\text{s.t.} \quad s_{t+1} = s_t + x_t - d_t \quad \forall 1 \leq t \leq T \tag{15.4b}$$

$$x_t \leq M z_t \quad \forall 1 \leq t \leq T \tag{15.4c}$$

$$x_t \in \mathbb{Z}_0^+ \quad \forall 1 \leq t \leq T \tag{15.4d}$$

$$s_t \in \mathbb{Z}_0^+ \quad \forall 1 \leq t \leq T \tag{15.4e}$$

$$z_t \in \{0, 1\} \quad \forall 1 \leq t \leq T \tag{15.4f}$$

Constraints (15.4b) define consistent stock values, whereas constraints (15.4c) model the logical condition $z_t = 1 \Leftrightarrow x_t > 0$. This model will serve as the basis for the following chapters, in which it will be extended in two ways. First, we model more complicated costs. In particular, non-linear costs per part, known as *scales*, are incorporated. Second, we develop a *robust* MIP model taking uncertainties in the demand forecast into account.

15.5.1 Modeling Non-Linear Costs

The objective from Eq. (15.4a) models storage costs per part as well as fixed production cost, but does not include production costs *per part*. This is due to the fact that production costs per part are assumed to be constant. Moreover, the total production quantity $\sum_{t=1}^{T} x_t$ is fixed by the All-Time Buy from Eq. (15.1). Due to storage costs, it will never be favorable to produce more parts in total than the All-Time Buy. As a result, total production costs are simply a constant and thus are neglected in the model. However, the assumption of constant costs per part is too simplistic for many real scenarios. Usually, the production or ordering costs per part decrease with the produced quantity. This leads to a non-linear scaling of total production costs with the production quantity, as sketched in Figure 15.5.

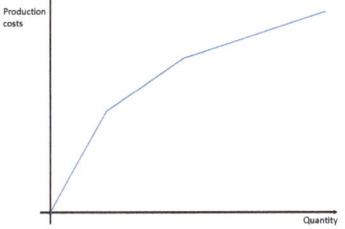
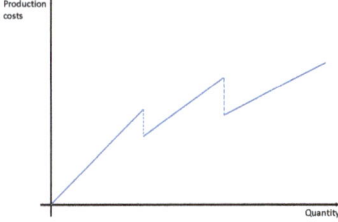

(a) Continuous, piecewise-linear costs.

(b) Piecewise-linear costs with discontinuities.

Fig. 15.5: Sketch for piecewise-linear production costs. In (a), the production costs per part for the first, say, 100 parts are 1 € per part while for the second 100 parts production costs are only 0.9 €. Also in (b), the production costs for 100 parts are 1 € per part, but the production costs for a total production of 200 parts are 0.9 € per part, even for the fist 100 parts. This leads to discontinuities.

In the following, we focus on modeling a piecewise linear cost scaling. Many equivalent formulations for modeling piecewise linear functions exist in literature, see e.g. [22] for a survey. The formulation used here is called *multiple choice model*. Let $y = f(x)$ be the costs for producing x parts, where f is a (continuous or non-continuous) piecewise linear function. Our goal is to derive a linear description of $y = f(x)$ by introducing additional auxiliary variables. To this end, let $(B_j)_{j=1...J+1}$ be the sequence of x-values that separate linear segments of f, e.g., the production quantities at which a cheaper price per part starts in Figure 15.5. Now, f is linear in each interval $[B_j, B_{j+1}]$. Thus, there exists $q_j \in \mathbb{R}$ and $n_j \in \mathbb{R}$ such that $f(x) = q_j x + n_j$ for each $j = 1 \ldots J$. We introduce two sets of auxiliary variables. First, the binary variables $b_j \in \{0, 1\}$, $k = j \dots J$ indicate in which interval x is located. That is, we want to ensure the implication $x \in [B_j, B_{j+1}] \Rightarrow b_j = 1, b_k = 0 \, \forall k \neq j$. Second, continuous variables $w_j \in \mathbb{R}_0^+$, $j = 1 \dots J$, are introduced which are used to model the quantity x for $x \in [B_j, B_{j+1}]$. More specifically, we enforce the implication $x \in [B_j, B_{j+1}] \Rightarrow w_j = x, w_k = 0 \, \forall k \neq j$. Both desired implications can be modeled via the linear constraints

$$\sum_{j=1}^{J} b_j \leq 1 \tag{15.5}$$

$$B_j b_j \leq w_j \leq B_{j+1} b_j \quad \forall 1 \leq j \leq J \tag{15.6}$$

$$x = \sum_{j=1}^{J} w_j . \tag{15.7}$$

Finally, we require

$$y = \sum_{j=1}^{J} (q_j w_j + n_j) . \tag{15.8}$$

It follows that $y = f(x)$ as desired. This multiple choice linearization technique needs to be applied for every time interval, introducing variables $y^t \in \mathbb{R}$, $b_j^t \in \{0, 1\}$ and $w_j^t \in \mathbb{R}_0^+$ for $t = 1 \dots T$.

As a side effect, the above linearization technique allows to model the logical on-off condition $z_t = 1 \Leftrightarrow x_t > 0$, enforced by Eq. (15.4c), in a more efficient way. The constraints from Eqs. (15.4c) and (15.5) can be replaced by the single equation

$$\sum_{j=1}^{J} b_j = z \tag{15.9}$$

without changing the feasible space of model. However, the resulting model has the desirable property of being *locally ideal*, c. f. [22], which means that the linear relaxation of Eqs. (15.6)-(15.9) satisfies all integrality conditions. That is, for all vertices of the relaxation polytope it holds $b_j \in \{0, 1\}$, $j = 1 \dots J$. Additionally, the local ideal formulation requires no large constant M. Both properties of the local ideal formulation significantly increase the computational performance.

15.5.2 Robust Optimization

In the basic model, it is assumed that the demand d_t is known a priori. However, the demand forecast comes with some uncertainty. In fact, only a probability distribution for the cumulative demand $\sum_{i=1}^{t} d_i$ is known instead of a fixed demand value d_t. Lot sizing under uncertainty is a vivid field of ongoing research, for further reading we refer to [21, 2, 24, 18, 28]. In the basic model from Eq. (15.4), the requirement that the inventory does not run out of stock is implicitly enforced by the non-negativity of the stock variables s_t. When d_t is a random variable with known distribution the requirement of non-negative stock can be replaced by the weaker requirement that the probability of running out of stock in time interval t is at most $\epsilon > 0$,

$$P\left(\sum_{j=1}^{t} x_j \geq \sum_{j=1}^{t} d_j\right) \geq 1 - \epsilon \quad \forall 1 \leq t \leq T. \tag{15.10}$$

Note, that we do not enforce the probability for running out of stock in *any* time interval t is at most ϵ, which would be a much stronger requirement. We only require that the probability in each individual time interval t is at most ϵ. In other words, we require that the inventory is non-empty in each individual time interval t with probability at least $\alpha = 1 - \epsilon$. Therefore, this technique is often called α-*service-level*, see also [24]. The cumulative distribution function of the cumulative demand $\sum_{k=1}^{t} d_k$ is known from the prediction and denoted by F_t. Thus, we may rewrite Eq. (15.10) as

$$F_t\left(\sum_{j=1}^{t} x_j\right) \geq 1 - \epsilon \quad \forall 1 \leq t \leq T, \tag{15.11}$$

which is equivalent to

$$\sum_{j=1}^{t} x_j \geq D_{1-\epsilon}^{t} \quad \forall 1 \leq t \leq T, \tag{15.12}$$

where $D_{1-\epsilon}^{t}$ denotes the $(1 - \epsilon)$-quantile of F_t.

Apart from considering probabilities for running out of stock, also storage costs need to be adapted when incorporating uncertain demands. With unknown demand, also the stock amount is implicitly uncertain. To overcome this issue, we replace the stock amount s_t simply by the expected amount

$$s_t = \sum_{j=1}^{t} x_j - \mathbb{E}\left[\sum_{j=1}^{t} d_j\right] \quad \forall 1 \leq t \leq T. \tag{15.13}$$

Including out-of-stock probabilities and expected demand, the final MIP model reads

$$\min \quad \sum_{t=1}^{T} (c_{\text{fix}} \cdot z_t + c_{\text{storage}} \cdot s_t + y_t) \tag{15.14a}$$

$$\text{s.t.} \quad z = \sum_{k=1}^{J} b_k \qquad \forall 1 \leq t \leq T \tag{15.14b}$$

$$s_t = \sum_{i=1}^{t} x_i - \mathbb{E}\left(\sum_{i=1}^{t} d_i\right) \quad \forall 1 \leq t \leq T \tag{15.14c}$$

$$x_t = \sum_{j=1}^{J} w_j^t \qquad \forall 1 \leq t \leq T \tag{15.14d}$$

$$y_t = \sum_{j=1}^{J} q_j w_j^t + n_j b_j^t \qquad \forall 1 \leq t \leq T \tag{15.14e}$$

$$B_j b_j^t \leq w_j^t \leq B_{j+1} b_j^t \qquad \forall j \leq t \leq J \, \forall 1 \leq t \leq T \tag{15.14f}$$

$$\sum_{j=1}^{t} x_j \geq D_{1-\epsilon}^t \qquad \forall 1 \leq t \leq T \tag{15.14g}$$

$$x_t \in \mathbb{Z}_0^+ \qquad \forall 1 \leq t \leq T \tag{15.14h}$$

$$s_t, w_j^t, y^t \in \mathbb{R} \qquad \forall j \leq t \leq J \ \forall 1 \leq t \leq T \tag{15.14i}$$

$$z_t, b_j^t \in \{0, 1\} \qquad \forall 1 \leq j \leq J \, \forall 1 \leq t \leq T \tag{15.14j}$$

To summarize, we developed a MIP model which determines a cost-optimal ordering policy. The model incorporates non-linear ordering costs and is robust against uncertainties in the demand forecast.

15.6 Pooling

As we have seen, forecasting demands of various kinds can be embedded into planning models that address subsequent steps in the supply chain. One such instance in the supply chain is the homogenous mixing process of raw materials into final blends through blending intermediate materials that can be stored and in turn used in the mixing process. This can be modeled by the Pooling Problem as described in Section 7.3.3.

The application of pooling in practice requires various adjustments like the inclusion of recipes, for instance when mixing yogurt. Each yogurt has specific ingredients. The substances measured may be lactose, fat content, concentration of food colors, or traces of bacteria from the soil where the fruit for the yogurt was grown. By mixing the same yogurt with different fat contents together the desired fat content can be reached. Following the notation of Section 7.3.3, the recipes can be included as follows. The raw and intermediate materials for the mixture are saved in multiple batches. Let M be the set of raw materials and let $I = \dot{\bigcup}_{m \in M} I_m$ be the available containers in which these materials are stored, where I_m is the set of batches that contain material $m \in M$. Then a blending recipe describing how much of each material is in the mix can be specified by the fraction values $\sigma_{ml} \in [0, 1]$, $\sigma_{mj} \in [0, 1]$ for all materials $m \in M$, pools $l \in L$, and outputs $j \in J$. The recipe can be followed by adding the following constraints:

$$\sum_{i \in I_m, l \in I_L} q_{il} = \sigma_{ml} \forall l \in L, \quad \sum_{i \in I_m, l \in I_J} q_{ij} = \sigma_{mj}, \forall j \in J$$

Since ingredients of various kinds are usually procured and processed in a similar manner with some regularity, forecasts can be used to anticipate the measured values of the substance and use them as input for the optimization. Alternatively, forecasting can be leveraged to predict arrival dates of new batches for the mixing process. Demand forecasting provides information on which blends should be produced. This and other integrated approaches open up new possibilities on how to produce a more sustainable inventory management.

15.7 Conclusion and Outlook

This chapter explored an application of prescriptive analytics, which reduces the complexity of an analytics problem by separating it into two more manageable steps. First, a forecasting algorithm captures uncertainty over the environment based on observations. Second, the forecast is passed to an optimization algorithm, which determines the optimal policy by treating any remaining uncertainty with robustness constraints. This separation of concerns allows focusing on the uncertainty aspect in the forecasting step and on the policy aspect in the optimization step. Each step can leverage domain expertise and build on existing solutions for the corresponding sub-problem. We demonstrate this principle at the task of finding cost-optimal order policies for the All-Time Buy of spare parts. The optimal policy minimizes the total cost of orders and storage while guaranteeing sufficient stock to satisfy the demand. To this end, we first forecast the demand curve of a spare part based on observed past demands. Specifically, we model the demand as a growth curve in a hierarchical Bayesian model. The growth curves capture domain expertise about the natural shape of demand curves and the hierarchy enables information sharing between similar spare parts. The output of this Bayesian model is a posterior distribution over future demand curves, which includes a measure of uncertainty. For the second step, we formulate an optimization problem for finding the cost-optimal policy. We start with an existing solution for the single-item uncapacitated lot sizing problem, which assumes that the demand is known. Next, we extend this model to more realistic, non-linear production costs. Using the multiple-choice model for the production cost, we can derive a locally ideal, linear mixed-integer program. To make the model robust against the uncertain demand, we adapt the storage availability and cost related constraints. Finally, we connect optimization and forecasting by using the forecasted demand as input to the optimization problem. Lastly, we outline the possibility of using forecasting in another use case, where a production of blended goods is modeled based on a pooling problem. In this production context, forecasting could be applied to predict the future demand of the goods or predict the assessment of quality parameters in blending ingredients or their availability. This information can then be funneled into the optimization model. Our focus is the case, where goods are produced according to specified recipes. To adhere to these ingredient ratios, we modify the standard pooling problem by including additional constraints.

Future aspects include the generalization of the forecasting model, such that further predictors can be included in the model. Additionally, the optimization algorithm currently does not make use of the forecasting model's uncertainty quantification, leaving room for further efficiency gains regarding the optimal stocking policy. This gives room for theoretical work on how to properly incorporate uncertainty in the stocking policy algorithm, as well as empirical work on the gains that such a combination of forecasting and optimization yields.

References

1. M. J. Betancourt and M. Girolami. Hamiltonian monte carlo for hierarchical models, 2013.
2. J. H. Bookbinder and J.-Y. Tan. Strategies for the probabilistic lot-sizing problem with service-level constraints. *Management Science*, 34(9):1096–1108, 1988.
3. U. Dombrowski, C. Malorny, R. Kolshorn, and T. Blöcker. Teiledienst. In *After Sales Service*, pages 35–124. Springer Berlin Heidelberg, 2020.
4. U. Dombrowski and S. Schulze. Lebenszyklusorientiertes Ersatzteilmanagement — Neue Herausforderungen durch innovationsstarke Bauteile in langlebigen Primärprodukten. In *Beiträge zu einer Theorie der Logistik*, pages 439–462. Springer Berlin Heidelberg, 2008.
5. L. Fortuin. The all-time requirement of spare parts for service after sales—theoretical analysis and practical results. *International Journal of Operations & Production Management*, 1980.
6. J. S. Hong, H.-Y. Koo, C.-S. Lee, and J. Ahn. Forecasting service parts demand for a discontinued product. *IIE Transactions*, 40(7):640–649, 2008.
7. Q. Hu, J. E. Boylan, H. Chen, and A. Labib. OR in spare parts management: A review. *European Journal of Operational Research*, 266(2):395–414, apr 2018.
8. Z. Huang. Extensions to the k-means algorithm for clustering large data sets with categorical values. *Data Mining and Knowledge Discovery*, 2(3):283–304, 1998.
9. R. A. Hughes, I. R. White, S. R. Seaman, J. R. Carpenter, K. Tilling, and J. A. Sterne. Joint modelling rationale for chained equations. *BMC medical research methodology*, 14:1–10, 2014.
10. T. Jónás, Z. E. Tóth, and J. Dombi. A knowledge discovery based approach to long-term forecasting of demand for electronic spare parts. In *2015 16th IEEE International Symposium on Computational Intelligence and Informatics (CINTI)*, pages 291–296, 2015.
11. G. Loukmidis and H. Luczak. Lebenszyklusorientierte Planungsstrategien für den Ersatzteilbedarf. In *Erfolgreich mit After Sales Services*, pages 251–270. Springer Berlin Heidelberg, 2008.
12. C. Menden, J. Mehringer, A. Martin, and M. Amberg. Vorhersage von Ersatzteilbedarfen mit Hilfe von Clusteringverfahren. *HMD Praxis der Wirtschaftsinformatik*, 56(5):1000–1016, 2019.
13. J. R. Moore. Forecasting and scheduling for past-model replacement parts. *Management Science*, 18(4):B200–B213, 1971.

14. B. Nicenboim, D. Schad, and S. Vasishth. An introduction to bayesian data analysis for cognitive science. *Forthcoming*, 2023.
15. S. Paun, B. Carpenter, J. Chamberlain, D. Hovy, U. Kruschwitz, and M. Poesio. Comparing bayesian models of annotation. *Transactions of the Association for Computational Linguistics*, 6:571–585, dec 2018.
16. Y. PochetWolsey and L. A. Wolsey. *Production Planning by Mixed Integer Programming*. SPringer, 1 edition, 2006.
17. A. Ramírez-Hassan and S. Montoya-Blandón. Forecasting from others' experience: Bayesian estimation of the generalized bass model. *International Journal of Forecasting*, 36(2):442–465, apr 2020.
18. R. Rossi, O. A. Kilic, and S. A. Tarim. Piecewise linear approximations for the static–dynamic uncertainty strategy in stochastic lot-sizing. *Omega*, 50:126–140, 2015.
19. J. L. Schafer. *Analysis of Incomplete Multivariate Data (Monographs on Statistics & Applied Probability)*. Chapman & Hall/CRC, 1997.
20. E. Schubert and P. J. Rousseeuw. Fast and eager runtime improvement of the PAM, CLARA, and CLARANS algorithms. *Information Systems*, 101:101804, nov 2021.
21. E. Silver. Inventory control under a probabilistic time-varying, demand pattern. *A I I E Transactions*, 10(4):371–379, 1978.
22. S. Sridhar, J. Linderoth, and J. Luedtke. Locally ideal formulations for piecewise linear functions with indicator variables. *Operations Research Letters*, 41(6):627–632, 2013.
23. D. Steuer, V. Hutterer, P. Korevaar, and H. Fromm. A similarity-based approach for the all-time demand prediction of new automotive spare parts. In *Proceedings of the 51st Hawaii International Conference on System Sciences*, 2018.
24. H. Tempelmeier. Stochastic Lot Sizing Problems. In J. M. Smith and B. Tan, editors, *Handbook of Stochastic Models and Analysis of Manufacturing System Operations*, International Series in Operations Research & Management Science, chapter 10, pages 313–344. Springer, March 2013.
25. R. H. Teunter and L. Fortuin. End-of-life service: A case study. *European Journal of Operational Research*, 107(1):19–34, 1998.
26. K. Tjørve and E. Tjørve. The use of gompertz models in growth analyses, and new gompertz-model approach: An addition to the unified-richards family. *PLOS ONE*, 12(6):e0178691, 06 2017.
27. L. A. W. Wolsey. *Integer Programming, 2nd Edition*. Wiley, 2 edition, 2020.
28. M. Xiang, R. Rossi, B. Martin-Barragan, and S. A. Tarim. Computing non-stationary (s, s) policies using mixed integer linear programming. *European Journal of Operational Research*, 271(2):490–500, dec 2018.
29. P. M. Yelland. Bayesian forecasting of parts demand. *International Journal of Forecasting*, 26(2):374–396, apr 2010.

Chapter 16
Optimization of Underground Train Systems

Lukas Hager[1], Tobias Kuen[2]

Abstract This chapter presents two approaches for enhancing the sustainability and efficiency of underground train systems. The first approach focuses on the optimization of DC railway power systems, employing a novel Mixed-Integer Quadratically Constrained Quadratic Program (MIQCQP) to control substation feed-in voltages effectively. By minimizing energy losses, this optimization approach demonstrates substantial potential for cost and emission reduction, contributing to a more energy-efficient underground train network. Validation results confirm the accuracy of the proposed model, and realistic instances reveal significant energy savings. The second approach deals with energy-efficient timetabling, a critical aspect in reducing the environmental impact of railway operations. The presented approach seeks to minimize energy consumption through the implementation of two key strategies: promoting energy-efficient driving patterns and optimizing recuperated energy from braking. Leveraging operational data, including power consumption profiles and travel time distributions, the optimization methods demonstrate remarkable potential in reducing energy consumption, subsequently leading to lower electricity costs and environmental benefits. This chapter is largely based on previous work of Hager and Koop on optimization of DC railway power systems and of Bärmann et al. [1] on energy-efficient timetabling.

Key words: Braking loss minimization, energy efficiency, metro systems, mixed-integer quadratic programming, traction power supply systems, transmission loss minimization, clique problem with multiple-choice constraints, combinatorial optimization.

[1]Friedrich-Alexander-Universität Erlangen-Nürnberg, FAU, Erlangen, Germany
[2] Fraunhofer Institute for Integrated Circuits IIS, Fraunhofer IIS, Nuremberg, Germany

Corresponding author: Tobias Kuen
e-mail: tobias.kuen@iis.fraunhofer.de

© The Author(s) 2024   303
C. Mutschler et al. (eds.), *Unlocking Artificial Intelligence*,
https://doi.org/10.1007/978-3-031-64832-8_16

16.1 Optimization of DC Railway Power Systems

16.1.1 Introduction

Due to rising energy prices and ecological challenges, energy efficiency in railway transport is a crucial research area. Prior studies mainly focused on reducing energy consumption from the consumer's side, such as energy-efficient timetabling and velocity profile optimization. In contrast, this work concentrates on the supplier's side, which has limited literature. Existing research highlights potential savings in the railway system but overlooks electrical infrastructure operations.

This work proposes a novel Mixed-Integer Quadratically Constrained Quadratic Program (MIQCQP) global optimization approach for controlling feed-in voltages at substations to minimize power losses. Direct current (DC) grids are considered, where regenerative energy utilization poses challenges.

While previous works explored feed-in voltage optimization, they did not consider recuperation losses. The proposed MIQCQP model, based on the optimal power flow model, addresses this issue. Its novelty lies in achieving global energy loss minimization, illustrated through a computational study using a realistic network structure.

16.1.2 Optimal Power Flow and mathematical MIQCQP model

The following MIQCQP is based on the optimal power flow (OPF) model. OPF is widely used to determine optimal power generation patterns in electric networks. In [7], the authors describe a DC version of OPF used as a basis for the presented model. For a better reading, see Table 16.5 where we list all parameters, variables, and symbols. The electric network is modeled as a directed graph $G = (X, E)$ with nodes X representing loads and sources, and arcs E representing current flow paths. The DC-OPF model can be stated as follows:

Minimize: Provision costs or power losses by controlling feed-in voltages

$$\text{s.t. Power equation: } P_i = U_i \cdot I_i \quad \forall i \in X,$$

$$\text{Kirchhoff's law: } I_i = \sum_{j \in \delta^+(i)} I_{ij} - \sum_{k \in \delta^-(i)} I_{ki} \quad \forall i \in X,$$

$$\text{Ohm's law: } I_{ij} = y_{ij}(U_i - U_j) \quad \forall (i, j) \in E.$$

The constraints in this model represent well-known relations from electrical engineering. The power equations describe the non-linear relationship between power generation or consumption $P_i \in \mathbb{R}$ at each node $i \in X$ as a product of voltage $U_i \in \mathbb{R}$ – which is a control variable if node i is feeding traction power system – and current

$I_i \in \mathbb{R}$. Kirchhoff's current law specifies that at each node, the sum of the directed currents adds up to zero. Here, $I_{ij} \in \mathbb{R}$ is the current flow on the arc $(i, j) \in E$, while $I_i \in \mathbb{R}$ models the demand at node $i \in X$. Ohm's law describes the voltage drop $U_i - U_j$ along an arc $(i, j) \in E$ depending on the given admittance $y_{ij} \in \mathbb{R}_+$.

16.1.2.1 Snapshot Model

Next, the DC-OPF is applied to railway traction power supply systems. The presented optimization model considers a snapshot in time of the daily operation, where trains have fixed positions in the railway network represented by an admittance matrix. The power demands of the trains are also fixed. The optimization goal is to minimize energy losses.

The directed graph $G = (S \cup N \cup Z, E)$ consists of three subsets: power conversion units N of feeding traction power substations, additional busbar nodes S, one for each substation, and the set of trains Z. Each busbar node in S is directly connected to its corresponding feed-in node in N, and the admittance on the connecting arc models the internal voltage drop of the substation. The set of arcs E is divided into arcs E_{NS} and arcs E_{SZ}, i.e., $E = E_{\text{NS}} \dot\cup E_{\text{SZ}}$. arcs in E_{NS} always point from the power conversion unit to the corresponding busbar, while transmission losses occur on arcs in E_{SZ}. The static railway DC-OPF (SR-DC-OPF) can be stated as follows:

$$\min_{U_i, i \in N} \sum_{(i,j) \in E_{\text{SZ}}} y_{ij} I_{ij}^2 + \sum_{i \in Z} P_{\text{B},i} + \sum_{i \in N} (A + BI_i + CI_i^2),$$

s.t. *Power equation for substations: Eqs.* (16.1), (16.2),

Power equation for trains: Eqs. (16.3), (16.4), (16.5), (16.6), (16.7),

Ohm's law for each line: Eqs. (16.8), (16.9), (16.10), (16.11), (16.12),

Kirchhoff's current law for each node: Eq. (16.13),

Voltage restrictions of traction power supply: Eqs. (16.14).

The above model will be explained in detail, beginning with the power losses modeled in the objective and then passing to constraints Eqs. (16.1)–(16.14).

The Objective Function contains transmission losses in lines, braking losses in trains and conversion losses in substations. The transmission losses on an arc $(i, j) \in E_{\text{SZ}}$ can be derived via Ohm's law and depend on the arc admittance y_{ij} and the current flow I_{ij}. Braking resistance losses can appear at every braking train $i \in Z$ when the feed-in voltages are chosen such that a part of the braking energy, namely $P_{\text{B},i}$, cannot be fed back to other trains. The losses at any substation $i \in N$ are approximated with a quadratic function in the feed-in current I_i using parameters A, B and C, as proposed in [2]. While the transmission and substation losses generally decrease with higher feed-in voltages, the braking losses increase with increasing feed-in voltage. A detailed discussion on these effects can be found in [10].

Power Equation for Substations. At the feed-in nodes $i \in N$, the feed-in power is the product of the control variable feed-in voltage and the corresponding current:

$$P_i = U_i \cdot I_i \qquad\qquad \forall i \in N. \qquad (16.1)$$

However, since the power P_i at the feed-in node $i \in N$ does neither appear in the objective function nor in other constraints of the model, its modeling can simply be neglected at feed-in nodes.

At the busbar nodes $i \in S$, representing the phyiscal busbars in substations, power is neither consumed nor generated. Therefore, they are modeled as pure passing nodes. The power equation will be neglected and the variables are fixed

$$P_i, I_i = 0 \qquad\qquad \forall i \in S. \qquad (16.2)$$

The Power Equation for Trains $i \in Z$ is more complex. The current splits into two parts:

$$I_i = I_{\text{trac,DC},i} + I_{\text{aux},i} \qquad\qquad \forall i \in Z. \qquad (16.3)$$

Here, $I_{\text{trac,DC}}$ is the current needed for traction and additional parts of the auxiliaries. The current I_{aux} is also needed for auxiliaries, however, it does not underlie current limitation characteristics of trains as $I_{\text{trac,DC}}$ does. Therefore, the train's power equation splits into the current underlying limitation characteristics and into the one that is not:

$$U_i \cdot I_{\text{trac,DC},i} = P_{\text{trac},i} + P_{\text{aux,DC},i} + P_{\text{B},i} \qquad\qquad \forall i \in Z, \qquad (16.4)$$
$$U_i \cdot I_{\text{aux},i} = P_{\text{aux},i} \qquad\qquad \forall i \in Z. \qquad (16.5)$$

Here, $P_{\text{trac},i}$ is the traction power of the train, $P_{\text{aux,DC},i}$ is the auxiliary power subject to current limitations, $P_{\text{aux},i}$ is the auxiliary power that is not subject to current limitations, and $P_{\text{B},i}$ is used to model the residual part of traction power that is converted into heat in a braking resistor. While $P_{\text{trac},i}$, $P_{\text{aux,DC},i}$ and $P_{\text{aux},i}$ are input parameters to the model, $P_{\text{B},i}$ is variable and depends on the power flows in the network and therefore implicitly on the feed-in voltages. Note, the braking resistance power is only relevant for braking trains ($P_{\text{trac}} \geq 0$), i.e. so

$$P_{\text{B},i} = 0, \text{ if } P_{\text{trac},i} \geq 0 \qquad\qquad \forall i \in Z. \qquad (16.6)$$

The limitations on $I_{\text{trac,DC},i}$ are described by a piecewise non-linear function in the voltage of the train. These limitations result from an intersubsection of piecewise linear boundaries on $I_{\text{trac,DC}}$ and the power equation from Eq. (16.4). For a non braking train $i \in Z$, $P_B = 0$ holds. Thus, the voltage-current combination ($U_i, I_{\text{trac,DC},i}$) must satisfy the equation $U_i \cdot I_{\text{trac,DC},i} = P_{\text{trac},i} + P_{\text{aux,DC},i}$ and therefore, the piecewise linear current limitation leads to a narrowing of the feasible voltage interval, see the upper part of Figure 16.1. If a train $i \in Z$ is braking, ideally, all traction power that is not needed for auxiliaries is fed back into the system. However, since technically there is a piecewise linear limitation on $I_{\text{trac,DC},i}$, it might be the case that some residual part of traction power fis not fed back. This is converted into heat in a braking resistor and modeled in the variable $P_{\text{B},i}$. Voltage-current combinations

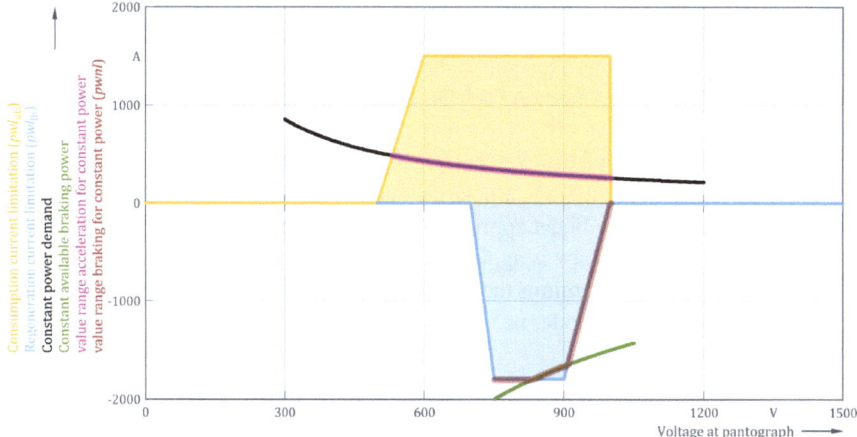

Fig. 16.1: Example for feasible voltage-current combinations of a train, with constant power for acceleration (black) and braking (green) including current limitations (yellow, blue). It results in a piecewise non-linear value range during braking (red) and a quadratic value range during consumption (pink).

$(U_i, I_{\text{trac,DC},i})$, where this is the case, lie on the piecewise linear function outside of the intersubsections with the hyperbola $(P_{\text{trac}} + P_{\text{aux,DC}})/U$, see the lower part of Figure 16.1. The feasible voltage-current combinations in braking trains are modeled via the piecewise non-linear function $\text{pwnl}_{\text{lb}}(U_i)$,

$$I_{\text{trac,DC},i} = \text{pwnl}_{\text{lb}}(U_i) \qquad\qquad \forall i \in Z. \qquad (16.7)$$

Ohm's Law is used to model for each arc $(i, j) \in E$ in the network the voltage drop. For arcs E_{SZ}, the modeling is straightforward via

$$I_{ij} = (U_i - U_j) \cdot y_{ij} \qquad\qquad \forall (i, j) \in E_{\text{SZ}}. \qquad (16.8)$$

For arcs E_{NS} Ohm's law is only applied in the direction of supply. This is due to the physics of rectifiers at substations. Since diodes, installed in rectifiers, have a junction they only conduct in a single direction. Thus, Ohm's law is only active if the voltage at the busbar is lower than the feed-in voltage. Otherwise, Ohm's law does not apply and current cannot flow back to the feed-in node, although the voltage drop would imply this. This is modeled in the following:

$$I_{ij} = (U_i - U_j) \cdot y_{ij} + \Delta_{ij} \qquad \forall (i, j) \in E_{\text{NS}}, \qquad (16.9)$$
$$\Delta_{ij} \leq M_\Delta(1 - z_{ij}) \qquad \forall (i, j) \in E_{\text{NS}}, \qquad (16.10)$$
$$I_{ij} \leq M_I z_{ij} \qquad \forall (i, j) \in E_{\text{NS}}, \qquad (16.11)$$
$$z_{ij} \in \{0, 1\}, \Delta_{ij} \in \mathbb{R}_+, I_{ij} \in \mathbb{R}_+ \qquad \forall (i, j) \in E_{\text{NS}}. \qquad (16.12)$$

The continuous auxiliary variable Δ_{ij} causes Eq. (16.9) to be trivially satisfied whenever the corresponding binary auxiliary variable $z_{ij} \in \{0, 1\}$ fulfils $z_{ij} = 0$. Indeed, exactly then Δ_{ij} has a feasible range of $[0, M_\Delta]$, where M_Δ is an upper bound for every Δ_{ij}. The input parameter M_I is some upper bound on maximum current flow on arcs in the network. If $z_{ij} = 1$ then $\Delta_{ij} = 0$ and Eqs. (16.9)–(16.12) reduce to the standard power equation $I_{ij} = (U_i - U_j) \cdot y_{ij}$. Finally, $z_{ij} = 1$ is enforced by Eq. (16.11) whenever there is a positive current flow from i to j. This modeling technique is known as the bigM approach in discrete optimization.

Kirchhoff's Current Law is applied at each node $i \in N \cup S \cup Z$ which says that the inflow, $\sum_{(i,j) \in \delta^{\text{in}}(i)} I_{ij}$, minus the outflow, $\sum_{(i,j) \in \delta^{\text{out}}(i)} I_{ij}$, equals the consumed or fed-in current I_i at that node, i.e.

$$I_i = \sum_{(i,j) \in \delta^{\text{in}}(i)} I_{ij} - \sum_{(i,j) \in \delta^{\text{out}}(i)} I_{ij} \quad \forall i \in N \cup S \cup Z. \qquad (16.13)$$

Note, currents at substation nodes will be zero, see Eq. (16.2), while currents at feed-in nodes will be negative or zero.

Voltage Restrictions are given because as the trains are mobile loads in the railway network, variable voltage drops inevitably occur. In order to deal with these voltage drops, the minimum and maximum voltages for substations and trains are standardized, e.g. in Europe this is done in the standard EN 50163 [6]. This results in node-specific bounds on the variables in the model:

$$U_{i,\min} \leq U_i \leq U_{i,\max} \qquad \forall i \in N \cup S \cup Z. \qquad (16.14)$$

16.1.2.2 Time Span Model

Next, snapshot modeling is extended to time spans. For a given time span $[0, T]$, a discretization $\mathcal{T} := \{0 = t_0, t_1, \ldots, t_n = T\}$ is introduced. The granularity of this discretization determines the size of the model, since a duplication of the constraints from Eqs. (16.1)–(16.14) from the SR-DC-OPF model is performed for each time step. It is assumed that the power demands of the trains as well as their positions, modeled implicitly by the admittances, are known a priori for each discretization time step t_i. Most DC traction power supply systems are fed by diode rectifiers which cannot be controlled actively in operations. Thus, it is required to optimize over a time interval and determine a constant feed-in voltage.

Since the trains will move during the considered time span, their neighboring relations, i.e. the arc set of the graph will change. Consequently, a separate graph $G^t = (N^t \cup S^t \cup Z^t, E^t)$ for each time step $t \in T$ is introduced.

The complete time-expanded DC railway optimal power flow model (TR-DC-OPF) is stated as follows:

$$
\min_{U_i^t,\, i \in N^t} \sum_{t=0}^{T} \left(\sum_{(i,j) \in E_{SZ}^t} y_{ij}^t (I_{ij}^t)^2 + \sum_{i \in Z^t} P_{B,i}^t + \sum_{i \in N^t} (A + BI_i^t + C(I_i^t)^2) \right)
$$

s.t. *Power equation for substations: Eqs.* (16.1), (16.2) $\qquad \forall t \in 0, \ldots T,$

Power equation for trains: Eqs. (16.3) – (16.7) $\qquad \forall t \in 0, \ldots T,$

Ohm's law: Eqs. (16.8) – (16.12) $\qquad \forall t \in 0, \ldots T,$

Kirchhoff's current law: Eq. (16.13) $\qquad \forall t \in 0, \ldots T,$

Voltage restrictions: Eq. (16.14) $\qquad \forall t \in 0, \ldots T,$

Constant voltage control: *Eq.* (16.15) $\qquad \forall t \in 0, \ldots T-1.$

The TR-DC-OPF model now includes the constraints of $T+1$ network snapshots with corresponding sets of variables and parameters. Additionally, the feed-in voltages are coupled via the constant voltage control constraints in TR-DC-OPF, as stated:

$$
U_i^t = U_i^{t+1} \qquad \forall i \in N^t, \forall t \in 0, \ldots T - 1. \tag{16.15}
$$

16.1.3 Case Studies

In this section, the potential benefit of an optimal voltage control at substations is demonstrated. Therefore the feed-in voltages of two exemplary railway networks are optimized.

16.1.3.1 Optimization of time stamps in a small network

First, the feed-in voltages are optimized for a small network consisting of two substations and two trains, see Figure 16.2. The two timestamps $t_1 = 00{:}03{:}10$ and $t_2 = 00{:}03{:}30$ are optimized. According to EN 50163, the feed-in voltages can only take values between 500 V to 900 V. In Table 16.2, the properties of the corresponding SR-DC-OPF models are shown.

In t_1 both trains in the network are accelerating and the objective consists purely of transmission and substation losses. Therefore, optimal feed-in voltages are 900 V, see Table 16.3. In snapshot t_2, an optimal solution is expected to allow the Train2 to recuperate a large portion of its braking energy to Train1. In Table 16.3 it is shown that the optimal feed-in voltages are 668.4 V at Substation A and 900.0 V at Substation B.

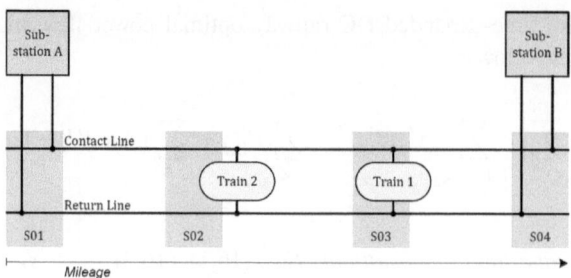

Fig. 16.2: Schematic diagram of the small network.

Table 16.1: Electrical parameters in the small network.

	$t_1 = 00:03:10$	$t_2 = 00:30:30$
P_{Train1}	799 000 W	-1 244 361 W
$P_{aux,Train1}$	100 000 W	100 000 W
P_{Train2}	1 111 477 W	788 956 W
$P_{aux,Train2}$	100 000 W	100 000 W
y_{SubA}	64.18 S	64.18 S
y_{SubB}	64.17 S	64.17 S
$y_{SubA,Train2}$	66.22 S	19.88 S
$y_{Train2,Train1}$	4.85 S	4.68 S
$y_{Train1,SubB}$	14.55 S	38.31 S

Table 16.2: Model parameters and results for t_1 and t_2 in the small network.

	$t_1 = 00:03:10$	$t_2 = 00:30:30$
continuous variables	24	25
binary variables	2	2
quadratic objective terms	5	5
U_{SubA} (optimized)	900.0 V	668.4 V
U_{SubB} (optimized)	900.0 V	900.0 V

16.1.3.2 Optimization of a realistic entire line

In this example a more realistic network, an airport rail link system including a realistic timetable, is optimized. In Figure 16.3 a visualization of the network configuration is given. Again, the nominal voltage is 750 V DC and the feeding is double-sided. Further, the trains have auxiliary loads of 100 kW each.

The headway period of 3 min is discretized with an accuracy of 1s. The properties of the TR-DC-OPF model and a comparison of optimized and non-optimized configurations are given in Table 16.4.

The total savings are 4.0%. The optimal feed-in voltages of 856.4 V, 840.6 V and 844.8 V are higher than the non-optimized voltages. Again, the same effects as in the previously shown example can be observed. The increase in feed voltages reduced

Table 16.3: Detailed power losses for the small network in two snapshots – comparison of non-opt. and optimized feed-in voltages as SR-DC-OPF.

| | $t_1 = 00{:}03{:}10$ | | $t_2 = 00{:}30{:}30$ | |
	non-opt.	optimized	non-opt.	optimized
U_{SubA}	820.0 V	900.0 V	820.0 V	668.4 V
U_{SubB}	820.0 V	900.0 V	820.0 V	900.0 V
Substation losses	13 635 W	11 281 W	2 037 W	1 656 W
Transmission losses	129 452 W	103 403 W	124 035 W	326 816 W
Braking losses	0 W	0 W	438 630 W	0 W
Total losses	143 087 W	114 684 W	564 702 W	328 472 W

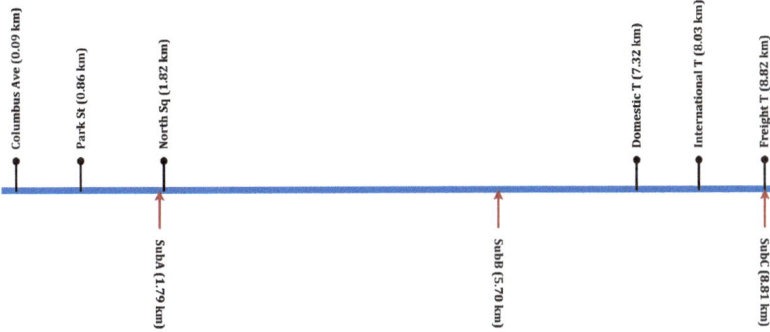

Fig. 16.3: Schematic diagram of the realistic network, with stations (black) and traction power supply substations (red); from Sitras Sidytrac Designer.

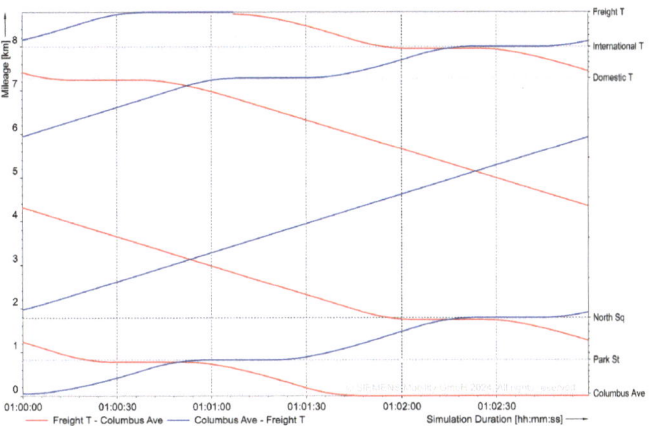

Fig. 16.4: Schedule of the realistic network; from Sitras Sidytrac Designer.

substation losses by 6.1% and transmission losses by 10.8%, but resulted in an increase in braking losses by 19.0%. In total, however, the savings in the transmission

Table 16.4: Model parameters and results for the realistic network – detailed power losses comparison of non-opt. and optimized feed-in voltages.

	continuous variables		10 790
	binary variables		540
	quadratic objective terms		4 035
	non-opt.	optimized	rel. difference
U_{SubA}	820.0 V	856.4 V	-
U_{SubB}	820.0 V	840.6 V	-
U_{SubC}	820.0 V	844.8 V	-
Substation losses	0.821 kWh	0.770 kWh	-6.1%
Transmission losses	15.125 kWh	13.491 kWh	-10.8%
Braking losses	4.604 kWh	5.477 kWh	+19.0%
Total losses	20.550 kWh	19.738 kWh	-4.0%

losses compensate for the increased braking losses.

Table 16.5: List of symbols used in the optimizaton models

Element	Name	Unit	Description
Graph	T		Length of the planning horizon
	S		Set of substations
	Z		Set of trains
	N		Set of no-load substations
	E		Lines
Variables	P	Watt (W)	Power
	P_B	Watt (W)	Non-recuperated braking Power
	U	Volt (V)	Voltage
	I	Ampere (V)	Current at node or arc
	$I_{\text{trac,DC}}$	Ampere (V)	Bounded Auxiliary current
	I_{aux}	Ampere (V)	Auxiliary current
	Δ	Ampere (A)	Slack variable for current balance
	z		Binary variable
Constants	P_{aux}	Watt (W)	Auxiliary power
	$P_{\text{DC,aux}}$	Watt (W)	Auxiliary DC-relevant power
	y	Siemens (S)	Admittance on a given line
	M_Δ	Ampere (A)	bigM for Δ
	M_I	Ampere (A)	bigM for I
	A, B, C		Substation loss parameters

16.2 Energy-Efficient Timetabling applied to a German Underground System

16.2.1 Industrial Challenge and Motivation

Traction energy consumption is among the most important cost factors in the electricity bill of a railway undertaking. It is significantly influenced by the manner in which the trains are driven. Thus, a significant reduction in energy consumption can be achieved by choosing energy-efficient velocity profiles. This includes making use of pure rolling phases, the so-called coasting. As far as possible, as a train consumes no traction energy at all in this phase and, due to the low rolling friction, only slowly looses speed. In Figure 16.5, the effects of choosing between different driving modes of a train are shown schematically for a train in an underground network.

These data show that by slightly slowing down the fastest possible speed profile on a given track, the train may consume up to 1/3 less in energy. In this respect, it is especially beneficial to extend the coasting phases of the train as much as possible. Altogether, choosing the optimal velocity profile for each train on each leg (= timetabled run between two stations) with respect to given total line travel times entails a huge leverage for bringing down the consumption of the overall underground system. This finding motivated the joint research project of FAU Erlangen-Nürnberg, Fraunhofer IIS and VAG Verkehrs-Aktiengesellschaft, the local operator of public transport in the German city of Nürnberg. Its idea was to take a given timetable draft toward the end of the timetable planning phase and to use the remaining degrees of freedom to slightly shift train departures within fixed windows around their currently planned departure times. The aim is to create the necessary flexibility to enable choosing the best-possible velocity profile on each leg. At the same time, these shifts in the departures times allow for the better synchronization of departure and arrival events. This is important as a braking train is able to feed back recuperated energy to the grid. However, this energy can only be used if there is another train in the network which is accelerating at the same time, otherwise it is lost because there is no energy storage in the system. Overall, there is a considerable potential for cost saving, as we have demonstrated in our collaboration. In the following, we will elaborate on the mathematical approach and present our case study for the underground system of Nürnberg.

16.2.2 Mathematical Research

Based on a timetable draft created by expert planners, the studied task is to determine slight modifications in the train departure times as well as choosing velocity profiles for all trains in an energy-optimal way. However, these modifications shall retain the timetable structure established in the draft according to stated criteria, e.g. dwell times (= passenger interchange times) in the stations, minimum headway times (=

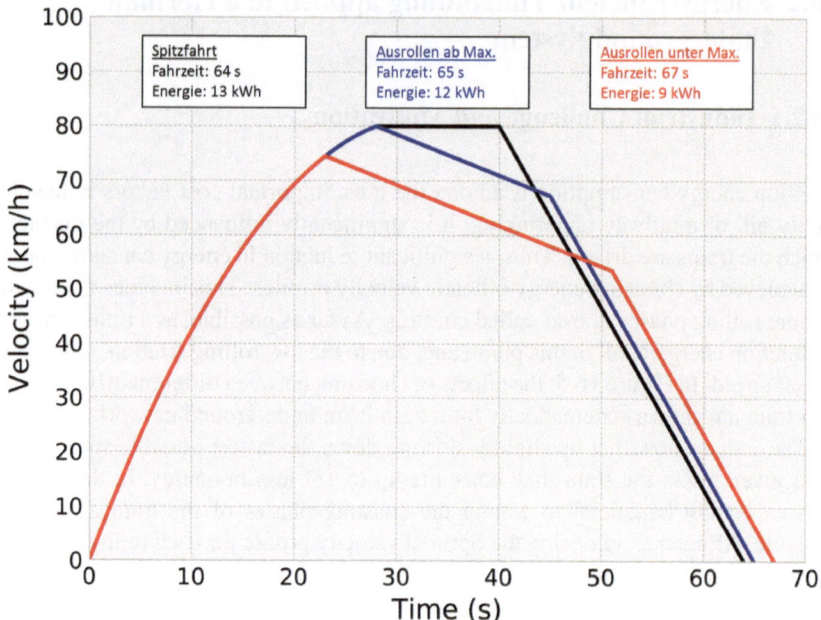

Fig. 16.5: Schematic representation of the effect of choosing between one of the three driving modes "always driving as fast as possible" (black), "accelerating to maximum velocity followed by coasting" (blue) and "accelerating to below maximum velocity followed by extended coasting" (red) on a sample underground leg

safety distances) between trains and desired connections between trains. In order to construct a mixed-integer programming (MIP) model for this task, we discretized the time horizon into time steps of e.g. 5 seconds each and determined a suitable discrete set of (e.g. 3) alternatives for the velocity profiles for each train on a given leg. The profiles were initially chosen as heuristic solutions to an optimal control problem; later we changed them against measured profiles of actual train runs in the network. Furthermore, we allowed departure time shifts up to a given amount, e.g. 15 seconds around the draft departure time for each leg – a change that is hardly noticeable by the passengers but that can still allow for significant energy savings as we were able to show. With allowed shifts of 15 seconds in increments of 5 seconds and 3 profiles to choose from, there are already $3 \cdot 7 = 21$ possible choices for the combination of departure time and velocity profile for each leg. Given that there are 24,000 legs to be served in the Nürnberg underground each day, this means there are $24,000^{21} \approx 10^{92}$ possible timetables adjustments to choose from. No company planner could hope to evaluate all of them manually in order to determine the most energy-efficient one. Via the techniques of discrete optimization we have developed over the course of this project, however, we are able to produce near-optimal timetables within one hour or less.

To this end, we came up with a model formulation for the set of the feasible timetable adjustments as a special case of the clique problem with multiple-choice constraints on an undirected graph G (see Chapter 7, Section 7.3.1.1 or for more details [[3], [4]]). Its nodes represent possible combinations of departure time and velocity profile for the legs to be scheduled, while the arcs model compatibilities between the departure configurations for different legs. Whenever the departure configurations for two specific legs do not violate any requirements for a feasible timetable, such as the above-mentioned ones, the corresponding nodes are connected by an arc. This results in an optimization model of the type

$$\min \sum_{t \in T} \max \left(P(x, t), 0 \right)$$

$$s.t. \, x \in X,$$

where $P(x, t)$ represents to total energy consumption at time step t, summed over all running trains, while X is the set of feasible timetable adjustments. Taking the maximum of $P(x, t)$ and 0 reflects that energy from a braking train can only be recuperated if it is used by other trains in the same time step. After linearizing the objective function with the help of additional auxiliary variables, the above model can be written as a MIP. We point out that all relevant types of timetabling constraints can indeed be expressed as pairwise node conflicts, which constitutes a very special structure. There are several ways to translate them into linear constraints. However, modeling the feasible region X in the most efficient way is very important as standard MIP solvers cannot solve the problem efficiently for real-world networks if a naive model formulation is used.

Our search for an adequate model formulation was inspired by the work of [9]. It was among the first to study the combined optimization of railway (or more precisely underground) timetables and energy consumption, giving a heuristic for reducing instantaneous power peaks. In [5], we took up their basic idea and studied the effects of optimal timetabling for small subnetworks of German railway traffic under different objective functions relating to power consumption patterns. During this work, we realized that the problem contains an interesting structure to be exploited in order to reduce solution times. The nodes of the compatibility graph can be partitioned by the legs they belong to, and within each partition V_l they can be sorted by departure time. For the special (but still NP-hard) case of a single energy profile available for each leg, the compatibility structure then allows for a totally unimodular description of the timetabling polytope. It could be improved to an even more efficient dual-flow formulation by using the canonical ordering of the departure times for each leg. This special structure also comprises problems in other application contexts, such as the piecewise linearization of path flows – e.g. of natural gas in a pipeline (see [8]). We generalized the core properties of the compatibility structure to the abstract notion of staircase compatibility in [3]. There, the resulting model formulations were successfully employed on much larger subnetworks of

Deutsche Bahn AG (DB), up to the Germany-wide network, for minimizing peak power consumption. When using our improved formulations, we observed significant savings in solution times (over a factor of 100 in several cases), which allowed us to solve the problem for the Germany-wide network within a couple of minutes, but took hours to solve beforehand. For multiple energy profiles per leg to choose from, the structure of the feasible set still tends to favour similar reformulations but cannot be perfectly described by staircase compatibility. We continued our polyhedral studies by considering a special case with respect to the dependency graph of the subsets in the node partition according to legs. This is the graph that encodes which pairs subsets of nodes directly impose any restrictions on each other. We showed that the feasible set in this case can be completely described by stable-set inequalities if the dependency graph is cycle-free. This leads to the following overall formulation:

$$\min \ \sum_{t \in T} \max \left(P(x, t), \ 0 \right)$$

$$s.t. \ \sum_{v \in V_l} x_v = 1 \quad \text{for all subsets } V_l,$$

$$\sum_{v \in S} x_v \leq 1 \quad \text{for all stable sets } S \text{ in } G,$$

$$x \geq 0,$$

where $G = (V, \ E)$ is the compatibility graph and the subsets V_l for each leg l form a partition of V. Altogether, we want to choose exactly one departure configuration for each leg, as modeled by the variables x_v for each node $v \in V$. The stable sets in G represent exactly the subsets of nodes which are in pairwise conflict with each other. Note that the total number of stable sets is potentially large and difficult to generate in general. However, only stable sets involving nodes from just two subsets are needed, which significantly reduces the enumeration effort and the size of the formulation – especially if the number of departure configurations choices for each leg is small in comparison to the number of legs in the timetable.

We used this improved formulation to greatly reduce solution time for optimizing the timetable in the Nürnberg underground network, see [4], also for more details on the aforementioned polyhedral results. In this preliminary computational study, an optimized shifted schedule of the longest line U1 in the system reduced the overall energy consumption by about 18% during the morning rush hour interval between 5 a.m. and 9 a.m. when compared to the actual 2018 schedule. From there on, we have undertaken great efforts to broaden the available database and to extend the results to all three Nürnberg underground lines over the whole day in order to see how much of these 18% in savings can be expected to be obtained in practice. These efforts and the findings we had will be described in the next section.

16.2.3 Implementation

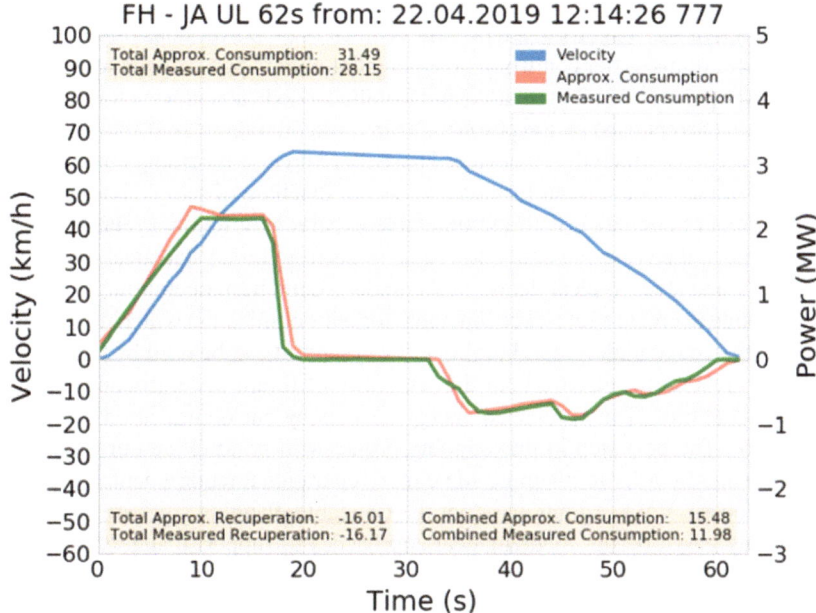

Fig. 16.6: A sample velocity profile (blue), approximated (red) and measured (green) power consumption profile on a given leg in the Nürnberg underground network

The initial spark for this project was a cooperation with DB in project E-Motion (2013–2016), funded by the German Ministry of Education and Research (BMBF). Its aim was to develop optimization algorithms to compute slight adjustments in the departure times of the trains to reduce peak power consumption for railway transport. The successful completion of this project led us to approach VAG in order to see if the same technique could be used to reduce peak consumption in the Nürnberg underground system. We soon learned that an even higher potential lies in reducing the overall power consumption by choosing energy-efficient driving patterns. So whereas in the project with DB we only adjusted departure times, the new task was to choose optimal travel times for each leg. After extending our timetabling model accordingly, we iteratively increased its performance. Firstly, the new degree of freedom, choosing travel times, added complexity to the mathematical model. An extensive study of its structure as described in [4] enabled us to give a more compact problem formulation that was much easier to solve and still respected all necessary constraints. Secondly, we incorporated additional timetabling constraints to make sure that our solutions are not just energy-efficient, but also real-world applicable. Note that these additions seamlessly fit into our new mathematical framework and therefore had no negative impact on solution times. Finally, we replaced the simulated velocity and power profiles, which were based on theoretical knowledge of train characteristics, and which we used at the beginning, with profiles that were based on actual measurements for train runs in the system. For some time, we used profiles approximated from velocity and acceleration measurements as recorded by the tachographs in the wagons together with characteristic power consumption curves of the power trains. After confirming the potential of our approach on these approximate power profiles, VAG purchased and installed a dedicated device for measuring the traction current used by the powertrains in a wagon. Figure 16.6 shows a sample profile recorded by the DL350 device. These more precise recordings improved the accuracy of the model output further, and, as our database of sample power profiles grew, we could cooperate with the timetable experts at VAG on a related topic as well. Namely, the collected data allowed us to perform broad statistical evaluations to identify and study typical delays in the underground train operation. As a result, we were able to create a reference timetable for our optimization which more closely matches the actual underground traffic in the system. It can be used by VAG during schedule creation to improve both the reliability of future underground timetables and the reliability of the projected energy savings by our optimization procedures in practice. The next step in this ongoing project will be to refine our timetabling model further in order to integrate some more operational requirements. At the end of this process, VAG is going to adopt the timetable planning software we are implementing based on our mathematical approaches to support planners in creating energy-efficient underground timetables.

16.3 Conclusion and Outlook

The findings of the project on the optimization of DC railway power systems underscore the untapped potential of tailored substation voltage adjustments in enhancing energy efficiency within underground train systems. This highlights the importance of customizing voltage regulation to optimize energy consumption based on individual operational dynamics and network characteristics.

Similarly, the insights from the study together with VAG in Nürnberg illuminate the substantial benefits of energy-efficient timetabling in minimizing energy consumption and operational costs of an underground train system. Synchronized schedules enable efficient energy recuperation and underscore the significance of optimizing train movements.

Looking ahead, an exciting avenue for future research lies in the integration of substation voltage adjustment and energy-efficient timetabling. This combined approach holds the promise of achieving even greater energy savings and operational efficiency.

Acknowledgements

The chapter is based on the work of Friedemann Koop (Siemens Mobility Gmbh) and Lukas Hager (FAU Erlangen) as well as the publication [1].

References

1. A. Bärmann, M. Merkert, P. Gemander, A. Martin, and F. Nöth. Energy-efficient timetabling in a german underground system. *Optimization Online*, 2020.
2. H. Biesenack, G. George, G. Hofmann, A. Schmieder, E. Braun, K. Girbert, R. C. Klinge, R. Puschmann, S. Röhlig, E. Schlechter, et al. *Energieversorgung elektrischer Bahnen.* Springer, 2006.
3. A. Bärmann, T. Gellermann, M. Merkert, and O. Schneider. Staircase compatibility and its applications in scheduling and piecewise linearization. *Discrete Optimization*, 29:111–132, 2018.
4. A. Bärmann, P. Gemander, and M. Merkert. The clique problem with multiple-choice constraints under a cycle-free dependency graph. *Discrete Applied Mathematics*, 277:1–16, 2020.
5. A. Bärmann, A. Martin, and O. Schneider. A comparison of performance metrics for balancing the power consumption of trains in a railway network by slight timetable adaptation. *Public Transport*, 9:95–113, 2017.
6. DKE German Commission for Electrical, Electronic & Information Technologies of DIN and VDE. Railway applications - supply voltages of traction systems; german version. Standard DIN EN 50163 ; VDE 0115-102:2005-07, 2005.
7. L. Gan and S. H. Low. Optimal power flow in direct current networks. *IEEE Transactions on Power Systems*, 29(6):2892–2904, 2014.
8. F. Liers and M. Merkert. Structural investigation of piecewise linearized network flow problems. *SIAM Journal on Optimization*, 26(4):2863–2886, 2016.
9. B. Sansó and P. Girard. Instantaneous power peak reduction and train scheduling desynchronization in subway systems. *Transportation science*, 31(4):312–323, 1997.
10. A. Spielmann. *Verbesserungsmaßnahmen zur Steigerung der Energieeffizienz von Metrosystemen.* Thesis, Georg-Simon-Ohm-Hochschule, Nürnberg, 2010.

Chapter 17
AI-assisted Condition Monitoring and Failure Analysis for Industrial Wireless Systems

Ulf Wetzker[1], Anna Richter[1], Vineeta Jain[1], Jakob Wicht[1]

Abstract With the increasing proliferation of wireless devices and Internet-of-Things (IoT) applications in various fields, such as patient monitoring, vehicle-to-everything (V2X) communication and industrial automation, there is a growing significance in developing robust methods and tools for evaluating and predicting link quality, monitoring information flow, as well as conducting failure analysis. This is particularly important in safety-critical industrial IoT (IIoT) environments such as smart factories, where challenging signal propagation conditions and interference from coexisting wireless technologies can severely impact network performance and application reliability. This contribution provides a comprehensive analysis of coexistence issues in industrial IIoT networks and highlights the complexities and challenges associated with performing failure analysis on a large scale. The necessity of using data-driven methods in the development of efficient and user-friendly failure analysis systems is discussed and the challenges regarding required datasets are highlighted.

Key words: data augmentation, object detection, Autoencoder, Dynamic Time Warping, Industrial IoT, failure analysis, coexistence problems, wireless monitoring systems.

17.1 Introduction

In the automation industry, wireless technology is increasingly used for machine-to-machine (M2M) communication. This trend is driven by substantially higher flexi-

[1]Fraunhofer Institute for Integrated Circuits IIS Division Engineering of Adaptive Systems EAS, Fraunhofer IIS/EAS, Dresden, Germany

Corresponding author: Ulf Wetzker
e-mail: ulf.wetzker@eas.iis.fraunhofer.de

321

bility and connectivity, which simplifies the reconfiguration of production processes and enables the introduction of advanced technologies such as autonomous robots, augmented reality (AR) or Human-Machine-Interfaces (HMIs). In order to meet the specific requirements of various industrial applications, a number of wireless communication systems specially adapted to this market [3] have been established. These systems use both unlicensed and licensed frequency bands or rely on cellular networks to ensure efficient and reliable communication in various industrial environments. Despite numerous advantages, the use of wireless communication poses many challenges, especially in the area of network operation and maintenance. Although many operational complications related to reception problems and interference with other wireless networks can be reduced or avoided through deliberate planning and design of the wireless application, this is not sufficient for long-term fault-free operation of the network. As the wireless environment is constantly changing, the state of the network must be continuously monitored and maintained. To prevent future disruptions and system failures, experts can evaluate the condition of the communication network and wireless connection to effectively and quickly find problems and their causes in the event of a malfunction. As the root causes of wireless communication failures are very difficult to identify and localize, on-site troubleshooting requires measurement equipment that can provide an insight into the entire network stack.

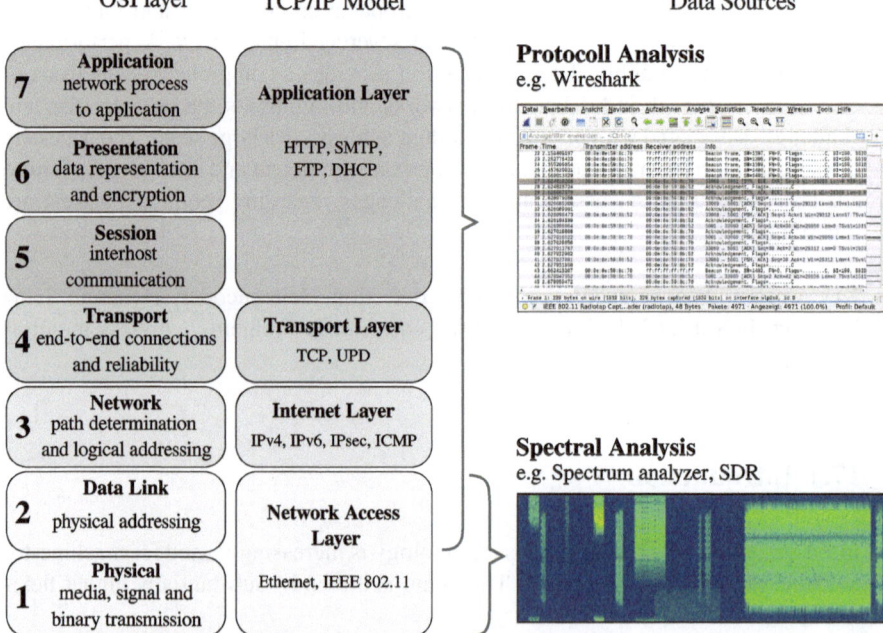

Fig. 17.1: Overview of data sources for a holistic analysis of all network layers considering the TCP/IP model as an example.

Spectral analysis provides comprehensive insights into the physical network layer and parts of the datalink layer as shown in Figure 17.1. Using this measurement method, it is possible to perform in-depth analyses of coexisting wireless applications in the same frequency band or other sources of interference. Protocol analysis is widely used because it covers almost all network layers, it presents the content of packets in an easily understandable form, and is also applied in wired networks. However, the evaluation and interpretation of the results obtained requires in-depth knowledge of radio standards, communication networks, network protocols and the associated applications, which is only available to specially trained technicians. Due to the increasing popularity of wireless communications, users are encountering interference problems more frequently. Controversially, more applications with high Quality of Service (QoS) and reliability requirements are being deployed that are affected by interference [8]. Due to the complexity of a network failure analysis, experts are increasingly needed for the prevention and elimination of particular costly or safety-critical failures. The need for user-friendly and highly automated solutions for the analysis of network data has therefore increased significantly. In the area of wired networks, this development has already been observed in the past few years. Automated analysis solutions are not only used for troubleshooting but also for intrusion detection. In both application areas, information is collected from the network, transport and application layers and analyzed offline or processed in real-time. In recent years, methods from the field of machine learning (ML) and AI have become more common, and are frequently used for classification, trend analysis, change point or anomaly detection as well as pattern recognition.

17.2 Verifying Data Source Accuracy in Protocol Analysis

There are two ways to capture the data exchanged between the devices within a wireless communication network. First, the data can be acquired directly at the sender or receiver of a frame. This method is particularly useful when it is known in detail which nodes are to be examined during a troubleshooting session. Since it is usually not possible to store data in large quantities on the embedded hardware of the network nodes, and the required effort for collecting and merging the resulting data is rather high, passive monitoring systems are used in the majority of all cases. In passive data capture, a dedicated monitoring node records all frames that are sent within reception range of the node [17]. This method is used often, since no special requirements are imposed on the network nodes and only the hardware and software of the monitoring node have to be extended accordingly. In addition, the monitoring system can be used flexibly or at several different locations. Commercial tools for passive monitoring such as [1] offer a comprehensive insight into the received frames and enable the user to perform further evaluations and root cause analyses.

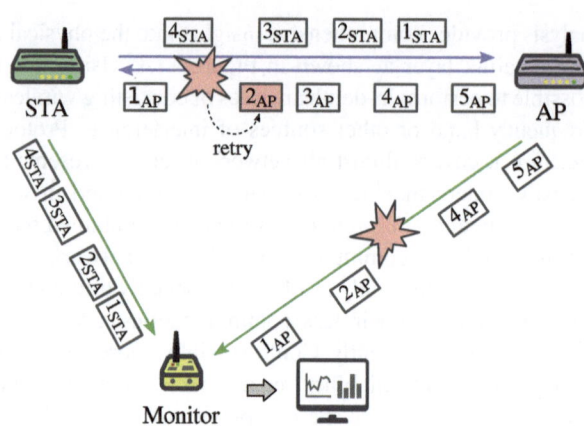

Fig. 17.2: Manifestation of reception differences during data acquisition with a passive monitor. Spatial differences between the original receiver of a frame and the monitor lead to different reception problems.

Despite the unquestionable advantages of passive network monitoring, some frames may not be received correctly due to transmission errors depending on the position of the monitor. This is caused by location differences between the actual participants in the data transmission and the monitoring node. These varying reception conditions inevitably lead to changes in reception strength and interference situations. Figure 17.2 shows such a monitoring scenario between a wireless access point (AP) and a station (STA). Due to the different transmission channels, the frames captured by the passive monitoring device differ from the frames received by the STA. Since a passive receiver is not involved in the actual data transmission, missed frames are not retransmitted and therefore not available as part of the dataset for further analysis. For the monitoring system, this results in spatial positions with good reception conditions at which it captures approximately the same number of frames as the actual STA. In contrast, insufficient capture conditions are characterized by positions of the monitoring device where significant deviations in the number of received frames can be observed. In non-static scenarios, these conditions fluctuate over time and can thus evolve continuously until a complete reception loss is reached, leaving the monitoring system blind.

Currently, there is no solution taking into account the quality of the captured data in relation to the propagation environment of the monitor. Using methods from the field of artificial intelligence (AI), however, it is possible to develop a system [16] for detecting anomalies in the data acquisition of passive receivers. In this work, the IEEE 802.11 protocol, which is widely known as Wi-Fi, was used as an example. Due to its widespread use and the resulting availability of commercial off-the-shelf (COTS) platforms, the IEEE 802.11 standard is used frequently in industrial environments and in particular for applications with increased QoS requirements in terms of data rate and latency. As the analysis method is data-driven, it can be easily retrained for other wireless communication protocols by applying a suitable dataset.

17.2.1 System Concept

For this purpose, the data stream of received frames can be considered as a multivariate time-series with irregular intervals, where the reception failures represent completely missing samples. Since network protocols, similar to natural language, are subject to a form of grammar, a logical sequence of samples exists within the time-series. This enables the detection of missing frames. Since some network protocols such as Wi-Fi have a high functional complexity and are continuously developed further, a simple rule-based detection of missing frames is time-consuming and cost-intensive. A more efficient approach is to use methods from the field of unsupervised learning to develop a system for assessing the data quality of network captures of wireless communication systems. Application-specific knowledge that needs to be refined continuously is not required in order to develop such a system; no in-depth knowledge of the function of the various communication protocols or expert knowledge in the field of communications engineering is needed. In addition, unsupervised learning also avoids the time-consuming process of labeling the training data. These approaches reduce the required expert knowledge and therefore increases the adaptability of the anomaly detection system.

If the absence of a sample within a sequence is considered abnormal behavior, data-driven methods such as autoencoders (AEs) can be used to determine the number of missing samples [13]. To implement this approach, COTS hardware can be used to capture the wireless frames. The acquired data is then subjected to classical feature engineering and pre-processing to be further processed by AE as a multivariate time-series.

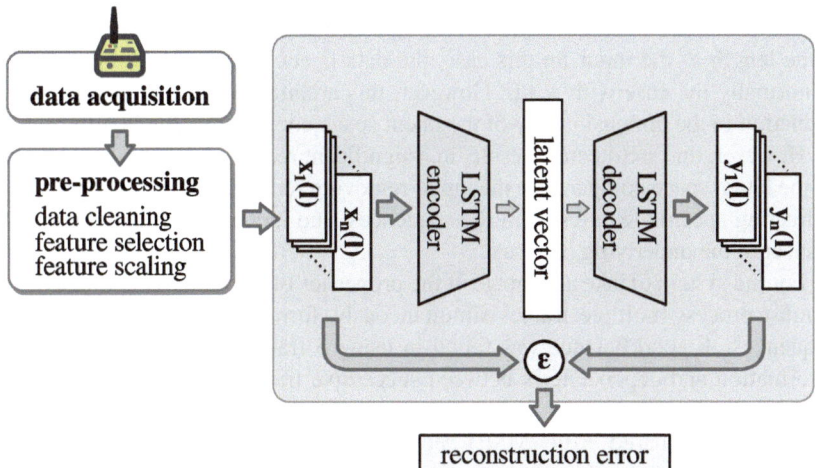

Fig. 17.3: Passive wireless monitoring pipeline including detection of position-related data acquisition gaps.

As shown in Figure 17.3, input data sequences of length l are then generated and passed to an AE for reconstruction. Missing samples within a sequence are reconstructed by the AE, causing a difference between the input and output sequence.

The calculated reconstruction error ε serves as a measure of whether the input sequence contained errors. If $\varepsilon > T$, where T is a predefined threshold depending on the application, the data cannot be reliably used for further analysis without further processing. Depending on the severity of the error, data imputation can be performed to correct the problem. Otherwise, the input sequence must be marked as incomplete before further analysis to avoid misinterpretation.

An AE is a type of neural network (NN) that tries to find an efficient encoding to represent the particular input data. The encoder creates an intermediate representation for the data, and the decoder reconstructs it. By training the network with error-free sequences, it learns its intrinsic properties and can process any further data efficiently and with minimal errors. When abnormal input sequences with different statistical properties are presented, the AE generates the most likely fault-free output sequence.

17.2.2 Autoencoder Architecture for Anomaly Detection

The long short-term memory (LSTM) units chosen as the basis for our AE are distinctive for capturing long-term dependencies between data points. A comparison between AEs with fully linked and 1D convolutional layers was conducted, where the LSTM-AE provided the highest accuracy in detecting missing samples in a given sequence. The encoder consists of only a single layer with 100 LSTM units and utilises the hidden states of the LSTM units for transforming its input to the latent representation. The decoder is comprised of one LSTM layer with the same number of units and an additional dense layer which forces the output y to have the same length as the input. In this case, the data is encoded but not compressed, as is normally the case with a AE. However, this architecture results in a significant reduction in the dimensionality of the latent space compared to the input.

However, this architecture results in a significant reduction in the dimensionality of the latent space compared to the input, removing redundancy from the input data, extracting specific features from the sequences, and increasing generalization with respect to the underlying patterns.

For the system to take advantage of the properties of the underlying transmission, it must process multiple frames simultaneously, forming what is called a "frame sequence". A good balance was found in using 5 frames, which provides enough information and dependencies between successive frames while keeping the computational effort low.

The mean squared error (MSE) between the input sequence x and the output sequence y constitutes the reconstruction error and was chosen as the loss function. Stochastic gradient descent, backpropagation algorithm, and Adam optimizer were used in training.

17.2.3 Dataset and Performance Evaluation

A wireless testbed was used to collect a representative dataset for training and testing. The testbed mimicked a typical industrial plant network consisting of multiple STAs connected to a central AP. A monitoring node captured wireless communications between all nodes. To obtain datasets with a large diversity, all hardware and software components were selected according to their adaptability. The final dataset contained 14.4 million frames in 20 different testbed configurations, randomly divided into 80% training data and 20% test data. To improve the test data, real data from the production line of an automotive factory was added as a validation dataset to provide a larger variance of traffic and loss patterns.

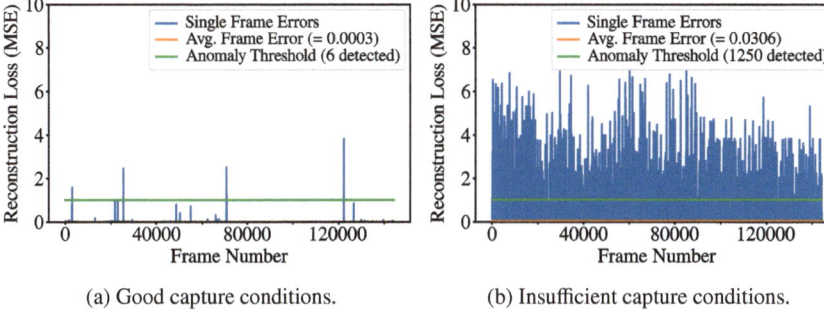

(a) Good capture conditions.　　　　　(b) Insufficient capture conditions.

Fig. 17.4: Reconstruction error per frame.

Since the problem is represented as an unsupervised learning task, feature selection was performed as a combination of expert knowledge from years of wireless network protocol analysis and experimentation with different configurations. Figure 17.4 shows the performance of the LSTM-AE using a validation dataset from the testbed. The majority of frames are decoded accurately under good capture conditions, but some show higher reconstruction losses due to previously unseen environments for the AE. A visual comparison of Figure 17.4a and Figure 17.4b shows a clear difference between good and insufficient capture conditions. As depicted in Figure 17.4b, the average frame error exhibits a significant increase under insufficient reception conditions in contrast to good conditions, as illustrated in Figure 17.4a.

A conclusive measure is obtained by examining the error distribution of the reconstruction errors shown in Figure 17.5. Figure 17.5a and Figure 17.5b illustrate the reconstruction error distribution for good and insufficient capture conditions, respectively. The dispersion index (DI), which represents the spread of the distribution, is almost four times higher for inadequate capture conditions. A high DI indicates an unreliable analysis by the monitor. Additionally, the model was validated on real-world data from an automotive factory, representing a different wireless environment than the dataset generated in the lab. Interestingly, even with unseen data, the DI remains much lower for good (Figure 17.5c) compared to poor (Figure 17.5d) acquisition conditions. Using a Wi-Fi dataset as an example, it could be shown that the informational quality of the traffic captured by a monitor node can be evaluated

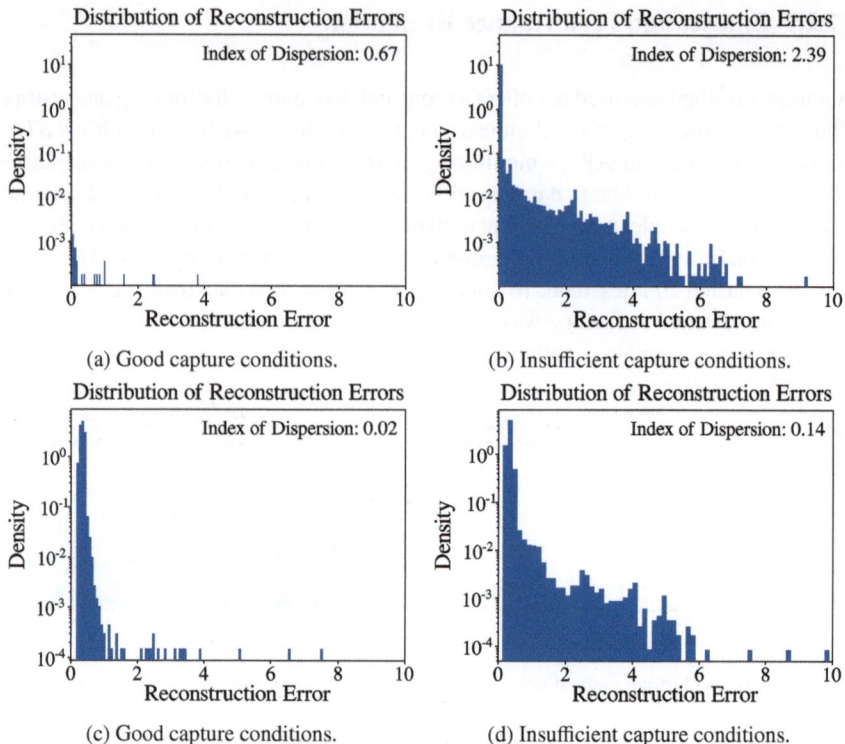

Fig. 17.5: Reconstruction error per frame on testbed data (a), (b), and data from previously unseen environments (c), (d).

using an unsupervised approach for anomaly detection. The described method is a significant improvement to the existing wireless network analysis systems on the market, as it increases the quality of the analysis results and allows non-experts to assess the validity of these results without any ambiguity.

17.3 Automated and User-friendly Spectral Analysis

There are various reasons for transmission failures in wireless communication systems. While the analysis of packet capture (pcap) files is well suited for investigating configuration and software problems, it cannot be used to draw direct conclusions about the physical causes of failures. To get a deep insight into the physical layer, we have to perform spectral analysis additionally. It enables the detection of individual frames, even under conditions with occurring collisions and bad signal propagation.

Commercially available real-time spectrum analyzers can be used as measurement instruments, but show strong limitations:

- They do not enable an automated detection of collisions and signals with low signal-to-noise ratio
- A detection and categorization of individual frames is not provided

To overcome these limitations by the development of a system for the gap-less and permanent real-time monitoring of radio channels, we have leveraged the benefits of spectral representation of signals as well as ML based image processing [18].

17.3.1 ML-based Spectrum Analysis

Spectrograms, usually presented as heat map images, visually represent the varying usage of frequency spectrum over time. Using advanced image recognition techniques, individual frames as well as collisions can be detected within a spectrogram. We have designed an algorithm to detect individual frames, even in case of collisions, as well as to detect collisions itself. Furthermore, frames showing low signal-to-noise ratio (SNR) or facing partial overlapping by adjacent channels can also be detected.

YOLOv4 [4], a single-stage object detector algorithm, was selected for the image processing, as it provides a high inference speed and sufficient accuracy. For our use case, we have adapted the model's hyper-parameters as well as the input resolution and the dimensions of the bounding boxes.

One of the major challenge in the development of the spectral surveillance system was the generation of high-quality data for training and testing our ML-model.

17.3.2 Generation of Training and Validation Data

To enable model training and testing on a variety of reproducible IEEE 802.11 frame sequences, a labeled dataset [19] was generated using a Rhode & Schwarz® SMBV100B vector signal generator [15] in combination with an Ettus Research® Universal Software Radio Peripheral (USRP) E320 software defined radio (SDR) [6]. Therefore, the following communication standards were used:

- IEEE 802.11 b/g,
- IEEE 802.11 n,
- IEEE 802.11 ac.

Additionally, we modified the following frame parameters:

- Payload length,
- modulation coding scheme (MCS),
- Data, beacon, trigger or sounding frames.

A signal processing tool-chain for the radio frequency (RF) signals received by the USRP was developed, using Python [7] and GNU Radio [2]. The steps of the synthesis of spectrograms from RF signals can be summarized as follows:

1. Threshold-based frame insulation,
2. introducing randomized channel models (incl. multi-path component effects, Doppler effect, attenuation per frame),
3. creating new random sequences of frames and collisions,
4. adding additive white Gaussian noise (AWGN) of varying magnitudes to emulate different SNR conditions,
5. applying fast Fourier transformation (FFT) for spectrogram generation and saving downsampled spectrograms (512×96 pixels) as image files, and
6. saving labels for frames and collisions.

Most of the parameters of the applied channel models, e.g., Doppler speed and time delay, have been randomized within constraints. The minimum and maximum number of randomly inserted frames, as well as the number of frame collisions, were parameterized. We optimized the spectrogram sample length for YOLOv4 algorithm's inference rate and set it to 4.5 ms. To handle fixed-length spectrogram samples, a $500 \mu s$ overlap was defined between successive segments to merge split frame labels.

Our training dataset contained 10,000 spectrograms, split into 80% training data and 20% validation data. The focus lay on 24 MHz and 40 MHz measurement bandwidths, corresponding to standard Wi-Fi channels.

17.3.3 Model Validation Using Artificial and Measurement Data

On the validation dataset, which is synthetically created to match the training data, the model achieved an average precision of approximately 96% when detecting Wi-Fi frames and around 60% average precision detecting collisions. The lower precision in collision detection appears due to challenging edge cases with low overlap, low SNR, and high signal power differences between frames. Performance metrics, such as true positives and false positives, were obtained by setting confidence thresholds for the classifier's inference at 85% and 50%, respectively, for Wi-Fi frame detection and collisions. The average precision is reported with a 50% confidence threshold to allow comparison with other work from the area of object recognition.

The model was further evaluated using real measurements. A relevant dataset was generated, considering the statistical distribution of frame parameters in real-world scenarios. A sequence of 4,000 collision-free IEEE 802.11ac frames was generated using a signal generator. In addition to the spectral measurements, Wi-Fi frames were captured in the pcap format using a commercial Wi-Fi module. During the quantitative validation, the system could detect frames with an accuracy between 98% and 97.6% depending on the transmission power as illustrated in 17.6 (a), detecting significantly more frames than the classical monitoring system. Problems in detection only occurred with dense sequences of very short frames and collisions. While the model properly detects the majority of frame collisions, as shown in Figure 17.6 (b), it occasionally happens that only one of the frames involved in the collision is recognized reliably.

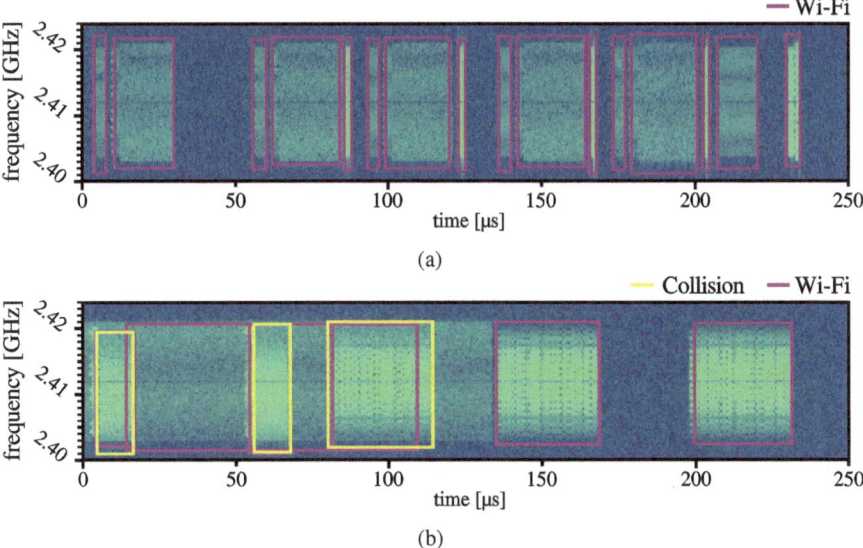

Fig. 17.6: Examples of successfully detected Wi-Fi frames and collisions in generated sequences (y-axis: frequency in GHz, x-axis: relative time in μs, color-scale: power ratio in dB).

17.3.4 System Architecture

The RF-band surveillance system's architecture for deployed inference is presented in Figure 17.7 and consists of three main components that process received radio signals in a pipeline and present the results to the user.

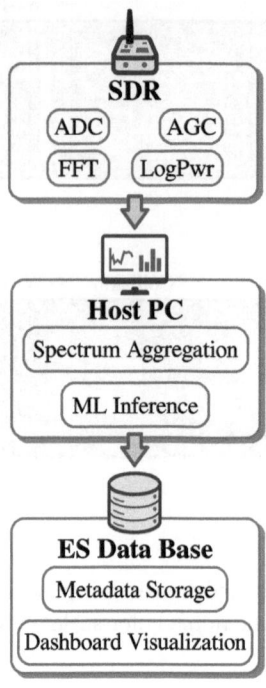

a) The **USRP E320 SDR** is responsible for signal reception. It implements automatic gain control (AGC) to optimize the dynamic range of the signal from the analog to digital converter (ADC). The USRP's digital signal processing capabilities efficiently compute the spectrum using FFT.

b) The **host computer** aggregates the spectrum data and converts them into spectrograms, which are directly fed into the ML inference model. The PC also calculates meta-information and statistics in parallel and fuses them with pcap information. All analysis results are stored in an elasticsearch database [5] and presented to the user on a dashboard.

c) To reduce data volume, **metadata extraction** is performed on the detected frames in the spectrogram. The metadata includes frame bandwidth, transmission time, center frequency, received signal strength, radio standard probability, and frame collisions. This information is used for communication standard-specific statistics on channel occupancy, frame rates, and detected interference.

Fig. 17.7: Overall system architecture.

The real-time acquisition of meta-information is essential for permanent analysis, as storing raw spectral information for extended periods would strain the storage system. To achieve high-performance inference of frame detection, the ML model was implemented in C++ using OpenCV and benchmarked on a mobile edge node with an Nvidia Turing graphics processing unit (GPU). This system can compute the algorithm in real-time and could even be implemented with higher energy efficiency in the future, using dedicated hardware.

17.4 Cross-layer Analysis

Although spectrum analysis provides intricate details about the physical layer, such as the spectral shape, power distribution, etc., and can detect co-existence issues and packet collisions, it has limitations in revealing information about higher protocol layers, such as source or destination addresses, flags within protocol headers, and payload content. So, even if interference or collisions are identified through spectrum analysis, the actual interferer remains unidentified. Therefore, in this section, we present a technique to improve automated spectrum analysis by supplementing protocol analysis information to it for a better root cause analysis of QoS degradation problems. This additional information enables quick identification and rectification of disruptions by taking measures such as reallocation of resources to non-interfering channels or triggering an alarm in case of abnormal or unusual traffic. Moreover, the monitoring and troubleshooting of industrial wireless networks benefit by combining information delivered from both protocol analysis and spectrum analysis.

Combining spectral and protocol domains, also known as **synchronization**, can be viewed as a time-series analysis problem, as both sequences are chronologically ordered. However, the task is complicated as there is no prior knowledge of the uncertainties and inaccuracies present in the traces due to missing frames, collisions and inaccurate temporal information. A possible solution is the use of dynamic time warping (DTW) to synchronize the traces. However, the DTW approach, which compares every frame of one sequence with every frame of the other sequence, incurs a high computational cost with a complexity of $O(n^2)$. This is infeasible in practical scenarios with millions of frames. To address this, global constraints were introduced, with the Sakoe-Chiba [14] and Itakura Parallelogram [10] being two popular DTW versions incorporating these constraints (shown in Figure 17.8). Nevertheless, in our scenario, global constraints are inadequate due to the unpredictable nature of the frame sequence, especially regarding frame rate and losses. We need local constraints that can adapt to the varying parameters of the sequences.

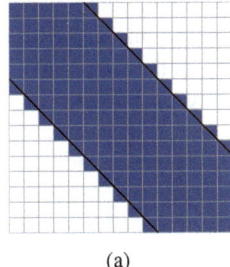 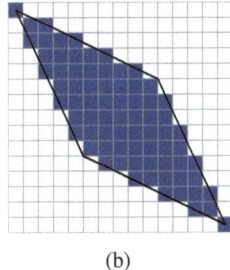

(a) (b)

Fig. 17.8: Enforcing Global Constraints in DTW: (a) Sakoe-Chiba and (b) Itakura from Vineeta Jain, licensed under CC BY 4.0. The spectral sequence is represented on the horizontal axis, while the protocol sequence is represented on the vertical axis. The optimal warp path, used to align the protocol sequence with the frames of the spectral sequence, is determined solely using the frames highlighted in blue (taken from [9], licensed under CC-BY 4.0).

17.4.1 Variable Adaptive Dynamic Time Warping: A Novel Approach

To address this, we propose a novel DTW-based approach called Variable Adaptive DTW (VADTW) [11]. This approach consists of coarse- and fine-grained synchronization levels. The coarse-grained level divides the spectral and protocol sequences into adaptive time bins, while the fine-grained level calculates variable window limits for each frame. These limits are then used to compute the warp path using DTW.

Coarse-Grained Synchronization level seeks to divide sequences into smaller subsequences based on their frame rate per unit of time (θ). Sequences are divided into bins of θ, with the number of frames within these bins counted. Afterwards, both sequences are cross-correlated based on frame rate to determine the displacement of one sequence relative to the other. This step aims to eliminate any sequence lag, optimizing the synchronization process. After lag removal, offline change point analysis is applied to the time bins to detect significant changes in the frame rate. A union of the change points from both sequences forms a single list of change points, from which adaptive time bins are determined. These adaptive bins, which adjust to the frame rate variation of the sequences and vary in time length, merge bins with a continuous similar frame rate and separate bins with significant variations.

The goal at **fine-grained synchronization** level is to find an optimal match between individual frames of both sequences. One approach is to apply DTW to all adaptive time bins independently and then merge the results to obtain an optimal warp path, a method we call **Adaptive DTW (ADTW)**. However, this approach may suffer from false positives and negatives because it does not consider the start and end points of a bin. An example can be seen in Figure 17.9(a) where the spectral sequence is represented on the horizontal axis, while the protocol sequence is represented on the vertical axis, and two large boxes represent the adaptive bins, while the green-colored boxes indicate the optimal warp path computed using ADTW. Due to the constraint of calculating warp path by exclusively using the frames within the time bins, the points marked in red are excluded from the warp path. Unfortunately, this exclusion can lead to a loss in precision and recall. To address this, VADTW introduces variable window lengths. Each frame in the first sequence is assigned a window limit based on the time length of its adaptive bin, and then it is matched with all frames in the second sequence that fall within this window limit. This approach allows matching to extend beyond the adaptive bin, overcoming the problem of ignoring the bin edges. An example of varying window constraints for synchronization is shown in Figure 17.9(b).

17.4.2 Experimental Results and Discussion

We conducted a comparative analysis between VADTW, ADTW, and DTW to prove that our proposed algorithms perform better than DTW when used for aligning sequences. For this purpose, we generated a synthetic dataset of 20, 000 IEEE 802.11n frames using a Rhode & Schwarz ® RSMBV100B vector signal generator. To demonstrate the capability of various algorithms to synchronize traces with missing frames, we artificially dropped frames using the Three-state Markov Chain model proposed in [12].

Table 17.1 displays the accuracy obtained for the synthetic dataset under various packet loss scenarios. In this table, "SS" denotes single-source packet drop, while "AS" represents all-source packet drop, involving the dropping of packets

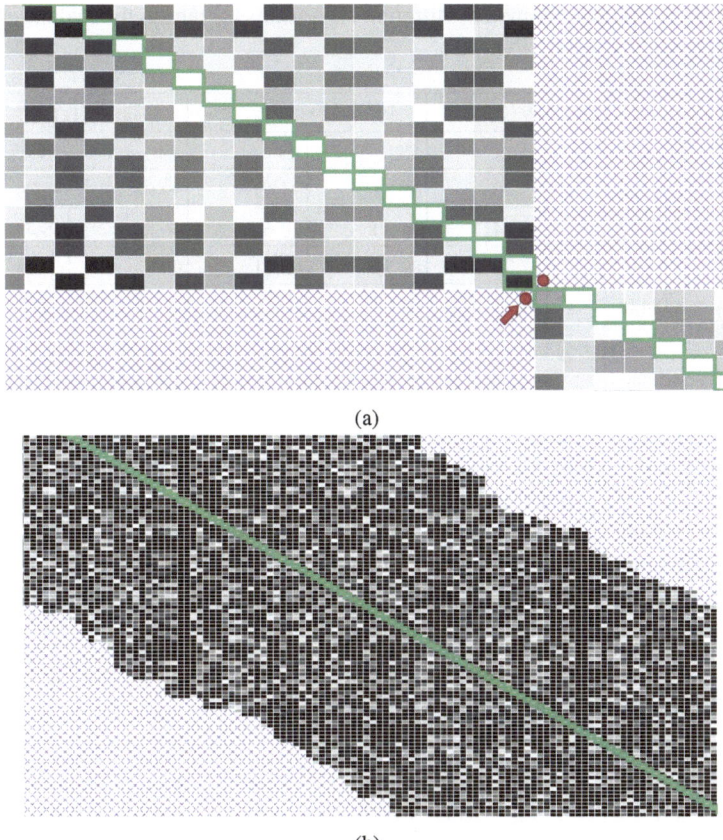

(a)

(b)

Fig. 17.9: A scenario depicting the problem arising due to not considering edge cases in (a) ADTW and how the issue is resolved by (b) VADTW from Vineeta Jain, licensed under CC BY 4.0

from both the spectral and protocol domains. From the results, it could be observed that VADTW achieves accurate alignment for more than 50% of pairs, even when subjected to a 50% random packet drop from both sources, which is noteworthy. Moreover, VADTW consistently outperforms both ADTW and DTW in all frame drop scenarios. DTW compares every frame of one sequence with every other frame of the other sequence to find a warp path. This leads to frames getting synchronized which are far apart in time, which is not possible in real-world, hence leading to false positives. In case of ADTW, the comparisons occur within adaptive time bins as explained in Fig 17.9(a). In other words, the protocol frame is exclusively compared to spectral frames falling within the range of the adaptive time bin to which the protocol frame belongs. This leads to false positives and false negatives. For VADTW, each protocol frame is associated with a variable window determined by its timestamp and

Table 17.1: Accuracy obtained by different algorithms for the synthetic dataset under various packet loss scenarios.

Packet drop (in %)	DTW		ADTW		VADTW	
	SS	*AS*	*SS*	*AS*	*SS*	*AS*
5	99.08	95.70	98.33	94.69	99.08	95.79
15	96.81	85.44	96.30	78.60	96.85	85.44
30	90.35	67.38	89.79	63.91	90.34	69.72
50	84.00	54.21	84.00	56.63	83.70	60.57

the length of the adaptive time bin to which it belongs (as depicted in Figure 17.9(b)). This removes the possibility of synchronizing frames which are very far in time and also, resolves the edge case issue of ADTW, leading to better accuracy than these algorithms. Further, with respect to execution time analysis, where DTW took 23.8 minutes to synchronize the dataset, VADTW achieved the same result in just 14.6 seconds. This marked improvement in execution time is of considerable significance given that spectral and protocol sequences can contain millions of frames.

17.4.3 Implications for Research and Beyond

The impact of this work extends beyond wireless networks. The two variants of DTW presented here - ADTW and VADTW differ mainly in their choice of local window constraints for synchronization. In ADTW, the window boundaries remain the same for all frames in a subsequence with similar sample properties, whereas in VADTW, window boundaries vary for each frame. These methods could be extremely useful for the research community in analyzing time-series with a large number of observations and many sudden change points. These methods could be particularly beneficial in cases where local window constraints are more accurate instead of global ones, such as predicting road traffic patterns. The proposed approaches can also be useful for problems that require online change point analysis, such as for automatic spectrum analysis in industries, as they are computationally less expensive and fast compared to DTW.

17.5 Conclusion and Outlook

In summary, the ubiquity of wireless communication systems in various areas of daily life and industry has made them an essential part of many people's digital lives. However, the exponential growth in the number of subscribers and devices poses a major challenge to the flawless operation of these systems. To meet the increasing demands on communication systems, there has been a rapid development of improved and new wireless standards competing for the limited frequency bands available. While these technical advances have brought significant benefits in terms of wireless network performance and robustness, the increasing number of competing subscribers has also led to an increase in network failures.

One approach to ensuring fault-free network operation is the use of monitoring systems. Continuous or immediate monitoring of the status of a wireless link, as well as analysis of faults and their causes, enables the early initiation of preventive measures or immediate countermeasures in the event of faults. In this section, the main techniques of network monitoring were presented and open questions for further improvement of these methods were addressed.

Given the ever-increasing diversity of network protocols and radio standards, conventional approaches for developing the necessary troubleshooting tools have significant disadvantages. The financial effort for expansion, adaptation and maintenance is significantly higher than using data-oriented approaches from the fields of ML and AI. Once a suitable ML/AI system has been developed, an appropriate dataset is needed for training to the respective standard. Obtaining these datasets can be done in different ways, as described in the three examples, and is often supported by data augmentation techniques.

In addition to the benefits mentioned above, AI-based systems offer a number of other advantages in the context of wireless communication systems. These include:

- Real-time adaptability: AI systems can adapt and learn from changing network conditions in real-time, enabling dynamic optimization and fault prediction.
- Anomaly detection: AI can efficiently detect unusual patterns and behaviors on the network, helping to identify potential security breaches or unauthorized access.
- Scalability: ML/AI-based solutions can scale efficiently to handle the huge amounts of data generated by the growing number of devices and subscribers in wireless networks.

Overall, the techniques we have developed to improve protocol analysis and spectral analysis have shown promising results, confirming the benefits of AI methods for fault analysis in wireless communication systems. Furthermore, we were able to link both domains together to develop a powerful and user-friendly comprehensive method for troubleshooting. The use of AI-driven solutions will be crucial in the future to ensure the smooth and reliable functioning of wireless communication systems.

References

1. NetAlly AirMagnet WiFi Analyzer Pro. 2021-01-21.
2. Gnu radio. http://www.gnuradio.org, Jan. 26th, 2021.
3. I. Aktas, A. Bentkus, F. Bonanati, et al. Position paper: Wireless technologies for industrie 4.0. Technical report, VDE, Tech. Rep, 2017.

4. A. Bochkovskiy, C.-Y. Wang, and H.-Y. M. Liao. Yolov4: Optimal speed and accuracy of object detection. https://arxiv.org/abs/2004.10934, 2020.
5. Elastic. Elasticsearch. https://www.elastic.com/, Jan. 25th, 2021.
6. Ettus Research, a National Instruments Brand. USRP B200. https://www.ettus.com/all-products/ub200-kit/, Jan. 8th, 2021.
7. P. S. Foundation. Python language reference, version 3.8. Available at http://www.python.org, Jan. 28th, 2021.
8. A. Frotzscher, U. Wetzker, M. Bauer, M. Rentschler, M. Beyer, S. Elspass, and H. Klessig. Requirements and current solutions of wireless communication in industrial automation. In *2014 IEEE International Conference on Communications Workshops (ICC)*, pages 67–72, 2014.
9. Z. Geler, V. Kurbalija, M. Ivanović, M. Radovanović, and W. Dai. Dynamic time warping: Itakura vs sakoe-chiba. In *IEEE Intl. Symp. Innovations in Intell. Syst. Appl. (INISTA)*, pages 1–6. IEEE, 2019.
10. F. Itakura. Minimum prediction residual principle applied to speech recognition. *IEEE Trans. Acoust., Speech, Signal Process.*, 23(1):67–72, 1975.
11. V. Jain, V. Fokow, J. Wicht, and U. Wetzker. A dynamic time warping based method to synchronize spectral and protocol domains for troubleshooting wireless communication. *IEEE Access*, 2023.
12. B. P. Milner and A. B. James. An analysis of packet loss models for distributed speech recognition. In *8th Int. Conf. Spoken Language Process.*, 2004.
13. P. Neuhaus, M. Henninger, A. Frotzscher, and U. Wetzker. Autoencoder-based characterisation of passive ieee 802.11 link level measurements. In *2021 Joint European Conference on Networks and Communications and 6G Summit (EuCNC/6G Summit)*, pages 294–299, 2021.
14. H. Sakoe and S. Chiba. Dynamic programming algorithm optimization for spoken word recognition. *IEEE Trans. Acoust., Speech, Signal Process.*, 26(1):43–49, 1978.
15. R. . Schwarz. R&S® SMBV100B. https://www.rohde-schwarz.com/product/smbv100b-productstartpage_63493-519808.html, Jan. 8th, 2021.
16. U. Wetzker, A. Frotzscher, and P. Neuhaus. Concept for the analysis of a radio communication system, Apr. 9 2020. US Patent App. 16/591,835.
17. U. Wetzker, I. Splitt, M. Zimmerling, C. A. Boano, and K. Römer. Troubleshooting wireless coexistence problems in the industrial internet of things. In *2016 IEEE Intl Conference on Computational Science and Engineering (CSE) and IEEE Intl Conference on Embedded and Ubiquitous Computing (EUC) and 15th Intl Symposium on Distributed Computing and Applications for Business Engineering (DCABES)*, pages 98–98, 2016.
18. J. Wicht, U. Wetzker, and A. Frotzscher. Deep learning based real-time spectrum analysis for wireless networks. In *European Wireless 2021; 26th European Wireless Conference*, pages 1–6, 2021.
19. J. Wicht, U. Wetzker, and V. Jain. Spectrogram data set for deep-learning-based rf frame detection. *Data*, 7(12), 2022.

Chapter 18
XXL-CT Dataset Segmentation

Roland Gruber[1], Steffen Rüger[1], Moritz Ottenweller[1], Norman Uhlmann[1], Stefan Gerth[1]

Abstract The objective of XXL-CT dataset segmentation is to use machine learning to virtually divide 3D volumes of complete vehicles, acquired through XXL computer tomography, into their individual components. Gathering labeled training data for this type of data is challenging. Previously, entity classification from XXL-CT data required significant manual effort involving over 120 employees for several months. This chapter shows how to develop entity segmentation procedures which significantly reduce the time from measurement to virtual analysis. The most time-consuming part of the data processing chain is currently the segmentation of individual assemblies, especially when dealing with overlapping metal sheets. We aim to demonstrate the transferability of trained networks to different XXL-CT vehicle data, taking into account the large shape variations of different metal sheets and their contact, weld or rivet points. Challenges include low data quality which is affected by acquisition and reconstruction artifacts, a dataset size of up to 1.7 terabytes, and the large number of individual instances to be segmented and their interrelationships which must be taken into account. This contribution considers three major aspects: The segmentation of CT datasets by means of neural networks, the development of solutions for the annotation of XXL-CT data, and the transferability of the trained network.

Key words: Instance Segmentation, Semantic Segmentation, XXL-CT

[1] Fraunhofer Institute for Integrated Circuits IIS, Fraunhofer IIS, Erlangen, Germany

Corresponding author: Roland Gruber
e-mail: `roland.gruber@iis.fraunhofer.de`

© The Author(s) 2024
C. Mutschler et al. (eds.), *Unlocking Artificial Intelligence*,
https://doi.org/10.1007/978-3-031-64832-8_18

18.1 Introduction

Machine learning in 3D XXL Computed Tomography (CT) datasets encounters various challenges. The unavailability of annotated datasets makes training accurate and robust models difficult. Preprocessing 3D data to handle noise, missing data, and outliers is a complex task. Furthermore, the large file size of the 3D datasets requires specially adapted algorithms which can efficiently scale with the demanding compute loads and memory requirements of 3D XXL-CT datasets. Generalizing trained models to new 3D data or different domains becomes challenging due to the variability and diversity in 3D shapes and structures.

Publicly accessible XXL-CT datasets for machine learning are scarce due to the considerable expenses involved in data acquisition and segmentation. The construction and maintenance of XXL-CT measurement systems require substantial financial investment and specialized expertise. Additionally, the manual nature of the data segmentation process adds to the overall cost. Error checking and proofreading are essential steps that further contribute to the financial burden associated with creating such datasets.

One of the major challenges for the segmentation of large volumes of 3D X-ray CT data is the lack of availability of annotated training data. For 2D image data there is a wide variety of different data collections available and also specialized pretrained networks [10, 14]. However for non-destructive testing (NDT), neither for X-ray radiography data nor for 3D CT data collections of annotated data is available. Further, there are only a few algorithms available for segmenting volumetric 3D datasets. The majority of these algorithms rely on slice by slice annotation referring to 2D trained segmentation networks. This approach loses the advantages of leveraging the interconnectivity of distinct objects in a 3D dataset.

When analyzing unique datasets such as the scan of the Me 163 airplane or the Honda vehicles (described later), the challenge of segmenting a large, interconnected 3D volume remains. In addition to the absence of pretrained networks, another hurdle is the lack of labeled data. This is primarily due to the unique nature of the scanned object, making it difficult to find existing annotations for training purposes.

In this work, we discuss the acquisition process of two comprehensive XXL-CT datasets, which comprise high-resolution scans of a historical Me 163 aircraft and a Honda Accord vehicle, procured at the Fraunhofer IIS facility.

We explore the complexities of our annotation strategies, encompassing both 3D instance and semantic labelling pipelines. Our instance labelling protocol details our manual annotation effort, enhanced by algorithmic support and refined through post-processing, with the goal of achieving high quality instance segmentations. In parallel, our semantic labelling approach adopts a hybrid model, integrating human annotator insights with automated machine learning systems to efficiently and consistently apply semantic labels to the CT data.

We further present the segmentation algorithms and their corresponding results, addressing the challenges associated with segmenting expansive volumetric datasets. Utilizing Flood Filling Networks and an adapted 3D U-Net framework, we present promising initial findings in the segmentation and localization of specific components within these intricate datasets. The paper concludes with a perspective on the future applications of these segmentation methodologies, particularly their capacity to refine manual annotation efforts and deepen the understanding of component relationships in the realm of non-destructive testing.

18.2 XXL-CT Dataset Acquisition

The XXL-CT facility of the Fraunhofer IIS in Fürth allows for the acquisition of volume datasets from large specimens [23]. This facility features a linear accelerator with energies up to 9 MeV, a line detector 4 m in length and allows an object size of up to 4 m in diameter. Using this state-of-the-art facility, we conducted scans on two distinct datasets.

18.2.0.1 Me 163 Airplane

For a first dataset, an Me 163 Second World War fighter airplane from the historic aircraft exhibition of the 'Deutsches Museum' in Munich, Germany was scanned in four consecutive CT scans, two for the fuselage (see Figure 18.1a) and two for the disassembled wings (see Figure 18.1b). The four radiologic datasets were subsequently reconstructed independently into volume datasets and then manually merged. The reconstructed volume datasets for the two fuselage scans (see Figure 18.2) span a spatial dimension of $6144 \times 9600 \times 5288$ 16-bit voxels, which is approximately 609 GB for the front part and $6144 \times 9600 \times 5186$ voxels (or 567 GB) for the rear part of the hull.

To acquire the X-ray projections, a linear accelerator with an acceleration voltage of 9 MeV and a line detector with a width of $w = 4$ m and a pixel spacing of 400 μm has been used. The distance between the X-ray source and the detector was set to $d_{S-D} = 12$ m, and the source-to-object distance was about $d_{S-O} = 10$ m. This resulted in a horizontal resolution of 9984 pixels. The X-ray projections has been acquired using a vertical stepping motor with a spatial resolution of 9984×5286 pixels. The magnification of 1.2 led to a horizontal voxel resolution of 330×330 μm^2 and a vertical sampling of 600 μm within the reconstructed volume.

The scanning process took approximately 17 days to complete.

18.2.0.2 Honda Accord Vehicle

The second dataset, a Honda Accord was measured in two consecutive XXL-CT scans (see Figure 18.3). One scan was taken of the front and one of the rear of the vehicle. Each scan was reconstructed independently into a volume dataset and then manually merged. The front had a spatial dimension of $4864 \times 4864 \times 2000$ 16-bit voxels, which is roughly 95 GB, while the rear part had a spatial dimension of

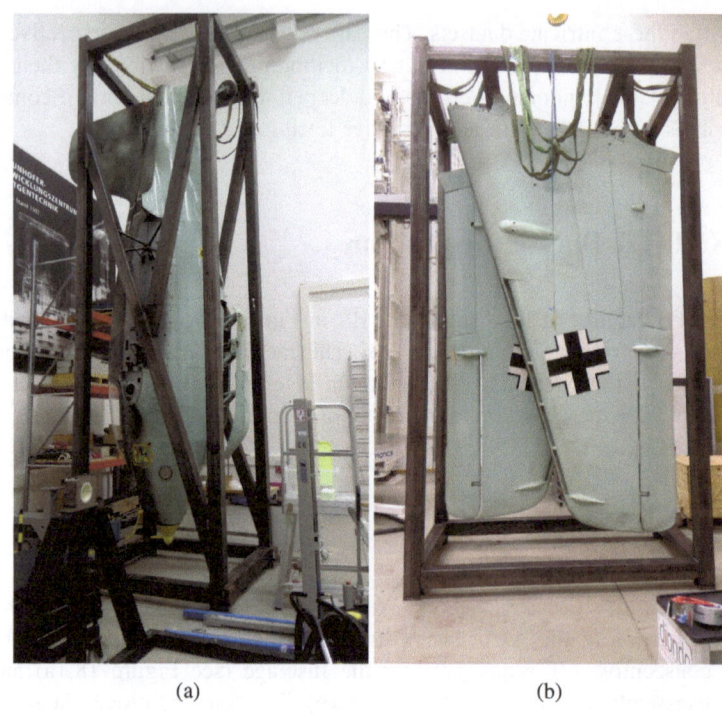

(a)   (b)

Fig. 18.1: Fuselage (Figure 18.1a) and wings (Figure 18.1b) of the Me 163 airplane inside the mounting brackets for the CT scan.

Fig. 18.2: Rendering of the CT scans of the front and back section of the Me 163 fuselage.

$4864 \times 4864 \times 2920$ 16-bit voxels (or approx. 138 GB). The X-ray projections were obtained using a linear accelerator with an acceleration voltage of 7.5 MeV and the same line detector used for the airplane scan. The distance between the X-ray source and the detector was set to $d_{S-D} = 11.758$ m, and the source-to-object distance was about $d_{S-O} = 9.913$ m, resulting in a horizontal resolution of 4870 pixels.

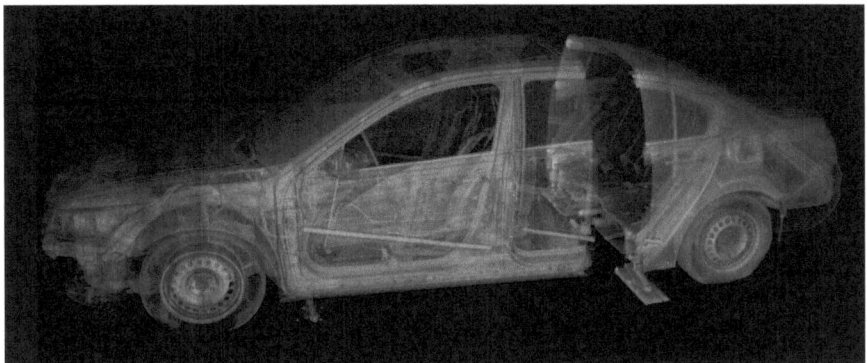

Fig. 18.3: Visualization of the front and rear XXL-CT scan of the Honda Accord.

18.3 Annotation Pipelines

Data annotation is a crucial step in the development and verification of machine learning applications. Accomplishing the segmentation task in the XXL-CT use case necessitates the availability of reliably and accurately annotated label data. For the two use cases discussed in this context, alternative annotation pipelines have been used. The first use case involves instance segmentation of materials along boundaries within a volume, while the second use case involves semantic segmentation or localization of familiar objects within a volume.

Semantic segmentation [3, 11, 15, 16, 20] entails assigning each voxel in the CT data to a specific label of interest (such as 'car tires'). *Instance segmentation* aims to label each individual entity in an image or volume with a unique identifier (often without knowledge of the entity's class). For example, each voxel of a 'screw' is assigned a unique ID, and each voxel of the neighbouring metal sheet is assigned another unique ID regardless of their class. *Semantic instance segmentation* [1, 13, 12], combines these two tasks by assigning a unique ID to each voxel of an entity and then assigning each entity a class such as 'tire' or 'wire'.

In the following we introduce two pipelines. First, the 3D instance labelling pipeline to tackle the instance segmentation task. Second, to address the semantic segmentation task the 3D semantic labelling pipeline.

18.3.1 3D Instance Labelling Pipeline

The purpose of instance segmentation is to assign the same ID to all voxels of an entity. Figure 18.4a shows a layer of the Me 163 reconstruction which serves as typical input for this use case. Figure 18.4b shows a possible corresponding partially manually created segmentation, using graphics tablets in conjunction with a gray value range bandpass algorithm and flood filling techniques to achieve the shown annotation. However, this annotation still exhibits some unpleasant properties and would not be desirable for the output of an instance segmentation. For instance, due to the sometimes low contrast between different metal sheets and rivets, a clear transition of the individual segments is not feasible. The initial segmentation depicted in Figure 18.4b contains numerous gaps and uneven edges. These are mainly due to the use of gray values to determine the segment boundaries and the elevated noise level of the volume dataset used. To achieve the desired annotation quality we utilized a post-processing step on the initial annotation. A layer of which is visible in Figure 18.4c, which contains a more uniform and desirable segmentation of entity edges.

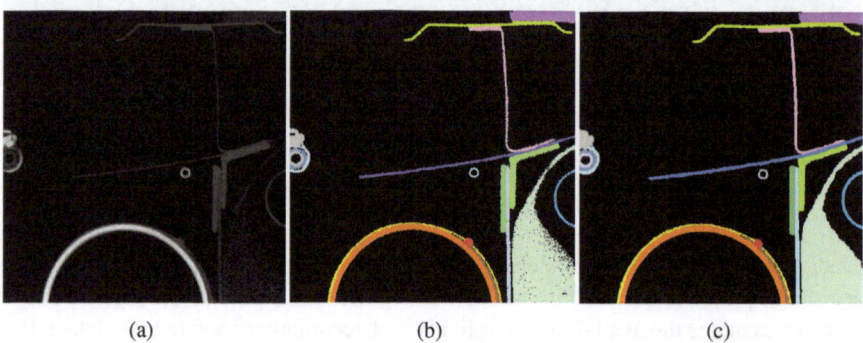

(a)   (b)   (c)

Fig. 18.4: Reconstructed slice of the Me 163 airplane Fuselage (Figure 18.4a) next to the result of the preliminary manual annotation (Figure 18.4b) and the post-processed annotation (Figure 18.4c) with closed defects and smoother contours.

To achieve this desired quality, we utilized a morphological closing filter [4] to post-process the manually annotated results. The aim was to close gaps between the quality of the manual annotation and the desired quality of the automatic segmentation, resulting in a semantically valid annotation with coherent boundaries and mostly uninterrupted contours, closely resembling the output of a potential human annotator. Specifically, a $3 \times 3 \times 3$ structure element was used to perform the morphological closing. The post-processing primarily resulted in sparse surface voxel alterations, but it also affected the surfaces of 'noisy' metal sheets, which are prone to more pronounced changes due to their 'noisy' nature as can be seen inside the pale green metal sheet in the lower right corner of the subfigures in Figure 18.4.

Finally, a connected component analysis with a chessboard metric [4] was performed to find separated segments. Small segments with less than 100 voxels were discarded to avoid over-segmentation. The threshold of 100 voxels was chosen using a heuristic approach.

Manual data annotation is often considered the most reliable method for accurately annotating complex image data (see Section 8.2.1). However, it requires substantial resources, including experienced personnel and time dedicated to annotation, even when specialized pipelines are used [19, 2]. Crowdsourcing approaches have been proposed to lower costs, but these methods also require specialized data management, annotation tools, and annotator skills.

In order to achieve a balance between expert and crowdsourcing approaches, we adopted a methodology where each sub-volume was initially annotated by one annotator and then reviewed and corrected by a second experienced annotator. This approach emulated partial perspectives, taking into account the possibility of unique biases, interpretations, or limitations in the annotations. By involving multiple an-notators, we aimed to capture a broader range of insights and mitigate the potential impact of individual annotator biases. This approach allowed us to consider different viewpoints and create a more comprehensive annotation that integrated the expertise and insights of multiple annotators. It is worth noting that the limited number of annotators involved was primarily due to cost constraints, thus avoiding the need for crowdsourcing.

The annotation process for the first two 512^3 sub-volumes took about 350 working hours each, while subsequent sub-volumes took between 10% to 50% of that time, depending on the complexity of the sub-volume.

The annotation guidelines provided to all annotators stipulated that the annotation should be based on the 'human interpreted reality' of the data, not the 'perceived visual representation.' An annotator should segment the 'probable' segment that they would like an automated annotation to generate, rather than focusing on the low-contrast voxels that they observe. This approach aimed to increase annotation uniformity and develop methods to separate all components meaningfully. After partial manual annotation, the individual segments required post-processing to close gaps between manual annotation quality and desired segmentation quality. As mentioned a morphological closing filter followed by an a connected component analysis was used for this purpose, resulting in simple surface voxel alterations and changes to the interior of 'noisy' metal sheets. The goal was to achieve semantically reasonable and visually pleasing segmentation results.

The dataset as well as a more complete description of the segmentation process and its challenges is described in [7, 6]

18.3.2 3D Semantic Labelling Pipeline

Semantic labels for a segmentation task refer to the assignment of regions or entities of interest in image data to certain class labels [24]. The semantic labels can be used by machine learning algorithms to identify and distinguish between them. For example, in the introduced XXL-CT dataset of the Honda Accord (see Section

18.2.0.2) semantic labels can belong to small entities of the car like different screws, springs, sheets, etc. but also to components and assemblies like gears, tires, brake discs, and so on.

The main goal of the 3D semantic labelling pipeline is to generate labeled datasets for optimization and evaluation of machine learning models which are trained to identify certain objects of interest. A 3-step hybrid approach for annotation is proposed:

1. Identification of voxels (xyz coordinates) of the object of interest in the XXL-CT data by a human annotator.
2. Those coordinates are fed into a weaker automated model for computation of a 3D binary mask.
3. Approval or further processing of the mask to the final semantic label.

Hybrid annotation approaches have two main advantages: efficiency and reproducibility.

After selecting relevant voxels the second step automates the annotation processes significantly by reduction of the time and costs that a human annotator would require for manually labelling each voxel by hand. Furthermore, improved accuracy and consistency of annotations are expected, as automated systems are not prone to human errors or bias. This allows reproducing the scheme for further semantic class labels of more objects that ultimately leads to more reliable annotations results.

For applying the proposed pipeline the introduced XXL-CT volume data of the Honda Accord of 95 GB are partitioned into front and rear. In the following, the front data will be referred to as XXL-CT data for semantic segmentation and contains areas from the bumper over the engine block to the B pillar (center of the vehicle). To account for the limitations of computer hardware resources and the faster processing time of human annotators, the front-facing CT data was divided into two sub-volumes, denoted as S_1 and S_2, which correspond to the right and left sides of the vehicle. Figure 18.5 shows both sub-volumes.

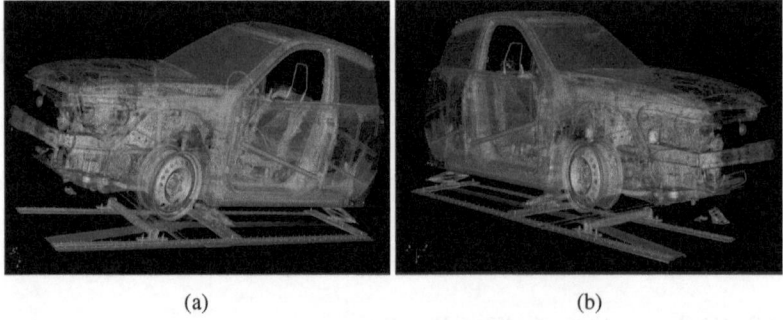

(a)   (b)

Fig. 18.5: Sub-volume S_1 (Figure 18.5a) and S_2 (Figure 18.5b) of the Honda front.

Within these datasets objects of interest were identified by a human annotator and processed with the proposed annotation pipeline. The final annotations of the objects of interest are presented in Table 18.1. Eight different object classes were distinguished in the annotated voxel dataset, which contained approximately 161.2 million voxels in total.

Object of Interest	Size	Origin Volume
Alternator	$153 \times 130 \times 106$	S_2
Brake Disc	$70 \times 285 \times 234$	S_1
Brake Disc	$85 \times 285 \times 215$	S_2
Crank Shaft	$76 \times 215 \times 133$	S_2
Flywheel	$25 \times 311 \times 252$	S_1
Frame Parts	$543 \times 131 \times 497$	S_1
Frame Parts	$623 \times 131 \times 497$	S_2
Gear	$33 \times 198 \times 166$	S_1
Spark Plug	$216 \times 70 \times 52$	S_2
Suspension Spring	$155 \times 249 \times 104$	S_1
Suspension Spring	$396 \times 472 \times 339$	S_2

Table 18.1: Annotation results of objects of interest in the XXL-CT dataset of the Honda Accord front.

18.4 Training Infrastructure and Segmentation Results

18.4.1 Instance Segmentation

Segmenting large volumetric datasets via instance segmentation can be challenging since it is often impractical or not feasible to fit the entire volume into the limited GPU RAM. To circumvent this obstacle, one approach is to downsample the dataset, while another is to work with sub-volumes or Fields of Views (FoV) of the whole dataset and then afterwards to combine the inference results of all sub-volumes. Downsampling, however, can lead to a significant loss of information which for example results in reduced contrast between previously distinguishable structures or the appearance of artifacts such as holes in thin structures. Where FoV-based approaches, due to the requirement of overlapping sub-volumes, often necessitate more computational resources and pose additional challenges when recombining the individually processed sub-volumes.

In our case, we opted for an FoV-based approach inspired by Flood Filling Networks (FFN) [9, 8]. This approach allows us to work with smaller sub-volumes, which than can be processed by the GPU within the available memory limits, while preserving the integrity of the original data without significant loss or the introduction of additional artifacts.

Flood Filling Networks belong to the family of supervised Convolutional Neural Networks (CNN) and are used for instance segmentation. The main component of FFN is a flood-filling algorithm which instructs the CNN to predict the likelihood of each voxel in the current FoV belonging to the segment occupying the center of the FoV. The FoV is usually a relatively small dimension, such as 48×48×48 voxels in our case. The predictions are stored in an accumulator volume which is identical in size to the input volume currently being processed. Initially, the accumulator is empty and then is successively filled by individual FoV prediction updates. If the predicted edge of the current segment extends beyond the current FoV, the corresponding neighboring FoV is added to a processing queue. For the next iteration the flood filling algorithm selects the next FoV from the queue and extracts the corresponding FoVs of the input volume and the values of the corresponding accumulator volume. These serve as input channels for the CNN. The CNN predicts an updated state of the accumulator FoV, which is then integrated into the whole accumulator volume. The iteration of a segment stops if the last FoV in its queue is consumed. Then another starting seed point can be selected to segment the next non-overlapping segment. The seed points can be chosen via multiple methods for example by user interaction or by a classical image processing algorithm proposed in conjunction with FFN. It is often possible to significantly speed up the computation by processing multiple FoVs in parallel.

The CNN attempts to detect segment edges and boundaries based on the grayscale values of a given sub-volume in the input dataset and the corresponding FoV of the current state of the accumulator volume. Since the FFN algorithm is entity-agnostic and does not have knowledge of any entity's class, including the one currently being segmented, it is feasible to train the CNN on one type of specimen and use it to infer on a different type of specimen [5].

The preliminary results of the inference run on a sub-volume of the Me 163 dataset, using a CNN trained on the same dataset, are shown in Figure 18.6. At the current stage of development of the FFN-based segmentation algorithm, the quality of the segmentation results is mixed. Although some segments have been adequately segmented, multiple segments have either been under-segmented or over-segmented. For example, the large central metal sheet in example Me 163_2 has been over-segmented into multiple segments. However, this type of error is acceptable since combining multiple correctly segmented segments into groups is a simple and time-efficient task for a human annotator during post-processing. Conversely, combining multiple segments into one segment, or under-segmenting, is challenging to correct.

Figure 18.7 shows the preliminary result of performing an inference on an subset of the 'car' dataset 18.2.0.2 with the same CNN exclusively trained on the 'airplane' dataset 18.2.0.1. While the segmentation results could be improved, it is worth noting that the proposed segmentation adequately represents most of the input data sub-volume. However, in some cases, several entities lack appropriate representation in the output segmentation, resulting in suboptimal segmentation results. Despite this, the early transition from a model trained solely on the Me 163 dataset to an

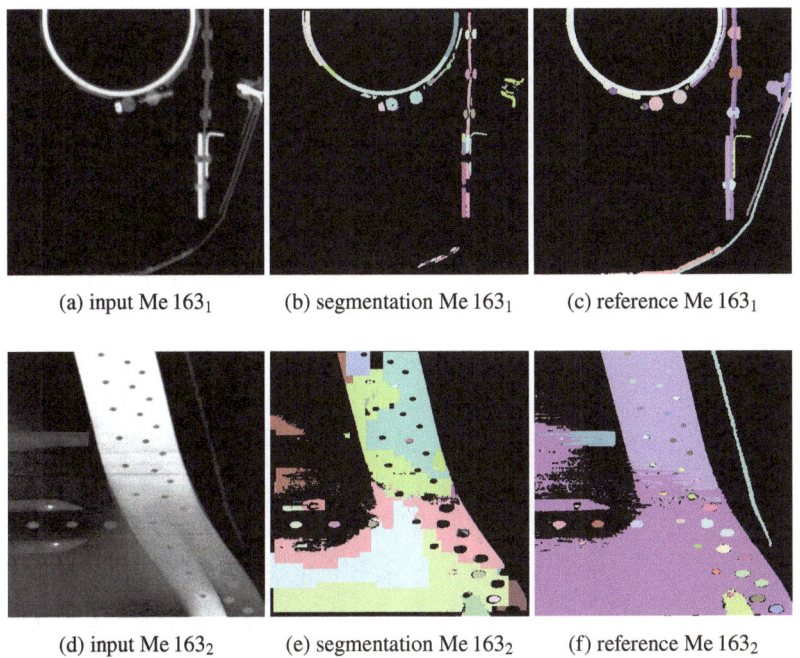

(a) input Me 163_1 (b) segmentation Me 163_1 (c) reference Me 163_1

(d) input Me 163_2 (e) segmentation Me 163_2 (f) reference Me 163_2

Fig. 18.6: Preliminary results of an FFN based instance segmentation trained and run on the Me 163 dataset.

inference run on the Honda dataset shows promise, with multiple thin metal sheets being adequately segmented.

18.4.2 Semantic Segmentation

A further approach is to directly assemble segments from input data. Therefore, an end-to-end machine learning infrastructure is introduced in the following. By using this infrastructure, the entire processing of XXL-CT data from preprocessing through feature extraction towards segment prediction is implemented and makes it possible to benefit from its holistic approach. This allows that relevant parameters e.g. for preprocessing for precise segmentation results can be learned directly from the input.

To realise the segmentation task for specific components in the XXL-CT data, an adapted U-Net [16] architecture has been applied. The U-Net was originally developed for semantic segmentation in the biomedical field and is characterized by its U-shape network topology (see Figure 18.8). The U-Net can be divided into two sub-architectures. The first is the contracting path on the left, which is also known

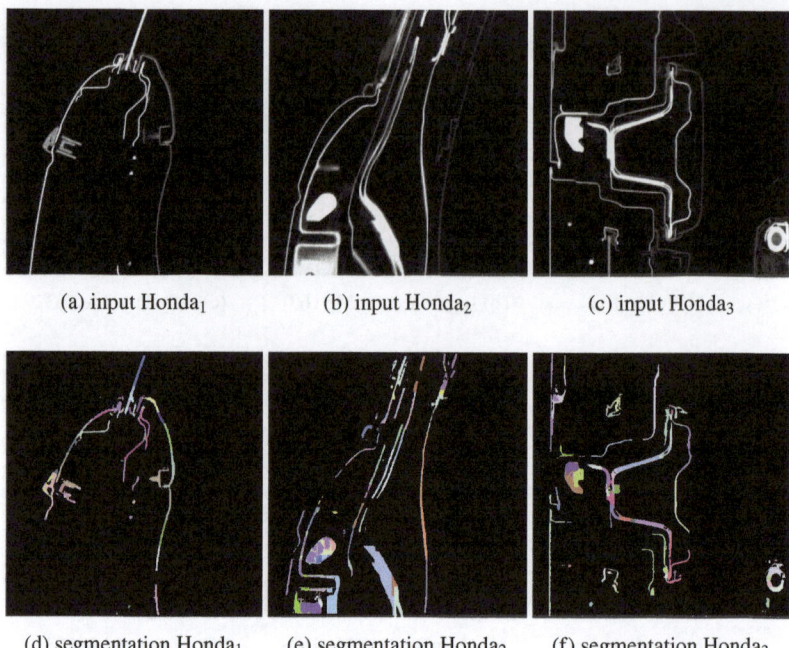

(a) input Honda₁   (b) input Honda₂   (c) input Honda₃

(d) segmentation Honda₁   (e) segmentation Honda₂   (f) segmentation Honda₃

Fig. 18.7: Preliminary results of an FFN based instance segmentation trained exclusively on the Me 163 dataset but applied on the Honda dataset.

as the encoder. This part offers the classification information. The contracting part is followed by the expansive part, which works as a decoder and completes the U-shape. The expansion part allows to learn localized classification information and in addition, increases the output resolution. The final convolutional layer then computes a fully segmented image. The skip connections between the corresponding stages of the contracting and the expansive part allow to propagate features of different granularities. Here, the networks task is to decide for each voxel in an XXL-volume whether it belongs to the target class.

To apply the classical U-Net on the XXL-volumes, the dimensions of all the operations marked by the arrows in Figure 18.8 as well as the feature maps have to be increased to fit the 3D data. The U-Net itself was designed to learn as many features as possible from a non-abundant amount of data. This strength of the architecture is used and its form reformulated to the 3D usecase that leads to three stages of up- and downsampling. The architecture, which finally has been applied, is shown in Figure 18.9.

The contracting path follows the typical structure of a convolutional network, which consists of repeated $3 \times 3 \times 3$ convolutions, each followed by a Rectified Linear Unit (ReLU) [25] and a $2 \times 2 \times 2$ max-pooling [18] with stride 2 for downsampling. With each downsampling, the count of feature channels doubles. Conversely, in the

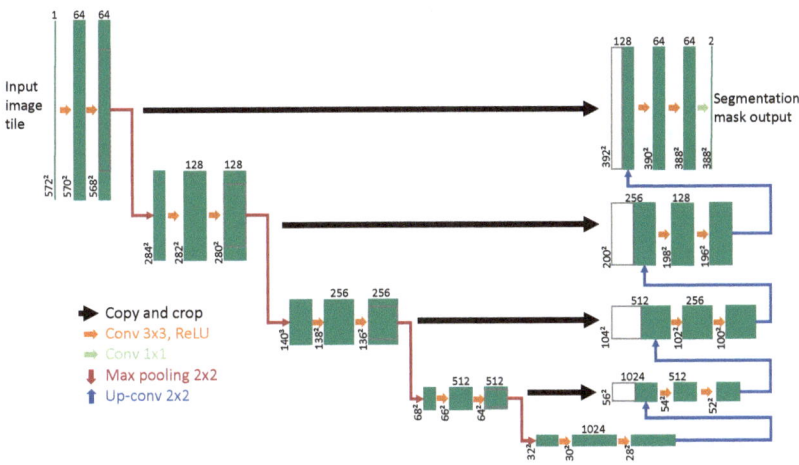

Fig. 18.8: Representation of the initial architecture of the U-Net. The green boxes correspond to the feature maps, where the numbers above gives the count of channels. The number at the bottom left describes the dimension of a feature map. The white boxes contain a copy of the dashed areas of the feature maps to the left of the arrows [21].

upsampling of the expansion path, the count of feature channels is halved again. In addition, at the beginning of each upsampling step, a concatenation with the corresponding feature map from the contracting path is applied.

To train the network Binary Cross-Entropy (BCE) [22] is used as the loss function (Equation 18.1), which is a combination of BCE and the sigmoid-function $\sigma(x) = \frac{1}{1+e^{-x}}$. The number of samples is represented by n, y_i denotes the label of the ground truth and $\hat{y}_i$ stands for the predicted label.

$$L(\hat{y}, y) = -\frac{1}{n} \sum_{i=1}^{n} y_i \cdot log(\sigma(\hat{y}_i)) + (1 - y_i) \cdot log(1 - \sigma(\hat{y}_i)) \qquad (18.1)$$

The following sections outline a concrete example of semantic segmentation of suspension springs inside the XXL-CT data. Basis of this is the above-described infrastructure. For evaluation of the semantic segmentation results, a suitable metric and evaluation scheme has to be selected. In addition to the introduced annotation results from Section 18.3.2 the 3D semantic labelling pipeline was applied to segment multiple suspension springs inside a XXL-CT volume dataset of a damaged Honda from a crash test. Consequential the results are two independent XXL-CT datasets here and as follows references as Honda$_{crashed}$ and Honda$_{uncrashed}$ with annotated

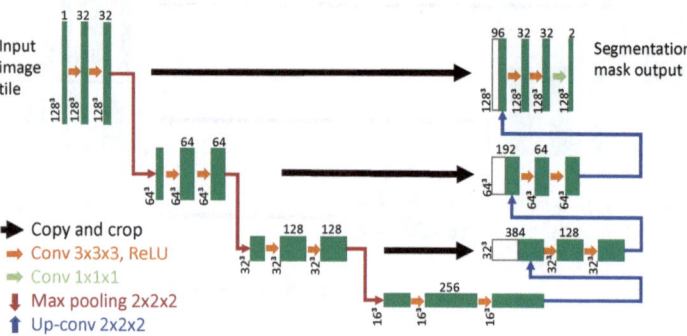

Fig. 18.9: Modified architecture of the applied U-Net. All the operations and feature maps have been extended to three dimensions. Compared to the original Figure 18.8, input and output do not differ in their dimensionality here.

suspension springs as a class label. Whereby the Honda$_{\mathrm{uncrashed}}$ is used for training and validation, the Honda$_{\mathrm{crashed}}$ is excluded and unseen during model optimization as test set.

The measurement of overlapping semantic segmentation predictions of the model and the annotated object regions of interest (ground truth) in 3D volume data was considered as a criterion for evaluation. In the following, a correct predicted voxel of a class label of interest is denoted as true positive (tp), where wrong predicted voxels of a class label of interest is denoted as false positive (fp). Further, a voxel of a class label of interest that is missed from the model prediction is denoted as false negative (fn). For measuring the model performance three scores were selected: precision (18.2), recall (18.3) and Dice score (18.4) [17].

Where precision gives indication of the model capability to predict correct voxels over the total amount of predicted voxels, recall measures the portion of correct predicted voxels to the total amount of voxels in the ground truth. The Dice score provides a balanced measure of model performance, taking into account both precision and recall.

$$\mathrm{precision} = \frac{\mathrm{tp}}{\mathrm{tp} + \mathrm{fp}} \tag{18.2}$$

$$\mathrm{recall} = \frac{\mathrm{tp}}{\mathrm{tp} + \mathrm{fn}} \tag{18.3}$$

$$Dice = \frac{2tp}{2tp + fn + fp} \tag{18.4}$$

To teach the model to recognize springs, the suspension spring-sub-volumes were extracted from the large volumes S_1 and S_2 according to Table 18.1 plus an extended region around the objects of interest to allow to split the volumes in cubic batches. One batch is set to the size of 128^3 voxels. The best results so far were achieved with a training of 50 epochs. To handle the 128^3 voxels size of the batches, one batch per epoch has been the maximum. A value of 66.1% was achieved for precision and 72.3% for recall. For the Dice score the result is 69.1%.

In order to illustrate how such a model looks in application, consider Figure 18.10. Each row shows a different perspective of a spring from the test set $Honda_{crashed}$. The left column of sub-figures always shows the input data, the second column the segmentation result and the rightcolumn the reference (ground truth). However, there are still some areas where voxels are incorrectly classified as spring. Considering the amount of data which was used for trained, the U-Net was still able to visibly segment an object correctly. The described training infrastructure for semantic segmentation can be applied to other components in 3D X-ray data if they are annotated according to the 3D Semantic Labelling Pipeline from Section 18.3.2.

18.5 Conclusion and Outlook

We developed two approaches for segmenting unique datasets and localizing specific components: Instance Segmentation using Floodfilling Networks and Semantic Segmentation using U-Nets.

Instance Segmentation using Floodfilling Networks approaches involves iteratively growing a region of interest by flooding voxels based on certain criteria. It starts with seed points and expands the region by considering neighboring voxels which meet specific conditions. Floodfilling Networks leverage machine learning techniques to learn the criteria for voxel selection and region expansion. This approach is particularly useful for segmenting objects with irregular shapes or when prior knowledge about the object's appearance or class is limited. Encouraging results are observed when applying the model to unseen data of unrelated specimen types.

Semantic Segmentation using U-Nets as type of convolutional neural network architecture have been widely used for image segmentation tasks. It consists of an encoder network and a decoder network. The encoder network captures high-level features by downsampling the input data, while the decoder network upsamples the features to produce a segmentation map. As we have shown, an adapted 3D U-Net architecture is well-suited for segmenting specific entities within volumetric CT data as it can capture both local and global context information. The transfer to data which was unseen during training shows promising results. Regarding the robustness of the 3D U-Net to identify objects of interest in unknown surroundings, we more

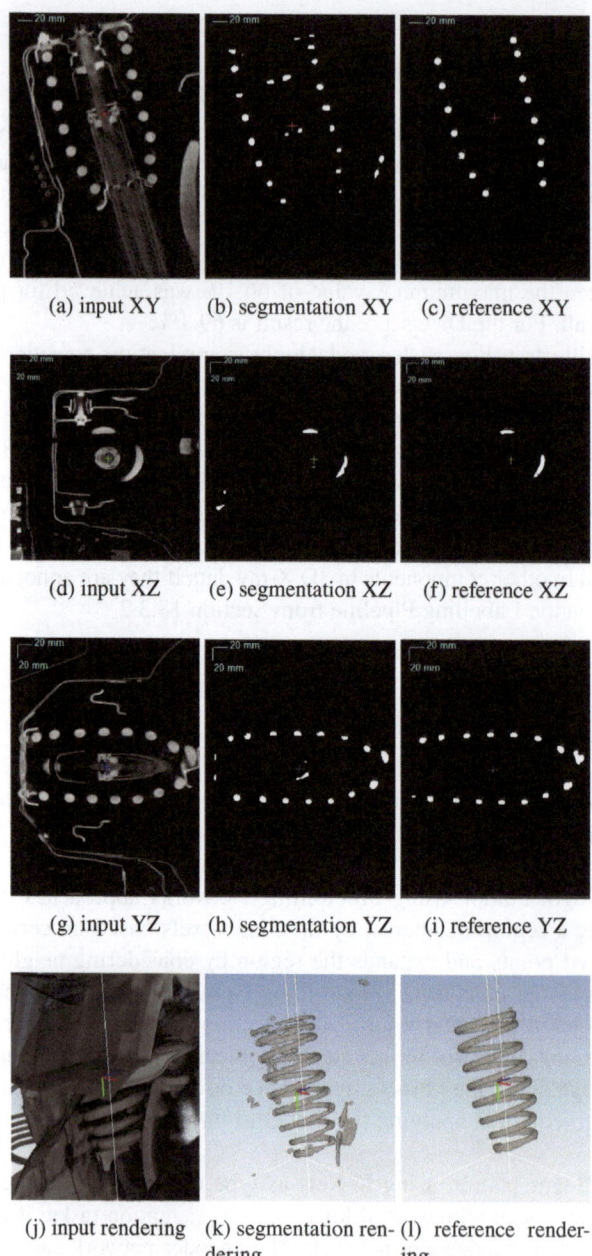

(a) input XY   (b) segmentation XY   (c) reference XY

(d) input XZ   (e) segmentation XZ   (f) reference XZ

(g) input YZ   (h) segmentation YZ   (i) reference YZ

(j) input rendering (k) segmentation ren- (l) reference render-
                    dering   ing

Fig. 18.10: Evaluation example of suspension spring semantic segmentation from
the Honda_{crashed}. From left to right in each row, the initial volume, segmentation
result and the ground truth is shown.

closely investigated the semantic segmentation of suspension springs in a vehicle. Therefore, we trained and optimized a 3D U-Net on CT data of an uncrashed Honda vehicle and evaluated the segmentation prediction on Honda crashed as test set that shows mechanical deformations of the vehicle, due to an intentionally performed crash before the CT scan.

Both approaches offer viable solutions for segmenting unique datasets and localizing specific components within 3D datasets. The choice between these approaches depends on the specific requirements and characteristics of the dataset and the application.

In the world of non-destructive testing, especially for unique scans or huge volumetric options segmenting data is still one of the major bottlenecks. For unique scans, the instance segmentation approach will be a topic for further developments. The focus would be how to integrate user input to retrain a network during the segmentation task. This could reduce the manual segmentation task dramatically. Contrasting for the semantic segmentation, future development will focus on the semantic connections of individual components depending on their location within the object. Additionally, managing labeled and segmented components to include them in the training of neuronal networks for specific tasks is still a open research question.

Acknowledgments

We want to thank the following colleagues for the manual annotation of XXL-CT data and their persevering in carrying out this work: Pooja Vineeth, Verena Malowaniec, Laura Heidner, and Kseniia Dudchenko. Additionally, we would like to thank Michael Salamon, Michael Böhnel, Thobald Fuchs and Nils Reims for the support at the data aquistion for the XXL-CT scans and Adrian Waldyra and Dimitri Prjamkov for the data processing and reconstruction support. Also we would like to acknowledge the work of Thomas Schäfer, who initiated the work on semantic segmantation using deep neuronal networks.

References

1. L. Chen, G. Papandreou, I. Kokkinos, K. Murphy, and A. L. Yuille. Deeplab: Semantic image segmentation with deep convolutional nets, atrous convolution, and fully connected crfs. *IEEE Transactions on Pattern Analysis and Machine Intelligence*, 40(4):834–848, 2018.
2. A. Fedorov, R. Beichel, J. Kalpathy-Cramer, J. Finet, J.-C. Fillion-Robin, S. Pujol, C. Bauer, D. Jennings, F. Fennessy, M. Sonka, J. Buatti, S. Aylward, J. V. Miller, S. Pieper, and R. Kikinis. 3d slicer as an image computing platform for the quantitative imaging network. *Magnetic Resonance Imaging*, 30(9):1323–1341, 2012. Quantitative Imaging in Cancer.
3. R. Girshick, J. Donahue, T. Darrell, and J. Malik. Rich feature hierarchies for accurate object detection and semantic segmentation. In *2014 IEEE Conference on Computer Vision and Pattern Recognition*. IEEE, 6 2014.

4. R. C. Gonzalez and R. E. Woods. *Digital Image Processing (3rd Edition)*. Prentice-Hall, Inc., USA, 2006.
5. R. Gruber, S. Gerth, J. Claußen, N. Wörlein, N. Uhlmann, and T. Wittenberg. Exploring Flood Filling Networks for Instance Segmentation of XXL-Volumetric and Bulk Material CT Data. *Journal of Nondestructive Evaluation*, 40, 2021.
6. R. Gruber, N. Reims, A. Hempfer, S. Gerth, M. Böhnel, T. Fuchs, M. Salamon, and T. Wittenberg. Fraunhofer ezrt xxl-ct instance segmentation me163, 2024.
7. R. Gruber, N. Reims, A. Hempfer, S. Gerth, M. Salamon, and T. Wittenberg. An annotated instance segmentation xxl-ct dataset from a historic airplane, 2022.
8. M. Januszewski, J. Kornfeld, P. H. Li, A. Pope, T. Blakely, L. Lindsey, J. Maitin-Shepard, M. Tyka, W. Denk, and V. Jain. High-precision automated reconstruction of neurons with flood-filling networks. *bioRxiv*, 10 2017.
9. M. Januszewski, J. Maitin-Shepard, P. Li, J. Kornfeld, W. Denk, and V. Jain. Flood-filling networks. *ArXiv*, abs/1611.00421, 2016.
10. A. Kirillov, E. Mintun, N. Ravi, H. Mao, C. Rolland, L. Gustafson, T. Xiao, S. Whitehead, A. C. Berg, W.-Y. Lo, P. Dollár, and R. Girshick. Segment anything, 2023.
11. P. Krähenbühl and V. Koltun. Efficient inference in fully connected crfs with gaussian edge potentials. In *Proceedings of the 24th International Conference on Neural Information Processing Systems*, NIPS'11, pages 109–117, USA, 2011. Curran Associates Inc.
12. Y. Li, X. Chen, Z. Zhu, L. Xie, G. Huang, D. Du, and X. Wang. Attention-guided unified network for panoptic segmentation, 2019.
13. C. Liang-Chieh, G. Papandreou, I. Kokkinos, k. murphy, and A. Yuille. Semantic Image Segmentation with Deep Convolutional Nets and Fully Connected CRFs. In *International Conference on Learning Representations*, San Diego, United States, May 2015. Institute of Electrical and Electronics Engineers (IEEE).
14. T.-Y. Lin, M. Maire, S. Belongie, J. Hays, P. Perona, D. Ramanan, P. Dollár, and C. L. Zitnick. Microsoft COCO: Common objects in context. In *Computer Vision – ECCV 2014*, pages 740–755. Springer International Publishing, 2014.
15. C. Liu, L.-C. Chen, F. Schroff, H. Adam, W. Hua, A. L. Yuille, and L. Fei-Fei. Auto-deeplab: Hierarchical neural architecture search for semantic image segmentation. *2019 IEEE/CVF Conference on Computer Vision and Pattern Recognition (CVPR)*, Jun 2019.
16. J. Long, E. Shelhamer, and T. Darrell. Fully convolutional networks for semantic segmentation. In *2015 IEEE Conference on Computer Vision and Pattern Recognition (CVPR)*, pages 3431–3440, 2015.
17. Y.-H. Nai, B. W. Teo, N. L. Tan, S. O'Doherty, M. C. Stephenson, Y. L. Thian, E. Chiong, and A. Reilhac. Comparison of metrics for the evaluation of medical segmentations using prostate mri dataset. *Computers in Biology and Medicine*, 134:104497, 2021.
18. K. O'Shea and R. Nash. An introduction to convolutional neural networks. *arXiv preprint arXiv:1511.08458*, 2015.
19. S. Pieper, M. Halle, and R. Kikinis. 3D Slicer. In *2nd IEEE Int. Symp. on Biomedical Imaging: Nano to Macro (IEEE Cat No. 04EX821)*, volume 1, pages 632–635 s, 2004.
20. O. Ronneberger, P. Fischer, and T. Brox. U-net: Convolutional networks for biomedical image segmentation. In *Lecture Notes in Computer Science*, pages 234–241. Springer International Publishing, 2015.
21. O. Ronneberger, P. Fischer, and T. Brox. U-net: Convolutional networks for biomedical image segmentation. In *Medical Image Computing and Computer-Assisted Intervention–MICCAI 2015: 18th International Conference, Munich, Germany, October 5-9, 2015, Proceedings, Part III 18*, pages 234–241. Springer, 2015.
22. U. Ruby and V. Yendapalli. Binary cross entropy with deep learning technique for image classification. *Int. J. Adv. Trends Comput. Sci. Eng*, 9(10), 2020.
23. M. Salamon, N. Reims, M. Böhnel, K. Zerbe, M. Schmitt, N. Uhlmann, and R. Hanke. XXL-CT capabilities for the inspection of modern electric vehicles. In *19th World Conference on Non-Destructive Testing*, 2016.
24. R. Sharma, M. Saqib, C.-T. Lin, and M. Blumenstein. A survey on object instance segmentation. *SN Computer Science*, 3(3):499, 2022.

25. S. Sharma, S. Sharma, and A. Athaiya. Activation functions in neural networks. *Towards Data Sci*, 6(12):310–316, 2017.

Chapter 19
Energy-Efficient AI on the Edge

Nicolas Witt[1,*], Mark Deutel[2,*], Jakob Schubert[1,*], Christopher Sobel[1,*], Philipp Woller[1,*]

Abstract This chapter shows methods for the resource-optimized design of AI functionality for edge devices powered by microprocessors or microcontrollers. The goal is to identify Pareto-optimal solutions that satisfy both resource restrictions (energy and memory) and AI performance. To accelerate the design of energy-efficient classical machine learning pipelines, an AutoML tool based on evolutionary algorithms is presented, which uses an energy prediction model from assembly instructions (prediction accuracy 3.1%) to integrate the energy demand into a multi-objective optimization approach. For the deployment of deep neural network-based AI models, deep compression methods are exploited in an efficient design space exploration technique based on reinforcement learning. The resulting DNNs can be executed with a self-developed runtime for embedded devices (dnnruntime), which is benchmarked using the MLPerf Tiny benchmark. The developed methods shall enable the fast development of AI functions for the edge by providing AutoML-like solutions for classical as well as for deep learning. The developed workflows shall narrow the gap between data scientist and hardware engineers to realize working applications. By iteratively applying the presented methods during the development process, edge AI systems could be realized with minimized project risks.

Key words: edge ai, tinyml, automl, classical machine learning, deep compression, design space exploration.

[1]Fraunhofer Institute for Integrated Circuits IIS, Fraunhofer IIS, 90411 Nuremberg, Germany
[2]Friedrich-Alexander-Universität Erlangen-Nürnberg, 91054 Erlangen, Germany
*equal contribution

Corresponding author: Nicolas Witt
e-mail: nicolas.witt@iis.fraunhofer.de

© The Author(s) 2024
C. Mutschler et al. (eds.), *Unlocking Artificial Intelligence*,
https://doi.org/10.1007/978-3-031-64832-8_19

19.1 AI on the Edge

Publicly known artificial intelligence (AI) functionality is primarily deployed in large cloud infrastructures, such as data centers. For instance, large-language models like ChatGPT serve millions of users. AI inference can run on desktop-sized machines similar to the personal computers used by content creators, such as AI-assisted editing. While some AI applications still rely on cloud computing, the rapid evolution of AI has led to more opportunities in the deployment of AI systems. Thus, the computing hardware targeted for AI inference can accommodate it. Taking advantage of the substantial computational resources and scalability of the cloud computing paradigm, complex AI functions have historically been located primarily in centralized data centers (the cloud). This conventional approach facilitated advanced AI applications but poses significant challenges, including latency, security, privacy, and data transmission overhead, especially in IoT scenarios.

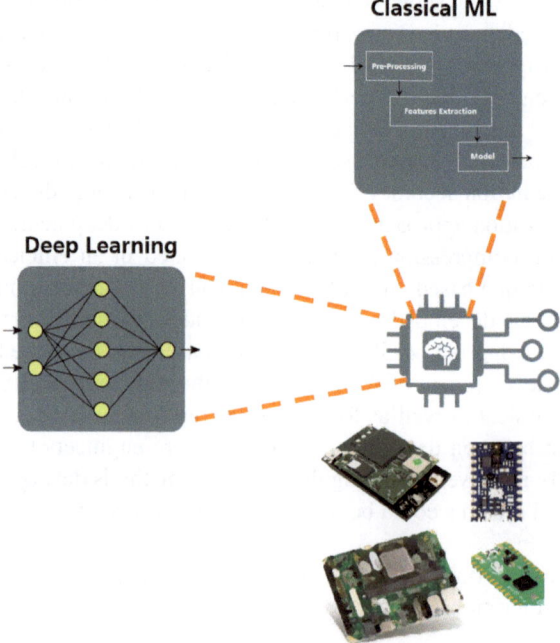

Fig. 19.1: Enabling AI functionality on the edge needs a resource-optimized approach of AI design.

Edge computing, a paradigm that emphasizes decentralized data processing and storage, has emerged as a solution to these challenges. The fusion of AI and edge computing has given rise to "Edge AI", a novel architectural framework that uses local processing capabilities to enable real-time, context-aware AI decision-making in close proximity to data sources. An overview of this emerging field can be found in [42]. It provides the reader with more detailed background and definitions that are beyond the scope of this article. In the absence of a clear definition of edge devices, our focus in this chapter is on resource-constrained devices such as drones, mobile robots, and IoT nodes (e.g., condition monitoring (CdM) sensor nodes), which typically rely on battery power. Processing units in these devices typically consist of microprocessors or microcontrollers with clock speeds of few hundred MHz and with only a few hundred kilobytes of memory (see Figure 19.1).

The advantages of designing AI systems for the edge are similar to those for edge computing, but due to the data-hungry nature of AI plays an even more important role. Here is a list of the most important advantages:

- **Data Privacy** - Processing data from private spaces (e.g., audio or video data in people's homes) on local devices follows the "Privacy by Design" paradigm and no private data has to leave the access realm of the user.
- **Low Latency** - Local processing of input data by AI models avoids the detour of raw data to distant data centers resulting in fast response times.
- **Small Size** - AI functions can even be realized in very small devices like IoT sensor nodes or hearing aids to name a few examples.
- **Low Communication Overhead** - As the raw video, audio, and vibration data can have a significant need for transmission bandwidth, extracting the essential information locally (e.g., person detection, keyword spotting, state information for CdM) leads to a significant reduction in communication. Instead of a video stream, only person counts, activation toggle, or OK / NOT OK state information has to be transmitted.
- **Energy Efficiency** - Also connected to the former advantage, edge devices can often exploit more efficient sleep modes or turn off frequently and are more fine-grained than servers in data centers, leading to less power consumption.

But besides these advantages, there are also several challenges in bringing AI to edge devices. The first major challenge is the lack of combined education in both AI and hardware development, resulting in a non-awareness of, e.g., data scientists on memory and energy demand of their trained models and hardware developers on the other hand not being trained to design high-performance AI models. This can lead to several opportunities in projects to miss requirements and/or financial and time restrictions. The main technical challenge to bringing AI to edge devices is the data and thus problem-dependent size of AI functions. An AI function can be realized by either classical machine learning pipelines (ML) consisting of several pipeline steps, usually feature extraction, scaling, dimensionality reduction, and the final ML model, or deep learning approaches where the used deep neuronal networks can have a large number of parameters to execute the intended functionality (see Figure 19.1).

The following sections show how the design space of AI functions can be explored to optimize not only AI performance (e.g., classification accuracy) but also optimize the resource footprint such as energy demand or memory consumption.

19.2 Energy-Efficient Classical Machine Learning

When searching for an energy-efficient machine learning solution suitable for deployment on the edge, there are several factors to consider. These factors include the processing of time series data that is collected by the on-board sensors, their sampling rate, and the computing power or memory required for the AI algorithm. Other factors to consider are the type of communication for data transmission and the specific inference frequency (i.e., how often the AI model is triggered) required for the use case. These factors cover a wide range of design options for potential AI solutions, which are generally narrowed down only through close interaction between hardware developers and data scientists. However, in practice, the limited overlap in competencies often leads to lengthier development cycles. The hardware developer might have difficulties gauging the probability of success for data-centric projects, while the software developer is often incapable of forecasting and ascertaining compliance with prerequisites like form factor or power requirements for the end system. The presented research aimed to narrow the gap between the two disciplines by designing AutoML approaches that consider both machine learning performance (for example, accuracy or F1 score) and, e.g., power consumption (measured in Watts) for inference on edge devices, as objectives for optimization.

19.2.1 Classification of Time Series Data

Efficient models of classical machine learning, such as decision trees or support vector machines (SVM), are often used for time series analysis. In this process, the time series data from the edge's onboard sensors, like an IMU (Inertial Measurement Units), are transformed using a sliding window approach [3]. Within each fixed-length window, various statistical or signal processing features are computed to summarize the data, such as mean, variance, spectral characteristics, and more. The extracted features provide a compact representation of the time series data that machine learning models can use. Further details are provided in Section 2.2.

The task of the developed AutoML system is to make a suitable selection or combination of the several hundred features. This requires selecting the appropriate machine learning model and optimizing its hyperparameters [44]. If the sensor sampling rate is adjustable or the optimal window size for the use case is unknown, these variables can also be incorporated into the approach and evaluated concerning the two optimization goals: achieving maximum AI performance and minimizing resource usage.

19.2.2 Multi-Objective Optimization

The AutoML approach involves utilizing multi-objective optimization (MOO) which is a strategy for tackling problems with multiple conflicting objectives. In this context, we consider two main objectives:

1. **Predictive Accuracy**: Maximizing the predictive accuracy of the machine learning model is crucial for ensuring its effectiveness in practical applications.
2. **Energy-Efficiency**: Minimizing the energy demand of the machine learning pipeline is vital for embedded devices that often have limited power resources.

To achieve these objectives, NSGA-II (Non-dominated Sorting Genetic Algorithm) [10] is employed. NSGA-II identifies a set of Pareto-optimal solutions, each representing a different trade-off between predictive accuracy and energy efficiency. These solutions provide diverse options that balance the two essential objectives. By exploring this set of trade-off solutions, decision-makers can choose the most suitable machine-learning pipeline for their specific embedded device application. A binary representation for encoding machine learning pipeline configurations expresses concrete choices and hyperparameters using binary values (individual). For example, if three different machine learning algorithms are considered, one allocates a specific binary sequence to each, such as '00' for algorithm A, '01' for algorithm B, and '10' for algorithm C. This binary string captures the choice of algorithm. The binary representation also extends to the feature selection or any other components of the pipeline such as additional preprocessing steps (e.g., scaling option, dimensionality reduction technique, such as PCA, etc.). For instance, '1' might indicate the inclusion of a particular feature, while '0' signifies its absence. Combining all these choices results in a single binary string that represents a specific combination of algorithm choice, hyperparameters, and pipeline structure (see Figure 19.2). During the evolutionary optimization process, crossover and mutation operations are applied to these binary strings to generate new pipeline configurations. Crossover involves exchanging bits between two parent configurations to produce offspring, while mutation entails flipping individual bits to introduce small changes.

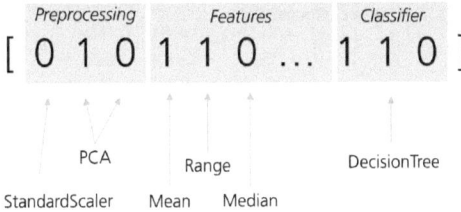

Fig. 19.2: Example of the binary representation of machine learning pipeline configurations.

19.2.3 Energy Prediction for Classical Machine Learning

To optimize the AI performance and energy requirements of machine learning pipelines across generations (see Section 19.2.2), information on the energy demand of each pipeline during the optimization process is needed. However, an actual deployment and execution of each pipeline on the hardware platform to measure its energy demand is not a practical solution in the large search space for optimization. Therefore, instead of taking measurements, prediction models were developed to estimate the energy demand for each ML pipeline to make the mulit-objective AutoML approach feasible.

The prediction of energy demand in prior research varies based on the level of abstraction used for estimating the energy demand. The level of granularity ranges from abstract Program Measurement Counters (PMC) [46], through functional block level [28] to an instruction set level [21, 22, 45, 1, 30] eventually forming a compromise between generalization and prediction accuracy. The two methods developed here are based on computational demand extracted from instructions in the source code and compiled assembly code, respectively. A dataset was formed by randomly sampling pipeline steps and recording their computational and energy demands as determined through physical measurements. Using this dataset a regression model is trained to predict the general energy demand of classic machine learning pipelines.

```
1  #define NUM_CLASSES 3
2  #define INPUT_SIZE 4
3
4  float instance[INPUT_SIZE + 1];
5  const float coef[NUM_CLASSES][INPUT_SIZE] = {...};
6  const float intercept[NUM_CLASSES] = {...};
7  const int8_t classes[NUM_CLASSES] = {0, 1, 2};
8
9
10 int classify(){
11   int indMax = 0;
12   float scores[NUM_CLASSES];
13   int i;
14   int j;
15   for (i = 0; i < NUM_CLASSES; i++){
16     scores[i] = intercept[i];
17     for (j = 0; j < INPUT_SIZE; j++){
18       scores[i] += (coef[i][j] * instance[j]);
19     }
20     if (scores[i] > scores[indMax]){
21       indMax = i;
22     }
23   }
24   return classes[indMax];
25 }
```

Fig. 19.3: Source code of a logistic regression classifier.

Approach 1: Program Code Based: In the first approach, the mathematical operations respectively write and read accesses to the data memory required for its calculation are taken from the implementation of the pipeline steps in C++, see Figure 19.3. The operations are categorized as either floating point or integer. A developer must provide a parametric description of the number of each operation, such as the number of features or classes. Consequently, during the optimization process, only the configuration of the pipeline step is needed as the feature vector for predicting energy demand. But in a program-level assessment, it is very hard to account for compiler optimizations and the true machine code running the AI inference. Therefore another approach was chosen.

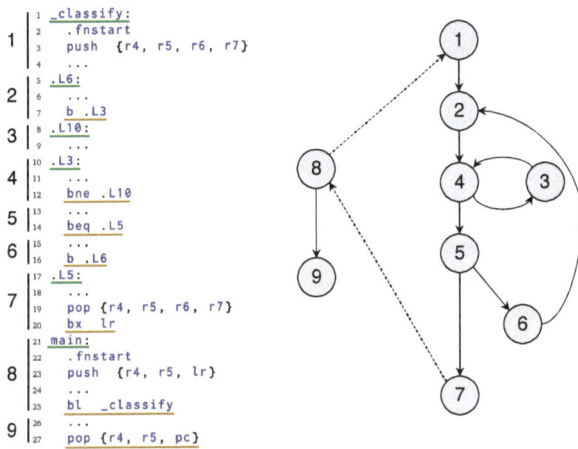

Fig. 19.4: Construction of the CFG from assembly code (Logistic Regression classifier). Green marked lines represent the end point and brown marked lines the starting point of a jump in the control flow. Instructions and directives not relevant for the control flow have been removed from the representation.

Approach 2: Assembly Code Based: A more precise and easier-to-automate method involves utilizing assembly code, which represents the program code at a lower level of abstraction and already incorporates the compiler optimizations. This representation is used in the second method of extracting computational demand but requires an initial analysis of the code's control flow to identify the assembly instructions executed on the platform. To accomplish this task, it is necessary to apply the Implicit Path Enumeration Technique (IPET) to determine the implicit control flow. At first, the assembly code must be used to extract a control flow graph (CFG), e.g., see Figure 19.4. A CFG is a graph that is directed and whose nodes consist of basic blocks that contain program instructions. The edges in the CFG represent the potential execution paths. The IPET calculates the frequency of execution for each node in the CFG. This method allows us to investigate the number of times each assembly instruction is executed. To minimize the feature space and thus the training dataset, the assembly instructions are grouped according to their required energy demand. Instructions with the same number of clock cycles required and using similar hardware devices in the processor are assumed to have similar power requirements. This creates a feature vector for use in either the training or test dataset. Like before, this feature vector is also parameterized. An example of the logistic regression classifier is presented in Table 19.1, based on the data illustrated in Figures 19.3 and CFG in 19.4.

G 1	G 2	G 3	G 4	G 5	G 6	G 7
$2C$	$3C+1$	$C\dot{M}+C$	$4C\dot{M}+6C-2$	0	$C\dot{M}+10C+30$	2

G 8	G 9	G 10	G 11	G 12	G 13	Sum
$C\dot{M}+4C+1$	0	78	0	1	0	$7C\dot{M}+26C+111$

Table 19.1: Resulting parametric description of the assembly computational demand in each group (G) of the CFG shown in in Figure 19.4. The parameter C represents the number of classes and M the length of the feature vector.

With **Approach 1**, the energy demand of each possible pipeline can be predicted with a mean relative deviation of 4.1 %. **Approach 2**, results in a better mean relative deviation of 3.1 % from the true energy demand and is better suited to integrate in an AutoML solution.

19.2.4 EA-AutoML Tool

The entire process of developing an energy-efficient machine learning solution for the Edge has been prototyped in an EA[1]-AutoML tool, offering a comprehensive workflow for fast and easy proof of concepts. After loading a labeled dataset of sensor data, the tool seamlessly performs feature extraction on windowed data. The optimization phase is then highly configurable, allowing the user to define the search space consisting of several preprocessing steps, feature calculations, and ML models. Upon the successful completion of the multi-objective optimization, the tool visualizes the obtained pipelines on a Pareto front (see Figure 19.5). This graphical representation offers a clear view of the trade-offs between the different objectives. In the example given, the damage to the bearings of an electric motor had to be classified via vibration data[2]. Between the two example candidates A and B, up to 80% of the energy demand can be saved with only a 4.5% loss in classification accuracy (see Figure 19.5).

Further comparing candidates in terms of the lifespan of a battery-powered edge device, taking into account parameters such as battery capacity, communication (e.g., Bluetooth or Wi-Fi), sleep mode power consumption, and the inference frequency. Figure 19.6 shows the comparison between solutions A and B. Selecting the more energy-effective pipeline B can significantly increase the potential lifespan of the edge device in this use case.

In the final step, the selected pipeline can be exported for embedded systems. The tool generates a C program that renders the pipelines executable on the target platform. This program can be seamlessly integrated into an existing project, ensuring a smooth transition from the prototype to practical application and in-field testing.

[1] *Evolutionary Algorithm*

[2] CWRU Bearing Dataset, see https://engineering.case.edu/bearingdatacenter/welcome

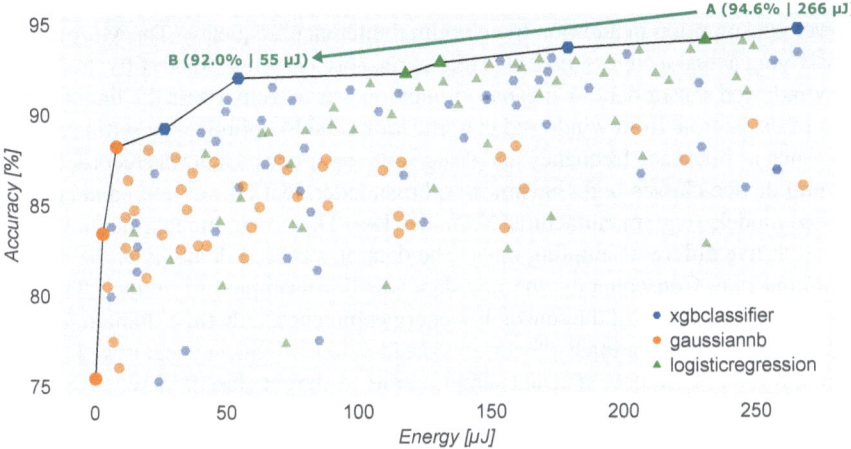

Fig. 19.5: Pareto-front featuring various machine learning pipelines that enable trade-offs between energy consumption and prediction accuracy.

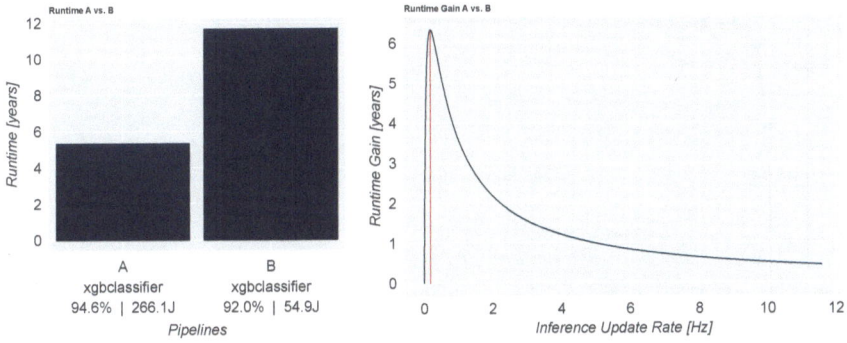

Fig. 19.6: Simulation of the life time and comparison of two machine learning pipelines for use on an Edge.

19.2.5 Application Example

The following section shows a concrete application, where AutoML was used to design an efficient AI pipeline for the edge. The use case is simplified due to confidentiality restrictions.

The objective of the application was to estimate the usage time of electrically or gasoline-powered appliances (e.g., garden appliances like lawnmowers) with an attached sensor tag using accelerometer data. The usage time is used to determine wear and tear, plan efficient maintenance schedules, and enable optimized distribution of fleet devices. The tags are battery-powered and cannot be recharged. Thus the precise estimation of usage time has to be as energy efficient as possible to prolong service intervals. This necessitates the implementation of an ML pipeline that has minimal power consumption in addition to an optimal inference frequency. The ML pipeline classifying actual *usage* vs. *no usage* (e.g., transport, carrying, stand still) should run on windowed sensor data. A lifespan simulation was used to assess the final estimation of usage time from windowed classifications and to optimize the settings of the tag, such as inference frequency and sleep/wake-up modes. Data was recorded from several device classes (e.g., lawnmower, brushcutter, leaf blower, etc.) and specific device models (e.g., manufacturer X, model 1A). The accelerometer on the tag can be set to five different sampling rates. The dataset was recorded only at the highest sampling rate. Consequently, the raw data was downsampled to mimic other supported sampling rates of the sensor. For energy efficiency, only time-domain features with minimal computational effort were considered (no frequency features). For each configuration of window size and sampling rate, different classifiers (e.g., Decision Tree, Support Vector Machine) were evaluated with 10-fold cross-validation. Deep learning methods were not considered due to the very resource-restricted target device. The models were evaluated and compared by computing the average F1-score, as well as Precision and Recall. The top-performing models were exported to C, enabling execution and testing on the target platform.

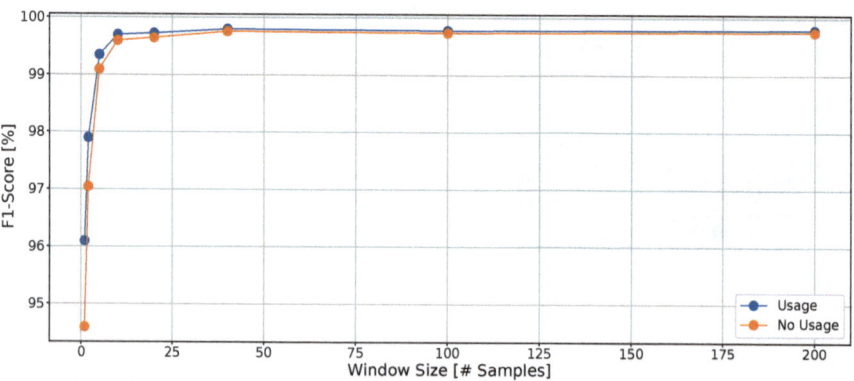

Fig. 19.7: Influence of different window sizes on performance. The example origins from models trained with 200 Hz data from one specific tool class.

As expected the F1-scores in Figure 19.7 decrease with decreasing window size, as there is less information in smaller windows. However, the performance hardly worsens from 200 samples down to 40. Even with only ten samples per window, the performance is still well beyond 99.5 %.

Finally, the task was to evaluate the effectiveness of the binary window-based classification pipeline in estimating the usage time. A simulation with a synthetic ground truth for usage times was used by including everyday usage profiles of each device class, i.e. typically daily usage time and single usage times. In addition to a consistent inference frequency (e.g., every 90 seconds), the tag was simulated with a wake-on-motion feature. The simulation yields, e.g., "Mean Absolute Percentage Error" (MAPE) of usage time estimation in the long run, e.g., over several days. The main influence on the final usage time estimate is the recall values for the *usage* and *no usage* classes of the pipelines operating in windowed sensor data, where a trade-off to resource-efficient pipelines (low sampling rate, low feature complexity) could be done with the help of the simulation. Figure 19.8 shows the simulation including predictions, inference calls, and the tag's battery status (energy needed for one inference heavily scaled up for visualization purposes).

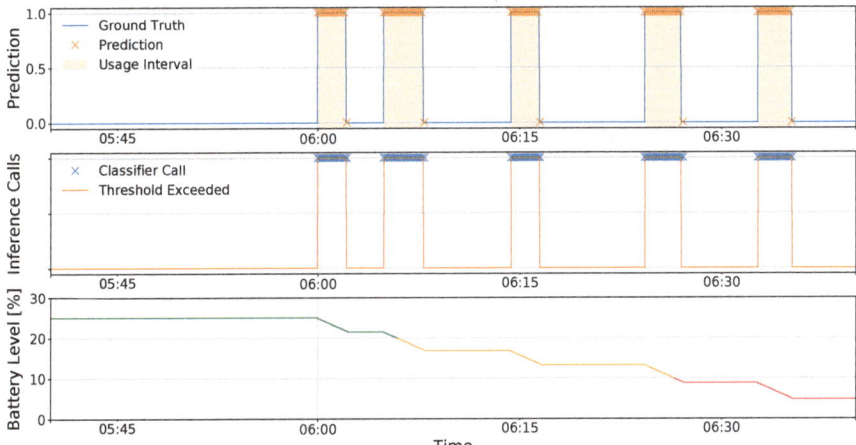

Fig. 19.8: Simulation visualization with predictions, inferences and battery level of the tag (with deliberately exaggerated energy demand per inference for visualization purposes).

Satisfactory results for usage time estimation and an optimized device lifespan could be achieved with classical ML classifiers, optimally selected features, and an optimized inference rate.

For more complex problems on edge devices, classical ML pipelines might not achieve the AI performance needed. The following section outlines the energy-efficient design of deep learning solutions for these problems.

19.3 Energy-Efficient Deep Learning

Deep Neural Networks (DNNs) have become dominant in various more complex applications involving autonomous decision-making, audio recognition [20], image classification [19, 29], or human activity monitoring [26]. DNNs offer a distinct advantage in their ability to learn and abstract correlations within high-dimensional, intricate data.

Nevertheless, the deployment of DNNs consumes substantial energy, resources, and time [43, 9]. In embedded scenarios where the trade-off between energy consumption, resource limitations, execution speed, and model accuracy are critical, DNNs often struggle to outperform classical machine learning approaches, particularly in light of their high energy demands [31]. However, given the ever-increasing amount of data to be processed on the edge, there is a growing demand for energy-efficient DNN execution on embedded devices.

Nonetheless, most DNN training and deployment frameworks primarily prioritize AI performance (e.g., accuracy) and do not explicitly address the critical aspects of energy efficiency and platform-specific constraints, including memory availability and processing speed. However, these factors are especially critical in most edge applications.

19.3.1 Deep Compression

Deep compression is an emerging area of research focused on the compression of Deep Neural Networks (DNNs). Notable techniques encompass DNN pruning [32] and weight quantization [25]. The primary objective is to decrease the resource footprint of a DNN on its designated platform, which includes its memory and energy requirements. This reduction is achieved by diminishing the number of trainable weights, subsequently reducing computational complexity, all while preserving the DNN's original accuracy.

Following these principles, various pipelines for DNN compression have been proposed. Han et al. [15] introduced a pipeline that combines network pruning, integer quantization, and Huffman coding. More recently Deutel et al. [13] proposed a pipeline combining different pruning and quantization methods. Other approaches concentrate on quantization during network training [25], while others emphasize structured pruning [34, 2], allowing for the immediate elimination of pruned weights. In the following, we discuss key aspects of DNN pruning and quantization and its use in the context of Edge AI.

19.3.1.1 Pruning

A commonly used method to compress DNNs for resource constrained edge systems is pruning. The technique is founded on the idea that some of a DNN's trained parameters can be removed without significantly compromising network accuracy. This is based on the observation that many DNNs tend to be overparameterized, harboring redundancy in their trained weights [11]. The concept of pruning has a long history, with it being initially proposed to enhance network generalization, reduce overfitting, and accelerate learning [17, 32, 18]. Today, pruning has evolved into one of the most popular techniques for DNN compression, often achieving significant reductions in size and computational complexity without sacrificing accuracy.

The simplest method for DNN pruning involves setting a subset of trainable parameters to zero during training, resulting in sparse parameter tensors. By nullifying these parameters, they are effectively excluded from the optimization process for training the DNN. Consequently, these removed parameters no longer impact the training of the network.

Granularity: Two other common techniques for DNN Pruning are *element pruning* and *structured pruning*, which differ in the granularity at which they introduce sparsity into a DNN.

Element pruning involves removing individual elements from parameter tensors, akin to early pruning methods by authors like [32, 18]. Structured pruning, a more recent approach, removes entire structures from parameter tensors, often focusing on filters or channels in convolutional layers as shown in [33, 35], but is extensible to rows or columns in linear layers as well. Structured pruning offers the advantage of complete removal of pruned structures from the parameter tensor, which element pruning cannot achieve since it creates tensors of arbitrary sparsity. However, structured pruning is more invasive and complex, requiring a global understanding of the DNN's structure and posing additional challenges in branching networks like residual networks [19].

Heuristics: A significant challenge in DNN pruning is determining which elements or structures have the least impact on a DNN's accuracy on the validations dataset when removed. The most accurate method, referred to as the "oracle criterion" [38], involves removing each network element or structure one by one and evaluating its impact on the loss. However, this approach is highly resource- and time-intensive, rendering it impractical in most cases.

As a result, alternative heuristics have been proposed in research to approximate optimal pruning more efficiently. This process of quantifying the importance of parameters in DNN parameter tensors is also known as "Sensitivity Analysis" [16]. Recent advances have focused on finding effective approximations, i.e., heuristics, for both element and structured pruning techniques. These include magnitude/threshold based heuristics [14, 16], L-norm based heuristics [33], gradient based heuristics [38], and heuristics based on the average percentage of zeros found in feature maps [23].

An alternative heuristic for pruning neural network structures is rooted in Layer-wise Relevance Propagation (LRP) [51, 50], which has its origins in the field of explainable AI (XAI). This technique assigns relevance scores to individual neurons within a neural network. However, although LRP-based heuristics can offer more informed decisions and generally lead to superior pruning outcomes, they often come at the expense of increased computation time.

Schedule: A *pruning schedule*, or *pruning recipe*, outlines when, how frequently, and to what extent a network undergoes pruning during training. One straightforward approach to scheduling pruning is known as *one-shot pruning* [32]. The general concept involves initially training a network to achieve a reasonable level of accuracy. Subsequently, the entire network is pruned using a specific heuristic to eliminate structures or elements with the lowest scores. Based on these scores, a certain number of them are then removed. Additionally, it is often beneficial to retrain the network after pruning.

An alternative type of pruning schedule is called *iterative pruning* [16, 33]. This schedule places a strong emphasis on the iterative process of pruning and retraining the network multiple times during training. As a result, not all parameters are removed at once, but over several pruning iterations. This gradual approach allows the network to adapt more effectively to the decreasing number of trainable parameters.

An extension to iterative pruning schedules is called *Automated Gradual Pruning (AGP)* [54]. The authors suggest a gradual increase in the number of pruned parameters instead of removing a constant number of parameters in each iteration. The algorithm automatically adjusts the number of pruned parameters in a DNN over a range of n pruning steps based on predefined parameters. AGP is based on the idea of initially pruning the network rapidly when redundant connections are abundant and gradually reducing the number of weights pruned in each subsequent iteration as fewer weights remain in the network.

19.3.1.2 Quantization

Quantization of DNNs revolves around the concept of reducing the precision at which a network's parameters are represented and processed. This strategy serves to reduce the memory footprint and inference time of a neural network. However, it is not without trade-offs, typically resulting in decreased accuracy. The primary objective of quantization is to strike a balance between accuracy, computational complexity, and memory footprint.

In general, quantization can be applied at two levels. The first is named *weight-only quantization*, where only the trained weight tensors of a DNN are quantized. The second is known as *full-network quantization*, which includes both weight and activation tensor quantization. Weight-only quantization is easy to implement and significantly reduces the memory footprint of the model. However, it increases the runtime overhead during inference, since all quantized tensors must be converted back to floating-point space before processing, and all multiply-accumulate operations also have to be processed in floating-point space. This conversion to floating-point space is not required in full-network quantization, where both weight and activation tensors are quantized, allowing most computations to be performed in integer space. However, full-network quantization requires the quantization of activation tensors at runtime, introducing additional computational overhead at runtime.

Quantization Scheme: A quantization scheme defines the mathematical connection between the representation of a DNN's initial floating-point values and their quantized equivalents. For edge systems, an 8-bit unsigned integer type is most commonly used to store the quantized values. Furthermore, the most common quantization schemes are based on a simple uniform affine mapping, as illustrated by Equation 19.1,

$$f(x) = \left\lfloor \frac{x}{s} \right\rfloor + zp, \; s = \frac{max_{data} - min_{data}}{255}, zp = -\frac{min_{data}}{s} \tag{19.1}$$

where the linear mapping is realized by the two parameters zero point zp and scale s, which are derived from the distribution of values in the full precision floating point tensor.

Applying Quantization: There are two methodologies on how to apply quantization to a DNN: *static post-training quantization (SPTQ)* and *quantization aware training (QAT)*.

SPTQ involves introducing quantization into a neural network after it completes its training [53, 6, 24]. It uses two key steps. First, quantization parameters are determined for all weight and activation tensors. Weight tensors' parameters are straightforward to calculate, as their values become constant after the network's training. However, deriving quantization parameters for activation tensors is more complex, as their actual values are only known during inference. A common practice is to sample from the network's test dataset to calculate these parameters. Second, quantization is applied to all weight and activation tensors based on the derived parameters and the selected quantization scheme. Weight tensors can be quantized immediately, replacing the original floating-point tensors with their quantized counterparts. For activation tensors, additional operators are added to the network to perform quantization and de-quantization during inference as they traverse the network.

QAT considers quantization during the training of the DNN [25]. Its main distinction from SPTQ is the inclusion of quantization parameters as additional trainable parameters during training. This involves augmenting the original floating-point model before training begins and integrating fake quantization operations into the network structure where tensors need to be downcasted to their quantized representation. The training process is then executed on this augmented network, where forward passes use fake quantized tensors and backward passes optimize the original floating-point weights. To facilitate backpropagation through the quantization operations, the *straight-through estimator (STE)* [7] is used. After training is completed, the quantized model can be directly exported without requiring any additional post-processing steps like SPTQ.

Some of the above-mentioned methods of pruning and quantization were integrated into a self developed workflow leading to a runtime environment for DNNs on embedded devices called dnnruntime [13].

19.3.2 Efficient Design Space Exploration

Deploying DNNs on embedded devices requires adhering to the constraints imposed by the target edge platform. These constraints present significant challenges when designing DNN models for such platforms. In essence, there is a need to reconcile conflicting goals and constraints that usually include aspects like memory availability, inference time, and power consumption of the deployed DNN model.

Despite extensive research in Design Space Exploration (DSE) for DNNs, i.e., Neural Architecture Search (NAS) and AutoML, see Chapter 1.1, there is still no definitive method that combines efficient design space exploration and robustness. In the literature, three prominent approaches to performing DSE for DNNs targeting edge systems are discussed. First, black-box Hyperparameter Optimization

(HPO) [55, 4, 40, 48, 12]. Second, differentiable NAS [36, 49]. Third, zero-cost NAS [41, 8].

Black-box HPO is the most reliable and consistent option of the three and can easily be extended to the multi-objective case, but it is typically time-consuming and sample-inefficient since it requires training and evaluation of multiple DNNs. However, it has been shown that the use of Bayesian optimization can improve sampling efficiency significantly [48, 12].

Differentiable NAS attempts to optimize the architecture as part of regular DNN training by relaxing the optimization problem. However, recent research has high-lighted stability issues and poor generalization [52].

Zero-Cost NAS is highly time-efficient because it does not train DNNs directly, but uses an empirical surrogate model. It can be used to quickly adapt DNN designs to different target platforms and resource constraints [41, 8]. However, since it provides only simple statistics from the surrogates, it does not provide precise performance information from actually trained DNNs [47].

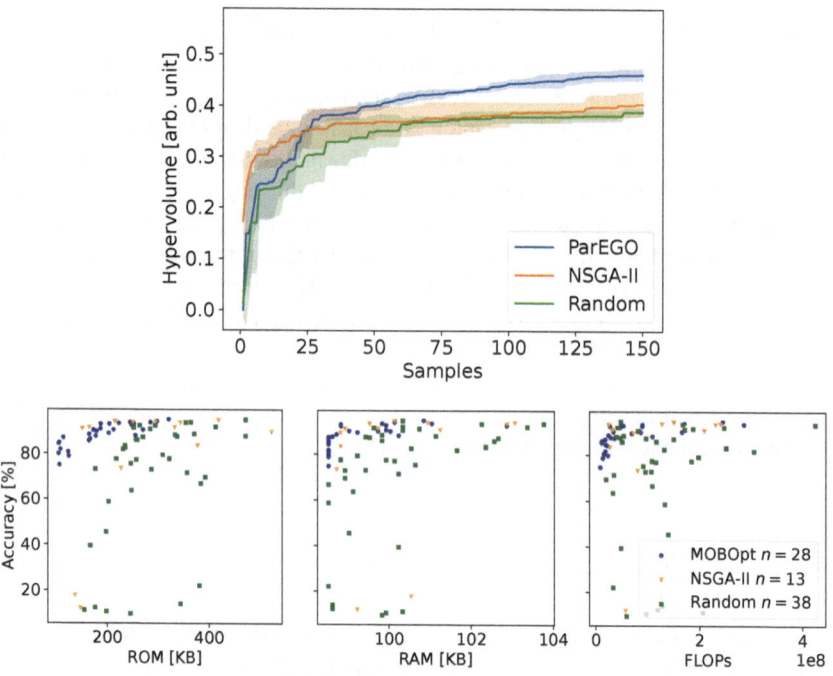

Fig. 19.9: Bayeisan optimization (blue) compared to evolutionary optimization (yellow) and random search (green, baseline) optimizing MobileNetv3 on DaLiAc. Top: Hypervolume improvement over the course of optimization, higher means better. Bottom: Pareto fronts resulting from the different optimizers considering accuracy, ROM, RAM, and FLOPs as objectives.

Bayesian optimization can help to significantly improve the time- and resource-efficiency of black-box optimization based NAS, see Figure 19.9. Expanding on this concept, Deutel et al. [12] implemented a time- and resource-efficient black-box optimization framework for DNN deployment on the edge, based on Bayesian optimization and reinforcement learning (RL). The novel solver employs an ensemble of locally parameterizable policies that compete with one another and are iteratively trained using Augmented Random Search (ARS) reinforcement learning agents [37] on the underlying Bayesian surrogate model. The network architectures proposed by our approach are readily deployable on common microcontrollers without the need for additional (re)training, also due to the self-developed runtime for DNNs dnnruntime.

19.3.3 Benchmarking Edge AI

The previous sections show the number of different ways that AI algorithms can be optimized to run on tiny edge devices with limited memory and processing power. This makes it difficult to directly compare different optimization methods. The MLPerf Tiny benchmark [5] shall provide a fair and reliable way to compare the resulting AI models. It is divided into the Open and the Closed Division. The division of the software stack is shown in Figure 19.10.

The Closed Division just allows varying hardware or deployment-specific components. It also allows to quantize the model weights as described in Section 19.3.1.2. This makes the different submitted solutions more comparable. In the Open Division, the test data is the only parameter that cannot be changed. Even the training data is allowed to be customized together with the training script and the model architecture. In the Open Division, it is also allowed to prune the used model weights.

From the four test scenarios, anomaly detection and image classification were chosen for evaluation. For anomaly detection, the ToyADMOS [27] and MIMII [39] datasets are used. In both cases, the sound of a toy car is recorded and mixed with noise from a factory environment. The goal is to classify if the machine is running normally or not. For image classification, the CIFAR-10 [29] dataset is used.

The characteristics measured in the MLPef Tiny benchmark are accuracy, latency, and energy demand. Accuracy is examined by evaluating the model on the corresponding test dataset, while latency and power consumption are measured with the help of an energy monitor. To get single inference values, it measures the execution time and the power consumption of multiple inferences for at least ten seconds.

Using the MLPerf Tiny benchmark the developed dnnruntime was benchmarked and compared to other submissions. The results are shown in Table 19.2. Although the achieved accuracy is the best result in three out of four scenarios, the required latency and energy consumption are almost always above the results of the compared submissions. Considering that the dnnruntime was developed as a research prototype by a single person, results similar to those of large frameworks with many developers cannot be expected. However, the results show that dnnruntime provides

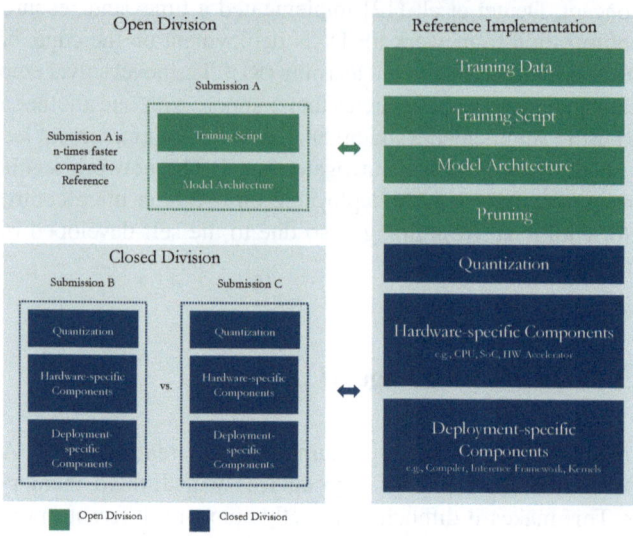

Fig. 19.10: Open and a Closed Division of the MLPerf Tiny benchmark (from [5]).

a good research foundation and is capable of delivering comparable results if more work is spent to optimize performance and efficiency.

19.4 Conclusion and Outlook

This chapter described different methods for the resource-optimized design of AI models and pipelines. Targeted edge devices are usually powered by microprocessors or microcontrollers with very restricted memory and computing power. All AutoML methods presented, use multi-objective optimization to find Pareto-optimal solutions for a trade-off between AI performance and resource demand (energy, memory).

Despite classical machine learning seems to be out fashioned these days, it can provide very energy-efficient data driven solutions. The respective chapter showed, how a fast automatic design was achieved by using a prediction model for energy demand based on assembly instructions inside an evolutionary algorithm for multi-objective optimization.

For the design of deep neuronal networks, deep compression methods were exploited in an efficient design space exploration technique based on reinforce-

	Anomaly Detection			Image Classification		
	AUC	Latency [ms]	Energy at 1.8V [μJ]	Accuracy [%]	Latency [ms]	Energy at 1.8V [μJ]
TFLiteMicro | Reference	0.86	10.523	417.7	87.5	664.01	27495.0
dnnruntime | v1	0.87	13.060	581.3	88.5	774.59	32899.5
dnnruntime | v2	0.87	12.965	574.8	88.5	774.59	33674.1
OctoML/microTVM/CMSIS-NN backend	0.86	8.6	443.2	87.5	389.5	20236.3
OctoML/microTVM/native codegen	0.86	11.7	663.7	87.5	389.5	21342.3
Plumerai/Inference Engine 2022.09	0.86	5.6		88.0	173.2	
STMicroelectronics/X-CUBE-AI v7.3.0	0.86	7.6	323.0	85.0	226.9	10681.6

	Keyword Spotting			Visual Wake Words		
	AUC	Latency [ms]	Energy at 1.8V [μJ]	Accuracy [%]	Latency [ms]	Energy at 1.8V [μJ]
TFLiteMicro | Reference	90.1	159.03	6608.0	82.8	545.85	22128.2
dnnruntime | v1	91.6	279.10	11853.4	82.8	1054.9	43907.3
dnnruntime | v2	91.6	278.86	11979.2	82.8	843.88	35893.4
OctoML/microTVM/CMSIS-NN backend	90.1	99.8	5230.3	85.8	301.2	15531.4
OctoML/microTVM/native codegen	90.2	144.0	5230.3	83.6	336.5	17131.6
Plumerai/Inference Engine 2022.09	90.2	71.7		85.2	208.6	
STMicroelectronics/X-CUBE-AI v7.3.0	90.2	75.1	3371.7	85.2	230.5	10066.6

Table 19.2: Detailed comparison of dnnruntime inference results with other submissions of the Closed Division.

ment learning. Optimizing different objectives was enabled through having a self-developed runtime for DNNs on embedded devices (dnnruntime), which can report expected FLOPS (floating point operations) and memory requirements during optimization. The dnnruntime was also benchmarked using the public MLPerf Tiny benchmark and showed promising results.

The developed methods showed to enable even less experienced persons in from each discipline to design energy-efficient AI solutions for edge devices. Thus bringing data scientists and hardware engineers closer together.

Further research will focus on improved compression techniques for very big models (e.g., large language models), hardware-aware optimizations, e.g., for AI accelerators, and making on-device learning available on microcontrollers for actual self-learning IoT systems. Overlapping research questions between data science and neuromorphic computing cover bioinspired neural architectures like spiking neuronal networks, as well as optimized hardware architectures for an analog execution of neuronal networks for ultra low power applications. To speed up the commercial availability of the proposed methods, future development targets the provision of web-based services for the independent development of hardware-optimized AI models, accompanied with certified courses for the professional expansion of know-how in the industry.

References

1. Power analysis of embedded software: A first step towards software power minimization. *IEEE Transactions on Very Large Scale Integration (VLSI) Systems*, 2(4):437–445, 1994.
2. S. Anwar, K. Hwang, and W. Sung. Structured pruning of deep convolutional neural networks. *J. Emerg. Technol. Comput. Syst.*, 13(3), 2017.
3. M. Bahri, A. Bifet, J. Gama, H. M. Gomes, and S. Maniu. Data stream analysis: Foundations, major tasks and tools. *WIREs Data Mining and Knowledge Discovery*, 11(3):e1405, 2021.
4. B. Baker, O. Gupta, N. Naik, and R. Raskar. Designing neural network architectures using reinforcement learning. In *International Conference on Learning Representations*, 2017.
5. C. Banbury, V. J. Reddi, P. Torelli, J. Holleman, N. Jeffries, C. Kiraly, P. Montino, D. Kanter, S. Ahmed, D. Pau, et al. Mlperf tiny benchmark. *arXiv preprint arXiv:2106.07597*, 2021.
6. R. Banner, Y. Nahshan, and D. Soudry. Post training 4-bit quantization of convolutional networks for rapid-deployment. *Advances in Neural Information Processing Systems*, 32, 2019.
7. Y. Bengio, N. Léonard, and A. Courville. Estimating or propagating gradients through stochastic neurons for conditional computation. *arXiv preprint arXiv:1308.3432*, 2013.
8. H. Cai, C. Gan, T. Wang, Z. Zhang, and S. Han. Once-for-all: Train one network and specialize it for efficient deployment. In *International Conference on Learning Representations*, 2020.
9. A. Canziani, A. Paszke, and E. Culurciello. An Analysis of Deep Neural Network Models for Practical Applications. *arXiv:1605.07678 [cs]*, 2017.
10. K. Deb, A. Pratap, S. Agarwal, and T. Meyarivan. A fast and elitist multiobjective genetic algorithm: Nsga-ii. *IEEE Transactions on Evolutionary Computation*, 6(2):182–197, 2002.
11. M. Denil, B. Shakibi, L. Dinh, M. Ranzato, and N. de Freitas. Predicting parameters in deep learning. *Advances in Neural Information Processing Systems*, 26, 2013.
12. M. Deutel, G. Kontes, C. Mutschler, and J. Teich. Augmented random search for multi-objective bayesian optimization of neural networks, 2023.
13. M. Deutel, P. Woller, C. Mutschler, and J. Teich. Energy-efficient deployment of deep learning applications on cortex-m based microcontrollers using deep compression. In *MBMV 2023; 26th Workshop*, pages 1–12. VDE, 2023.
14. S. Han, H. Mao, and W. J. Dally. Deep compression: Compressing deep neural networks with pruning, trained quantization and huffman coding. *arXiv preprint arXiv:1510.00149*, 2015.
15. S. Han, H. Mao, and W. J. Dally. Deep compression: Compressing deep neural networks with pruning, trained quantization and huffman coding. *International Conference on Learning Representations (ICLR)*, 2016.
16. S. Han, J. Pool, J. Tran, and W. Dally. Learning both weights and connections for efficient neural network. *Advances in neural information processing systems*, 28, 2015.
17. S. Hanson and L. Pratt. Comparing biases for minimal network construction with back-propagation. *Advances in Neural Information Processing Systems*, 1, 1988.
18. B. Hassibi and D. Stork. Second order derivatives for network pruning: Optimal brain surgeon. *Advances in neural information processing systems*, 5, 1992.
19. K. He, X. Zhang, S. Ren, and J. Sun. Deep residual learning for image recognition. *Conference on Computer Vision and Pattern Recognition*, 2016.
20. S. Hershey, S. Chaudhuri, D. P. W. Ellis, J. F. Gemmeke, A. Jansen, R. C. Moore, M. Plakal, D. Platt, R. A. Saurous, B. Seybold, M. Slaney, R. J. Weiss, and K. Wilson. CNN architectures for large-scale audio classification. In *2017 IEEE International Conference on Acoustics, Speech and Signal Processing (ICASSP)*, pages 131–135, 2017.
21. B. Herzog, S. Reif, J. Hemp, T. Hönig, and W. Schröder-Preikschat. Resource-demand estimation for edge tensor processing units. *ACM Transactions on Embedded Computing Systems (TECS)*, 21(5):1–24, 2022.
22. T. Hönig, B. Herzog, and W. Schröder-Preikschat. Energy-demand estimation of embedded devices using deep artificial neural networks. In *Proceedings of the 34th ACM/SIGAPP Symposium on Applied Computing*, pages 617–624, 2019.

23. H. Hu, R. Peng, Y.-W. Tai, and C.-K. Tang. Network trimming: A data-driven neuron pruning approach towards efficient deep architectures. *arXiv preprint arXiv:1607.03250*, 2016.
24. I. Hubara, Y. Nahshan, Y. Hanani, R. Banner, and D. Soudry. Improving post training neural quantization: Layer-wise calibration and integer programming. *arXiv preprint arXiv:2006.10518*, 2020.
25. B. Jacob, S. Kligys, B. Chen, M. Zhu, M. Tang, A. Howard, H. Adam, and D. Kalenichenko. Quantization and training of neural networks for efficient integer-arithmetic-only inference. In *Proceedings of the IEEE Conference on Computer Vision and Pattern Recognition (CVPR)*, 2018.
26. T. Kautz, B. H. Groh, J. Hannink, U. Jensen, H. Strubberg, and B. M. Eskofier. Activity recognition in beach volleyball using a Deep Convolutional Neural Network. *Data Mining and Knowledge Discovery*, 31(6):1678–1705, 2017.
27. Y. Koizumi, S. Saito, H. Uematsu, N. Harada, and K. Imoto. Toyadmos: A dataset of miniature-machine operating sounds for anomalous sound detection. In *2019 IEEE Workshop on Applications of Signal Processing to Audio and Acoustics (WASPAA)*, pages 313–317. IEEE, 2019.
28. V. Konstantakos, A. Chatzigeorgiou, S. Nikolaidis, and T. Laopoulos. Energy consumption estimation in embedded systems. *IEEE Transactions on instrumentation and measurement*, 57(4):797–804, 2008.
29. A. Krizhevsky, G. Hinton, et al. Learning multiple layers of features from tiny images. 2009.
30. M. Kumar, X. Zhang, L. Liu, Y. Wang, and W. Shi. Energy-efficient machine learning on the edges. In *2020 IEEE international parallel and distributed processing symposium Workshops (IPDPSW)*, pages 912–921. IEEE, 2020.
31. N. D. Lane, S. Bhattacharya, A. Mathur, P. Georgiev, C. Forlivesi, and F. Kawsar. Squeezing Deep Learning into Mobile and Embedded Devices. *IEEE Pervasive Computing*, 16(3):82–88, 2017.
32. Y. LeCun, J. Denker, and S. Solla. Optimal brain damage. *Advances in neural information processing systems*, 2, 1989.
33. H. Li, A. Kadav, I. Durdanovic, H. Samet, and H. P. Graf. Pruning filters for efficient convnets. In *International Conference on Learning Representations*, 2016.
34. H. Li, A. Kadav, I. Durdanovic, H. Samet, and H. P. Graf. Pruning Filters for Efficient ConvNets. *arXiv:1608.08710 [cs]*, 2017.
35. M. Lin, R. Ji, Y. Zhang, B. Zhang, Y. Wu, and Y. Tian. Channel pruning via automatic structure search. In *Proceedings of the Twenty-Ninth International Conference on International Joint Conferences on Artificial Intelligence*, pages 673–679, 2021.
36. H. Liu, K. Simonyan, and Y. Yang. DARTS: Differentiable architecture search. In *International Conference on Learning Representations*, 2019.
37. H. Mania, A. Guy, and B. Recht. Simple random search of static linear policies is competitive for reinforcement learning. In *Advances in Neural Information Processing Systems (NeurIPS)*, 2018.
38. P. Molchanov, S. Tyree, T. Karras, T. Aila, and J. Kautz. Pruning convolutional neural networks for resource efficient inference. In *5th International Conference on Learning Representations, ICLR 2017-Conference Track Proceedings*, 2019.
39. H. Purohit, R. Tanabe, K. Ichige, T. Endo, Y. Nikaido, K. Suefusa, and Y. Kawaguchi. Mimii dataset: Sound dataset for malfunctioning industrial machine investigation and inspection. *arXiv preprint arXiv:1909.09347*, 2019.
40. E. Real, A. Aggarwal, Y. Huang, and Q. V. Le. Regularized evolution for image classifier architecture search. In *Proceedings of the aaai conference on artificial intelligence*, 2019.
41. X. Shen, Y. Wang, M. Lin, Y. Huang, H. Tang, X. Sun, and Y. Wang. Deepmad: Mathematical architecture design for deep convolutional neural network. In *Proceedings of the IEEE/CVF Conference on Computer Vision and Pattern Recognition (CVPR)*, 2023.
42. R. Singh and S. S. Gill. Edge ai: A survey. *Internet of Things and Cyber-Physical Systems*, 2023.
43. V. Sze, Y.-H. Chen, T.-J. Yang, and J. S. Emer. Efficient Processing of Deep Neural Networks: A Tutorial and Survey. *Proceedings of the IEEE*, 105(12):2295–2329, 2017.

44. A. Truong, A. Walters, J. Goodsitt, K. Hines, C. B. Bruss, and R. Farivar. Towards automated machine learning: Evaluation and comparison of automl approaches and tools. In *2019 IEEE 31st international conference on tools with artificial intelligence (ICTAI)*, pages 1471–1479. IEEE, 2019.

45. P. Wägemann, C. Dietrich, T. Distler, P. Ulbrich, and W. Schröder-Preikschat. Whole-system worst-case energy-consumption analysis for energy-constrained real-time systems. *Leibniz international proceedings in informatics: LIPIcs; 106*, 106:24, 2018.

46. M. J. Walker, S. Diestelhorst, A. Hansson, A. K. Das, S. Yang, B. M. Al-Hashimi, and G. V. Merrett. Accurate and stable run-time power modeling for mobile and embedded cpus. *IEEE Transactions on Computer-Aided Design of Integrated Circuits and Systems*, 36(1):106–119, 2016.

47. C. White, M. Khodak, R. Tu, S. Shah, S. Bubeck, and D. Dey. A deeper look at zero-cost proxies for lightweight nas. *ICLR Blog Track*, 2022.

48. C. White, W. Neiswanger, and Y. Savani. Bananas: Bayesian optimization with neural architectures for neural architecture search. In *Proceedings of the AAAI Conference on Artificial Intelligence*, 2021.

49. B. Wu, X. Dai, P. Zhang, Y. Wang, F. Sun, Y. Wu, Y. Tian, P. Vajda, Y. Jia, and K. Keutzer. Fbnet: Hardware-aware efficient convnet design via differentiable neural architecture search. In *IEEE Conference on Computer Vision and Pattern Recognition, CVPR*, 2019.

50. R. Xu, S. Luan, Z. Gu, Q. Zhao, and G. Chen. Lrp-based policy pruning and distillation of reinforcement learning agents for embedded systems. In *2022 IEEE 25th International Symposium On Real-Time Distributed Computing (ISORC)*, pages 1–8. IEEE, 2022.

51. S.-K. Yeom, P. Seegerer, S. Lapuschkin, A. Binder, S. Wiedemann, K.-R. Müller, and W. Samek. Pruning by explaining: A novel criterion for deep neural network pruning. *Pattern Recognition*, 115:107899, 2021.

52. A. Zela, T. Elsken, T. Saikia, Y. Marrakchi, T. Brox, and F. Hutter. Understanding and robustifying differentiable architecture search. In *International Conference on Learning Representations*, 2020.

53. R. Zhao, Y. Hu, J. Dotzel, C. De Sa, and Z. Zhang. Improving neural network quantization without retraining using outlier channel splitting. In *International conference on machine learning*, pages 7543–7552. PMLR, 2019.

54. M. Zhu and S. Gupta. To prune, or not to prune: exploring the efficacy of pruning for model compression. *arXiv preprint arXiv:1710.01878*, 2017.

55. B. Zoph and Q. Le. Neural architecture search with reinforcement learning. In *International Conference on Learning Representations*, 2017.